GENETICS
OF LIVESTOCK
IMPROVEMENT

Second Edition

GENETICS

OF LIVESTOCK

IMPROVEMENT

JOHN F. LASLEY

Department of Animal Husbandry
University of Missouri

Prentice-Hall, Inc.   Englewood Cliffs, New Jersey

Library of Congress Catalog Card Number: 70–171845

ISBN: 0–13–351189–8

10   9   8   7   6

Printed in the United States of America

PRENTICE-HALL INTERNATIONAL, INC., *London*
PRENTICE-HALL OF AUSTRALIA, PTY., LTD., *Sydney*
PRENTICE-HALL OF CANADA, LTD., *Toronto*
PRENTICE-HALL OF INDIA PRIVATE LTD., *New Delhi*
PRENTICE-HALL OF JAPAN, INC., *Tokyo*

Contents

Preface

Established fundamental principles of animal breeding do not change over a period of years. Information about these principles and methods of presenting and using them do change, however, with the publication of more research results and with the experience of the teacher, breeder, and author. In this second edition, we have tried to present the principles of animal breeding in as simple and practical a manner as possible for the agricultural student and the breeder of farm animals. This was the main objective of the first edition and it is still the main objective of the second.

In the first chapters of this book we have laid a foundation for animal breeding by presenting some of the fundamental concepts of animal genetics. We have included a discussion of chromosomes and chromosome abnormalities and of some of the most recent concepts of how genes function. Again, we have stressed the different kinds of phenotypic expression of genes as the base on which to build selection and mating systems because, basically, this is what animal genetics is all about. Chapters dealing with gene frequencies and selection and mating systems are included to present some of the basic principles involved in the improvement of quantitative traits and the genetics of populations. Finally, the last chapters of the book show how the fundamental principles of qualitative and quantitative genetics may be applied to the improvement of certain species of farm mammals.

I am grateful to graduate students, Charles Chrisman, David Crenshaw, S. N. Pani, and Stanley Starling for their suggestions for the improvement of several chapters of this book. I am also grateful to Dr. Robert Koch and

Dr. Earl Lasley for their ideas and suggestions for the improvement of Chapter 11, which deals with the subject of population genetics, and which is one of the most important and basic chapters in this book. Finally, I am especially grateful to hundreds of undergraduate students at the University of Missouri who have used the first edition in an animal breeding course and have given me the incentive to write this book and to continue to research methods of presentation of the fundamental principles included in the first and second editions of *Genetics of Livestock Improvement*.

JOHN F. LASLEY

Glossary of Abbreviations

xv

ColoAESB	Colorado Agricultural Experiment Station Bulletin
DTW	Deutsche Tierarztliche Wochenschrift
ECR	Experimental Cell Research
Faoun,EAAP	Food and Agricultural Organization of the United Nations, European Association of Animal Production
FlaAESB	Florida Agricultural Experiment Station Bulletin
FS	Fertility and Sterility
H	Hereditas
IowaAESRB	Iowa Agricultural Experiment Station Research Bulletin
ISGDR	Illinois Swine Grower's Day Report, Urbana, Ill.
JAgS	Journal of Agricultural Science
JAR	Journal of Agricultural Research
JAS	Journal of Animal Science
JAUPR	Journal of Agriculture of the University of Puerto Rico
JDR	Journal of Dairy Research
JDS	Journal of Dairy Science
JEB	Journal of Experimental Biology
JEZ	Journal of Experimental Zoology
JG	Journal of Genetics
JGP	Journal of Genetic Psychology
JH	Journal of Heredity
JMB	Journal of Molecular Biology
JLCM	Journal of Laboratory and Clinical Medicine
JRF	Journal of Reproduction and Fertility
MoAESRB	Missouri Agricultural Experiment Station Research Bulletin
MVP	Modern Veterinary Practice
NCRP	National Council on Research Publications
NHF	National Hog Farmer
N.J.AESB	New Jersey Agricultural Experiment Station Bulletin
NLP	National Livestock Producer
NV	Nordisk Veterinaermedicin
NV-T	Norsk Veterinaer-Tidsskrift
NZJST	New Zealand Journal of Science and Technology
NZJSTA	New Zealand Journal of Scientific and Technological Agriculture
OG	Obstetrics and Gynecology
OklaAESMP	Oklahoma Agriculture Experiment Station Miscellaneous Publication
PASAP	Proceedings of the American Society of Animal Production
PBRT	Poultry Breeders' Round Table
P6ICAH	Proceedings of the 6th International Congress of Animal Husbandry

P8ICG	Proceedings of the 8th International Congress of Genetics
PNAS	Proceedings of the National Academy of Sciences
PS	Poultry Science
PSB	Proceedings of the Society of Biology
PSCBC	Proceedings of the Scottish Cattle Breeding Conference
PSEBM	Proceedings of the Society of Experimental Biology and Medicine
QHJ	Quarter Horse Journal
S	Science
SA	Scientific American
SAg	Scientific Agriculture
SDAESC	South Dakota Agricultural Experiment Station Circular
SM	Scientific Monthly
TR	Thoroughbred Record
t.t.	Translated Title
TU	Tierärzliche Umschauung
USDAARAR	United States Department of Agriculture, Agricultural Research Administration Report
USDAC	United States Department of Agriculture Circular
USDAFB	United States Department of Agriculture Farmers Bulletin
USDIP	United States Department of Interior Publication
USYA	United States Yearbook of Agriculture
VR	Veterinary Record
WLJ	Western Livestock Journal
YJBM	Yale Journal of Biology and Medicine
Z	Zuchtungskunde
ZTZ	Zeitschrift für Tierzüchtung und Zuchtungsbiologie

1

Development
of the Livestock
Industry

1.1 NEED FOR ANIMAL PRODUCTS

The population explosion together with a poor distribution of food are among the world's greatest problems today. In many highly populated, poorly developed countries most children suffer from malnutrition in their early years. This may cause a high death loss or an adverse effect on the body that persists throughout the lifetime of affected individuals. For example, in one South American country, 82 children out of every 1000 born die before they reach one year of age, another 12.4 per 1000 die before their fourth birthday, and among those that survive many are mentally retarded. Some of this mental retardation is due to a severe protein deficiency during childhood, which appears to affect adversely the central nervous system, which does not have the ability to repair itself. These figures are similar to those from other parts of the world where the population has outgrown its food supply (see Fig. 1.1).

The food supply in these overpopulated areas consists mostly of starchy grains, which supply enough energy for life processes, but lack the protein

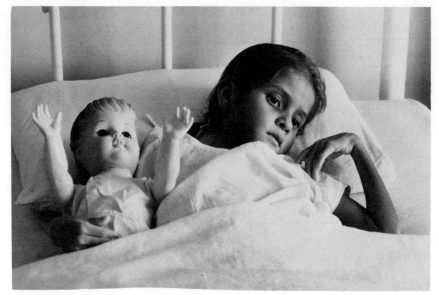

Fig. 1.1 This child is suffering from protein malnutrition. Many scientists recognize the necessity of supplying the proper kind and amount of protein to children, in some countries to prevent death and mental retardation among those who survive. Animals and animal products supply good quality proteins in the diets of young, growing children. (Courtesy of the Rockefeller Foundation.)

necessary for growth and repair of body tissues. Thus in these areas a severe lack of protein of good quality often results in the disease known as *kwashiorkor* in young children. It is characterized by retarded growth, swollen eyelids, lack of appetite, severe diarrhea, and abnormal color and texture of the hair. Death usually follows the appearance of such symptoms unless effective treatment is administered.

The discovery of high-lysine corn and its production in areas where corn is the major item in the human diet offers promise of helping relieve these protein shortages. Animal products such as meat, milk, and eggs, however, are the major sources of high-quality protein for man. This explains the need for the production of large quantities of animal products and their proper distribution to children in underdeveloped countries of the world.

The production of large quantities of animal products requires improved and efficient methods. Efficient production has many aspects, one of which is the application of effective breeding methods. The purpose of this text is to present fundamental genetic principles and how they may be applied to more efficient animal production.

1.2 DEVELOPMENT OF THE LIVESTOCK INDUSTRY IN THE UNITED STATES

Animal products have long been important in the diet of people in the United States. When the first settlers came to America from Europe, they brought animals to supply meat, eggs, milk, wool, feathers, leather and many other staples of life. The early animals were generally nondescript in appearance. Not until the work of Robert Bakewell and his contemporaries in England in the late 1700's do we find the beginning of breeds and the use of the show ring and of records as a means of developing farm animals along special use lines. The work of these men led to the establishment of pure breeds of farm animals.

Purebred animals were first introduced into the United States after the Revolutionary War. The first purebreds to be imported were dairy cattle and sheep, because milk and wool were two of the colonists' greatest needs. There were still enough wild animals in our woodlands and enough cull dairy cattle and sheep to supply meat. The need for more efficient meat-producing animals came later. Our present-day breeds of beef cattle and swine were imported shortly before the Civil War.

With the Industrial Revolution and the movement of people from farms into villages and cities, the need arose for more meat and wool to be used off the farm. This led to the realization that animals have to become more efficient in their production. The best method then known to increase the efficiency of production was to breed animals for purity of the desired characteristics. The use of this method contributed to the development of our pure breeds. The market for breeding animals was usually very close to home, and the breeders in a given area knew the breeding history of their neighbors' animals, so they knew where to get the types and breeds of animals they wanted for their herds or flocks. As the market for the sale of purebreds expanded and some breeders became better known than others, it became necessary to develop some means of verifying the ancestry of these animals. This led to the formation of breed associations during the last part of the 19th century.

The formation of breeds and their registry associations resulted in the recognition of the fact that purebred sires could bring about considerable improvement in livestock production. Numerous experiments were conducted by experiment stations prior to 1925 in which the offspring of purebred males were compared with those of nonpurebred males. In general, the comparisons favored purebred sires, and livestock men became interested in the use of purebred or registered sires. Actually the word *registered* came to mean that the animal was something special.

With the formation of many breeds, the question inevitably arose of which was best. Many livestock men had their own ideas, and many of them made comparisons between different breeds, as did some of the colleges and experiment stations. These comparisons showed that there was no single best breed and that breeds differ in many performance traits. Breeds superior in some traits of economic importance were inferior in others. If this had not been found to be true, we would now have only one breed, or at most only a very few different breeds, in each of the classes of farm animals.

The increased popularity of the show ring in this country stimulated studies of several classes of farm animals to determine the best types from the standpoint of production. The general conclusion was that the correlation between type and performance was very low. This meant that selecting for good type would not automatically give good performance. As far as the size of animals was concerned, however, an intermediate between large and small seemed to be the most desirable when all factors were considered.

Still later, interest was shown in crossbreeding both of cattle and of hogs, and many experiment stations compared crosses with purebreds. In general, the results favored the crosses, but in many cases the differences were small. These studies brought out the fact, however, that some breeds crossed better than others and that it was just as important in crossbreeding as in purebreeding to use good selection practices and a systematic mating system.

Recently, more and more emphasis has been placed on good records of performance of farm animals as a means of selection and improvement. Some new methods of testing, along with well-planned mating systems, show much promise in improving the efficiency of livestock production in the future.

1.3   IMPROVEMENTS HAVE BEEN MADE IN THE EFFICIENCY OF LIVESTOCK PRODUCTION

The application of improved methods of breeding, feeding, management, and disease control during the last few years has greatly increased the efficiency of livestock production. Estimates from agricultural statistics show that the production per animal unit in the United States has increased 30 percent in the last 35 years. Efficiency of production has increased in almost all classes of livestock. Since dairy-cattle testing began in 1906, for all cows milked, the average milk yield per cow per year has increased 1700 pounds and the butterfat production has increased 63 pounds. About 30 years ago, well-fed hogs required 8 to 9 months to attain 200 pounds' weight and required 400 pounds of feed per 100 pounds of gain. During the past 10 years, the age to attain 200 pounds has been reduced to 5 to $5\frac{1}{2}$ months, or even less, and only 250 to 300 pounds of feed are required per 100 pounds of gain. Ton litters at 6 months of age was the goal set up in some states 25 years ago. To-day, litters weighing 2000 to 3000 pounds at that age are fairly common, and a few litters have weighed more than 5000 pounds at 6 months.

Egg production per hen was increased about 84 percent in the last 30 years. The production of broilers has increased greatly and has become an important industry in the United States; the efficiency of production also has increased. In 1947, $12\frac{1}{2}$ weeks and 12.3 pounds of feed were required to produce a 3-pound broiler. In 1952, 10 weeks and 10.2 pounds of feed were required to produce a broiler of the same weight; more recently, even better results are being achieved on many specialized poultry farms.

Improved management, feeding, and disease control practices have been responsible for much of the improvement in the efficiency of livestock production. Improved selection and mating systems have also been important. Over 90 percent of the broilers reaching the market today come from the one-way cross of White Cornish males on White Plymouth Rock hens. This cross takes advantage of the meatiness and growing ability of the Cornish breed and the increased number of eggs for incubation along with the meatiness of White Plymouth Rock hens. Inbred lines within the egg-producing breeds have been developed and are crossed to give a linecross hen with superior egg-laying ability for commercial egg production.

Systematic crossbreeding systems have been utilized for more efficient production of market hogs. The Holstein breed of dairy cattle has become more and more popular because of its heavy milk production and a lower percentage of butterfat in the milk. Skim milk, or low butterfat milk, has become popular with consumers because of the high calorie content of the fat and the supposed correlation of animal fats in the diet with certain heart diseases. New breeds of cattle have been, and are still being, imported into the United States and are receiving a lot of attention because of their possible use in crossbreeding systems for fast and efficient gains and the production of high-quality beef with more lean and less fat. Yearling weight records in beef cattle have been moving upward in recent years, with a few instances in which bull calves weigh more than 1600 pounds at one year of age. Efficiency of gains in the feed lot also has been improving. Systematic crossbreeding systems in beef cattle are finding increasing use in the commercial production of market animals.

1.4   CHANGES NOW TAKING PLACE IN LIVESTOCK PRODUCTION

The past century has seen a great change in farming and in the livestock industry in the United States. Farming has developed from a small family operation to a large-scale, big-business operation. Specialization is now the keynote. The old village general store of a few years ago has given way to the large, specialized stores of today. The farmer has turned from general farming operations, in many instances, to specialization in certain crops or livestock. Furthermore, he has become more specialized as to the type of livestock he produces, whether they are sheep, dairy cattle, beef cattle, swine, or poultry. Or, to paraphrase the old saying, "farmers are tending to put more of their eggs in one basket."

Now that farming has become a big business, farmers are doing away with the small time system of mental bookkeeping. One cannot remember the milk production of an individual cow in the herd, even though some people think they can. Most likely to be remembered is that a particular cow produced a large pail of milk when she first freshened; but how quickly she slackened off in milk production is soon forgotten. The cow that gives an average amount of milk when she first freshens and maintains this level of production long into the lactation period is perhaps a better money-maker than one whose production starts at a high level and declines rapidly.

The livestock industry in America is now facing increased competition for the home markets from substitute and synthetic products. For instance, oleomargarine has supplanted butter on most dinner tables of the working class. Lard is almost unheard of around the kitchen, with the increasing popularity of vegetable shortenings. Vegetable oils are not only an excellent product, but they can be produced more cheaply than animal fats. Nylon, orlon, and other synthetic fibers have put the silkworm out of business and are crowding the sheep from the forefront as a producer of clothing fibers.

To meet this competition, farmers will have to use greater care in the selection and breeding of their livestock. They will have to keep good records on each individual and carefully study these records to continue to make a profit in a livestock enterprise. Farmers can no longer cull only those animals that are noticeably below average in performance. In the future it will become increasingly necessary to cull substandard animals in order to meet competition.

1.5 THE FUTURE OF LIVESTOCK PRODUCTION IN THE UNITED STATES

The production of livestock should continue to be as important to the national economy in the future as it has been in the past, but there are signs which point to the need for increased efficiency in the years to come. In addition, there are certain indications that the livestock producer will be forced to pay more attention to the quality of the product he offers for sale than he has in the past.

The population of the United States is increasing very rapidly although the birth rate is declining. If it continues to increase at the present rate, it may some day become so large that there will be a problem of producing enough food. This might cause competition between humans and farm animals for the yearly production of cereal grains. The numbers of swine and poultry would then decline, since these would be the chief animal competitors. Even if this problem should develop, more efficient production of livestock through improved methods of breeding, feeding, and management would help to solve it. Cattle and sheep would offer less competition for the grain production, for there will always be millions of acres of land that can be utilized only for grazing purposes.

One other fact points to a greater need for efficiency in livestock production. Most of the tillable land in the United States is now in production, except some areas where irrigation could be practiced if water were available. Furthermore, thousands of acres of fertile farming land are being taken out of production by the growth of cities and towns and by the expansion and enlargement of superhighways. Therefore, increases in the future food supply will have to be brought about by more efficient production and not by cultivating more new land.

Farmers should look to the future; they should not let temporary farm surpluses restrain progress in research. The development of new methods for the improvement of the efficiency of livestock production is a great challenge to research workers all over the world and should become increasingly important as time goes by.

REFERENCES

1. Craft, W. A. "Fifty Years of Progress in Swine Breeding," *JAS*, 17: 960, 1958.
2. Terrill, C. E. "Fifty Years of Progress in Sheep Breeding," *JAS*, 17: 944, 1958.

3. Warwick, E. J. "Fifty Years of Progress in Breeding Beef Cattle," *JAS*, 17: 922, 1958.

4. "After One Hundred Years," *U.S. Yearbook of Agriculture* (1962).

QUESTIONS AND PROBLEMS

1. Why were purebred registry associations formed?

2. What is kwashiorkor? In what countries does it appear with great frequency? Why?

3. Can a child have enough food to fill its stomach and still suffer from malnutrition? Explain.

4. Why are animal products important in the human diet?

5. Trace the history of animal breeding research and practices from the time the first animals were imported into the United States from abroad to the present time.

6. Which breed of livestock is best within a species such as beef cattle?

7. Discuss the importance of mating systems and selection in the effective production of livestock and animal products.

8. What changes have taken place in farming and livestock production within the past 20 years?

9. List several reasons why efficiency of production of livestock may become increasingly important in the future.

10. If the United States becomes overpopulated in the years to come, what class or classes of livestock may be decreased in numbers? Why?

11. Can livestock breeders take advantage of such specialized crosses as are used by the broiler industry? Explain.

2

Cells, Chromosomes, and Gametes

The *law of biogenesis* states that all living organisms come from other living organisms. This occurs through the process of reproduction in which the parents produce offspring and each of them transmits a sample one-half of their genes through the gametes to their offspring. It is the purpose of this chapter to discuss some of the basic materials and principles involved in the production of a new generation within the species.

2.1 THE CELL

The bodies of all animals are made up of microscopic "building blocks" called cells. The body contains millions of cells of different sizes and shapes. Most cells contain two major parts, the *cytoplasm* and the *nucleus*. The outer portion of the cell is the *cell membrane*, which serves as the framework and maintains the shape of the cell. Some cells, such as the red blood cells of mammals, lose their nucleus when they become specialized (differentiated) to perform a certain function. Other cells, such as the spermatozoa, lose much of their cytoplasm when they mature.

9

The *nucleus* is the oval-shaped body more or less near the center of the cell. The nucleus might be said to be the heart and brain of the cell because it carries the genetic material.

The material between the nucleus and the cell membrane is called the *cytoplasm*. Within the cytoplasm are many secretion products such as lipid droplets as well as others. The cytoplasm also contains certain highly specialized cellular elements called "organelles," which play important roles in cell functions. Some of these organelles within the cytoplasm that play an important part in cellular activity are the *Golgi apparatus, ribosomes, mitochondria,* and *lysosomes*. The Golgi apparatus is present in all cells and consists of one or more stacks of tiny flattened sacs known as saccules or cisternae. It was first observed by the Italian microscopist, Camillo Golgi, in the cytoplasm of nerve cells. For many years the function of the Golgi apparatus was a mystery, but recent research indicates that it is the primary site of the synthesis of large carbohydrate molecules which perform many vital functions within the body. The ribosomes are the organelles where the amino acids are assembled to form proteins. The mitochondria are the sites of chemical reactions that supply energy to the living cells. Lysosomes are small, usually spherical bodies that contain digestive enzymes which break down all of the major constituents of living organisms. Recent evidence suggests that they may play an important role in the occurrence of diseases such as cancer and gout.

2.2 THE CHROMOSOMES

Chromosomes are the threadlike bodies that can be seen within the nucleus of the cell when stained at the proper stages of cell division. One of the outstanding facts about chromosomes is that they are present in pairs in body cells and half pairs in the gametes. The members of each pair of chromosomes present in body cells might be called twins. Geneticists call these pairs *homologous chromosomes*, which is another way of saying that they are very similar in appearance (*homo* comes from the Greek word meaning equal or the same, and *logous* from the Greek word meaning proportion). Each body (somatic) cell normally contains one pair of chromosomes known as the *sex chromosomes* because they are involved in the determination of the sex of the individual. In mammals, the sex chromosomes are called the X and the Y, with the X chromosome being much larger and longer than the Y. The body cells of females are XX in composition, whereas those of males are XY. In poultry, the male is ZZ and the female is ZW. All chromosomes other than sex chromosomes are known as *autosomes*.

The central inner portion of each chromosome contains a long double helical structure called a deoxyribonucleic acid molecule (DNA for short).

This molecule resembles a ladder that has been twisted in opposite ways at the ends. The DNA molecule varies in length, depending upon the specific chromosome in which it is found.

New techniques for culturing cells in vitro have been developed within the past few years [3]. The principle used is to place certain cells such as lymphocytes (white blood cells) in a suitable culture medium which stimulates them to undergo further division. After a period of culture, colchicine is added to stop cell division at the metaphase stage. A balanced salt solution (Hank's) is then added, followed by one that is hypotonic (distilled water) to swell and disperse the chromosomes. The solution containing the cells is then placed on a glass slide, spread in a thin layer, dried, stained, and studied under the microscope. The stained slides usually show cells in the interphase stage (resting stage) and those that are mostly in the metaphase stage with the chromosomes clearly visible (Fig. 2.1). The chromosomes can then be counted or a photomicrograph can be taken to produce an enlarged picture so that the pairs of chromosomes can be arranged in a karyotype.

2.2.1 The Karyotype

Chromosomes vary in shape, size, and location of centromeres, both within and between species. The chromosome pairs, or homologous chromosomes, are alike in these respects. Much work has been done to characterize or group chromosomes according to their morphology. This grouping is called a karyotype.

Mitotic chromosomes in the metaphase can be characterized largely by their relative length and position of the centromere. In the metaphase each chromosome appears doubled (paired sister chromatids) and is united at the centromere (Fig. 2.2). The position of the centromere is used to classify these chromosomes. Chromosomes with a median centromere are called *metacentric* chromosomes. One with a submedian centromere is *submetacentric*. One with a terminal centromere is *telocentric*, whereas one with a nearly terminal centromere is *acrocentric*. The karyotypes for individuals of several species have been determined. These are shown for the most common domestic animals in Table 2.1.

2.2.2 Normal Chromosome Numbers

Each species of animals has a characteristic number of chromosomes, although in some cases two species may have the same number of chromosomes but may be distinguished by the shape of their chromosomes and the location of the centromere. Normal chromosomes occur in pairs in body cells (diploid number) and half pairs (haploid numbers) in the gametes. In the domestic horse all breeds and types appear to possess a diploid number

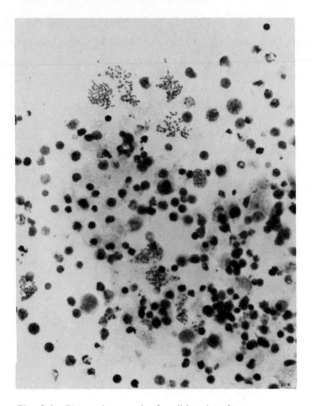

Fig. 2.1   Photomicrograph of a slide taken from a culture of lymphocytes showing cells at the interphase or inactive stage and many at the metaphase stage of cell division. At the metaphase the chromosomes appear doubled (sister chromatids) but are connected at the centromere.

Metacentric   Submetacentric   Acrocentric   Telocentric

Fig. 2.2   The doubling of the chromosomes (sister chromatids) at the metaphase of cell division. The points (dots) where the sister chromatids are attached are the centromeres. The point of attachment of the sister chromatids is a distinguishing feature of some chromosomes and helps identify certain pairs. The homologous pairs of chromosomes also differ in length.

of 64 chromosomes. However, the Mongolian wild horse (*Equus prezewal-skii*), which is assumed to be the ancestor of the domestic horse, has a diploid number of 66 chromosomes. This wild horse has two less metacentric and four more acrocentric chromosomes than the domestic horse, but hybrids between them are said to be fertile. Perhaps the two groups, domesticated and wild, have the same amount of chromosomal material occurring in pairs in somatic cells in spite of the difference in chromosomal numbers. Some scientists theorize that chromosome numbers can differ by the fusion of two acrocentric chromosomes to form a single larger metacentric chromosome [1]. It is also possible that two pairs of small acrocentric chromosomes may result from the separation of a large metacentric chromosome at the centromere.

The European wild boar (*Sus scrofa*) has one less pair of chromosomes than the domestic pig (*Sus scrofa*). It has been proposed that this difference in karyotypes exists because one submetacentric chromosome pair has divided in the wild pig to form two acrocentric pairs in the domestic pig [4].

The karyotype of European cattle (*Bos taurus*), Zebu (*Bos indicus*), and the American bison (*Bos bison*) are very similar in numbers and appearance, except that the Y chromosome is submetacentric in *Bos taurus* and acrocentric in *Bos indicus* [5] and *Bos bison* [2]. Fertile hybrids result from the cross of *Bos taurus* and *Bos indicus*, and no physiological and anatomical abnormalities are observed in the F_1 cross. However, crosses between *Bos taurus* and *Bos bison* usually result in sterile F_1 males, because the testicles are carried too close to the body and spermatogenesis cannot take place at this elevated temperature. The F_1 females lack mothering instinct, so many of their calves have to be raised on foster mothers. *Bos taurus* females mated with *Bos bison* males often abort the fetus because of an excessive production of amniotic fluid, indicating a mother-fetus incompatibility. Perhaps the same difficulties would result from a cross of the *Bos indicus* and *Bos bison*.

The mule and hinny are well-known hybrids from the reciprocal crosses of the donkey and the horse, but they are almost always sterile. This is assumed to be due to a dissimilarity of karyotypes of the two parent species with the horse possessing 64 chromosomes and the donkey 62. Many of the chromosomes are also different in morphology for the two parent species. The mule and hinny have 63 chromosomes in body cells, one of which does not have a homologous twin. These chromosomal differences in the mule and hinny are thought to result in improper synapsis during meiosis so that normal gametes are not formed.

2.2.3 Abnormal Chromosome Numbers

These are often called chromosomal aberrations (deviations from normal types). Since the development of simpler and more accurate methods of

Table 2.1 Chromosome numbers and types of Chromosomes as to the centromere attachment at the metaphase stage of cell division in some of the most common species of domestic animals

Species	Autosomes		Sex chromosomes		2N No.
	Metacentric, submetacentric, or subtelocentric	Acrocentric or telocentric	Metacentric or submetacentric	Acrocentric or telocentric	
Domestic horse (Equus caballus)	26	36	X	Y	64
Mongolian wild horse (Equus przewalskii)	24	40	X	Y	66
Persian wild ass, onager (Equus hemionus)	46	8	X	Y?	56
European wild boar (Sus scrofa)	26	8	X & Y		36
Domestic pig (Sus scrofa)	24	12	X & Y		38
Domestic cattle (Bos taurus)		58	X & Y		60

Species					
Domestic cattle, Zebu (Bos indicus)		58	X	Y	60
American bison (Bos bison)		58	X	Y	60
Domestic sheep (Ovis aries)	6	46	X & Y		54
Domestic dog (Canis familiaris)		76	X & Y		78
Domestic goat (Capra hircus)		58	Y	X?	60
Donkey (Equus asinus)	38	22	X	Y?	62
Domestic fowl* (Gallus domesticus)	32	22	Z	W?	78
Human		22	X & Y		46

*Some chromosomes are so small that it is difficult accurately to locate the centromere attachment in the metaphase. Note also that in the chicken the sex chromosomes are called Z (large) and W (small).

Attention is also called to the fact that not all reports agree on the point of attachment of the centromere.

determining chromosomal numbers and karyotypes in animals, this has been a very active field of research.

Chromosomal abnormalities include variations in numbers and structure as well as mixed populations of chromosomes in cells of the same individual [6]. Numerical chromosomal abnormalities occur because of errors in cell division (usually meiosis). These include polyploidy (euploidy) and aneuploidy. Body cells of mammals usually contain two sets ($2n$ or diploid) of chromosomes. A triploid individual would have three complete sets of chromosomes in body cells ($3n$ or triploid), whereas a tetraploid individual would possess four complete sets ($4n$ number). Aneuploidy is a chromosome abnormality in which only one or more chromosomes are duplicated or deleted. Thus, an aneuploid individual could have $2n + 1$ or $2n - 1$ chromosomes. Some individuals could also be $2n + 2$ or $2n - 2$ or even possess or lack a larger number of chromosomes. Extra chromosomes, however, usually are detrimental, whereas missing chromosomes usually cause death shortly after conception or early in embryonic life especially if the missing chromosome (or part of a chromosome) carries genes vital for life. Monosomics are also important; this refers to a condition where one member of a chromosome pair is missing, giving the $2n - 1$ number.

Structural chromosome abnormalities are of several types, including translocations, inversions, duplications, deletions, and isochromosomes. Translocations include those chromosome abnormalities in which there is an exchange of chromosome parts among chromosomes that are nonhomologous. Therefore, the chromosome that receives the new additional chromosomal material (usually at one end) carries more than its usual complement of genetic material. The chromosome that loses the translocated part possesses less than its normal amount of genetic material. This is known as a deletion. A duplication differs from a translocation in that there is an exchange of chromosome parts among two that are homologous so that one chromosome possesses duplicated chromosomal material or loci. Also, one of the homologous chromosomes involved in this exchange loses some of its genetic material. This is also called a deletion. Inversions include those abnormalities in which there is a rearrangement of the order of genes on the chromosome. Isochromosomes are those in which the sister chromatids at the metaphase do not split in a normal longitudinal manner giving two identical chromosomes. Instead, they divide at the centromere in a transverse fashion, giving two chromosomes each of which possesses duplicate arms but lacks the genetic information possessed by the chromosomes that split or divide in the normal longitudinal manner. This is shown in Fig. 2.3.

A mixture of cell types in the same individual results in types of chromosomal abnormalities known as *mosaicism* or *chimerism*. Mosaicism includes individuals derived from the same zygote which possess two or more distinct chromosome types, or genotypes. An example would be an individual with

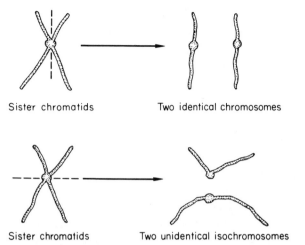

Sister chromatids   Two identical chromosomes

Sister chromatids   Two unidentical isochromosomes

Fig. 2.3 Isochromosomes are formed by a transverse separation of sister chromosomes rather than the normal longitudinal separation.

some body cells possessing some $2n$ and others $3n$ numbers of chromosomes. This may be distinguished from chimerism, which includes individuals with two or more chromosomal types from two or more zygotes. A good example would be the union of two separate fertilized eggs to form a single individual. This has been accomplished experimentally in the mouse.

2.2.4 Human Chromosomal Abnormalities

Chromosome abnormalities have been studied more completely in humans than in farm animals. As a result, most types of chromosomal abnormalities have been reported in the literature for humans. About twenty percent of aborted human fetuses have some type of chromosomal abnormality [7]. These include polyploidy and aneuploidy in addition to various structural abnormalities.

A few triploid humans have survived to birth, but these are mentally retarded and have certain congenital defects. All of these also possess some cells with the normal number of chromosomes as well as some which possess the triploid number, and therefore they are mosaics.

Many cases of trisomy and sex chromosomal abnormalities have been reported in humans after birth, some of which are also found in adults. Some type of chromosome abnormality occurs in about one out of each 240 children born. Autosomal abnormalities (trisomy) have been observed for many of the pairs of homologous chromosomes, but the one that has received the most attention is Downe's syndrome, or mongolism. This syndrome is due to an extra 21 chromosome (trisomic, or $2n + 1$), or to an extra portion of chromosome 21 due to a translocation. Klinefelter's syndrome involves the sex chromosomes and includes *XXY*, *XXXY*, or even *XXXXY* individuals. Such individuals are males but are mentally retarded and possess female characteristics. Turner's syndrome is found in women who have an *XO* sex chromosome composition. This means that she possesses only one *X* chromosome. Such individuals show webbing of the neck, low-set ears, wide-set eyes, underdeveloped breasts and sterility. It is thought that only one in 40 *XO* females survive to birth. No individuals of an *OY* chromosomal composition (having a *Y* but no *X* chromosome) have been observed, so this condition is probably lethal. This is understandable because the *X* chromosome is known to possess many genes necessary for the survival of the individual, whereas the *Y* chromosome may not possess so many vital genes. However, it probably carries some genes of importance. Men with an *XYY* chromosomal composition have been observed. They are usually tall, have a bad complexion, and in some cases are mentally retarded. Some reports indicate that more *XYY* individuals are found in mental institutions and institutions for the criminally insane than in the normal population.

2.2.5   Chromosome Abnormalities in Animals

Reports of chromosomal abnormalities in animals are not as numerous as those in humans. This is probably because chromosome numbers and abnormalities in animals have not been studied so intensively. Nevertheless, some abnormalities of interest have been reported in animals.

Polyploidy has been reported in swine by a French worker [8]. He reported triploidy in 21 percent of the ova recovered from sows mated 26 hours after estrus began. Probably all polyploid individuals die by mid-pregnancy. Polyploidy may result from the failure to form the haploid number of chromosomes (nondisjunction) in meiosis when gametes are formed. Thus, the gametes possess the diploid ($2n$) rather than the usual haploid ($1n$) number of chromosomes. Polyploidy may also result from the fertilization of the ovum by more than one sperm. This is called polyspermy, and in swine it often results when the ova are either immature or aged at the time of fertilization. Such ova apparently do not possess the ability to form the fertilization block that normally prevents the entrance of more than one spermatozoa into the cytoplasm. Failure to extrude the polar body in oogenesis may also cause polyploidy. Triploidy has been found to be associated with malformation syndromes in poultry (*Gallus domesticus*) and is also associated with a high embryonic death loss.

Aneuploidy has been reported in blastocysts of swine [9]. Many of the chromosomal karyotypes showed extra numbers of *X* and *Y* chromosomes, and some of these were already beginning to degenerate shortly after fertilization. It was estimated that about one-third of the ova loss in swine may be due to such chromosomal abnormalities. Occasionally pigs with chromosome abnormalities survive after birth, since intersex pigs have been reported with an abnormal sex chromosomal composition (XXY, etc.), giving symptoms of a syndrome similar to that of Klinefelter's syndrome in man.

A chromosomal abnormality involving a translocation and a deletion of chromosomal materials in swine has been reported [10]. A Swedish Landrace boar, when mated to 21 sows, sired litters that averaged 5.6 pigs as compared to 12.7 pigs per litter when the same were mated to other boars. Since the sire would transmit the deletion to one-half of his progeny, these progeny died during embryonic life, as the missing genes were vital to the life of the developing individual.

Many male pseudohermaphrodites have been observed in domestic animals. In the pig most intersexes appear to be genetic females with an XX sex chromosome composition, but they possess portions of the sex organs of the male. It has been postulated that this condition may be due to a portion of the Y chromosomes being translocated to the X chromosome. Another theory is that the X chromosome may contain certain male-determining genes that are homologous to those on the Y but are normally depressed when the chromosomal composition is XX. However, a mutation on the X chromosome may remove the depressing effect of the genes on the X chromosome, allowing the expression of the male-determining genes in such XX individuals. A third possible explanation is that male pseudohermaphrodites may be chimeric, possessing both XX and XY cells.

Livestock breeders have long been aware of the high probability of sterility in a heifer born twin with a bull calf. Romans in the first century B.C. called such a sterile cow a "taura" or a "female bull." Recently it has been discovered that sex chromosomal chimerism occurs in freemartin heifers and their male cotwins. That is, each twin, regardless of sex, contains both XX cells of the female and XY of the male. This is assumed to be due to a common blood supply of the twins during fetal life because they possess common fetal membranes. It has been postulated that testosterone, the male hormone, is produced by the male twin and inhibits the normal development of the female reproductive system early in embryonic life. More recently it has been postulated that the primary cause of freemartinism is the Y chromosome from the male in the blood of his female cotwin. A karyotype of chromosomes in both members of the twin pair will usually show a mixture of both male and female cells in either or both sexes. This is indicative of a common blood supply during fetal life and sterility in the female twin. The fertility of the male twin does not appear to be affected. The freemartin condition has also been observed in other species, but its occurrence appears to be infrequent.

The consumption of various drugs or other compounds (Lysergic acid diethylamide, or LSD, for example) has been reported to cause chromosome breakages and other abnormalities in lymphocytes of humans. Studies with mice indicate that the drug LSD can also cause changes of meiotic chromosomes in gametes. However, there is no complete agreement on the effects of this drug in some species. Much of the literature reporting induced chromosome breakage after LSD consumption was based on peripheral lymphocyte cultures, and these results may not be indicative of what has happened throughout the body but just in the bone marrow or blood cells.

A long list of compounds causing chromosome breakages in insects and small laboratory animals has been reported. Chromosome abnormalities have been reported in sheep with craniofacial malformations [12]. The condition results from the consumption of a poisonous range plant, *Veratrum californicum*, by the pregnant ewe on the fourteenth day of pregnancy. The main question concerning the effect of chromosomal abnormalities caused by the consumption of various compounds is whether or not such abnormalities also occur in the gametes and if they are transmitted to the next generation, where they may have an adverse effect.

Enough is now known about the occurrence of chromosomal abnormalities in animals and man clearly to show that they are related to certain congenital defects and reduced vigor in affected individuals. An increased knowledge of the cause and occurrence of chromosomal abnormalities may help scientists avoid some prebirth and birth losses in humans and domestic animals. It is also possible that scientists can discover something in the basic causes of such abnormalities that can be used to advantage in the improvement of the health and welfare of all living beings.

2.3 CELL DIVISION

Each living animal develops from a single cell, the fertilized egg. This single cell divides to form two cells, then these divide to form four, then eight, and so on. In the first divisions of the cells each mother cell and daughter cells are identical. Later the mother cells produce daughter cells that differentiate (or change) to form various tissues and organs of the body. The processes necessary for the production of the various body tissues and organs are not completely understood. They appear to be governed by a blueprint present in the nucleus of the fertilized egg. Although the mechanisms involved are not fully understood, they have been occurring over and over again with great accuracy for thousands of years. For example, a fertilized swine egg always produces a baby pig, not a calf. Although certain factors such as genetics and disease may cause a failure of production of some body tissues or parts, or may cause them to be defective, the new individual produced clearly belongs to the same species as its parents.

Two types of cell division are known. These are *mitosis*, responsible for the production of body cells, and *meiosis*, which is responsible for the production of gametes or sex cells.

2.3.1 Mitosis

Mitosis is a form of cell division in which the mother cells possess the diploid number of chromosomes (2*n*) and produce daughter cells possessing the same chromosome complement. Cells divide by mitosis mainly during embryonic growth and development. However, some cells divide in different stages of growth after birth, for the replacement of body cells such as in the skin and red blood cells, and in the repair of tissues when they are damaged. When cells are not dividing, they are in a stage called the interphase (*inter* means between). The different stages of cellular division include the prophase, metaphase, anaphase, and telophase.

The *prophase* is the first stage of cellular division. During this stage of division the cell appears to become more rigid and the tension of its surface membrane increases. As the prophase progresses, the chromosomes become visible and shorten and thicken. In the late prophase the nuclear membrane disappears and the spindle fibers form.

In the *metaphase* the chromosomes align on the equator of the cell (equatorial plate), but the chromosomes do not synapse as in meiosis. The spindle fibers make contact with the chromosomes at the centromere, as shown in Fig. 2.4. Each chromosome appears doubled with the duplicate parts (sister chromosomes) connected at the centromere. It is at this stage of cell division when each chromosome is doubled and is clearly visible after staining that photomicrographs are taken and the chromosomal karyotype is made in chromosome studies.

In the *anaphase* each doubled chromosome divides at the centromere with each half going to its respective mitotic center (centriole). Each of the two new daughter cells produced possesses each chromosome pair originally present in the nucleus of the mother cell.

The *telophase* is characterized by the chromosomes ending their movement to each mitotic center. The spindle fibers also disappear, the nuclear membrane reappears, the cytoplasm divides into two equal parts (cytokinesis) across the equatorial plate, and two daughter cells are produced that are identical with the original mother cell.

2.3.2 Meiosis

This is a type of cell division in which the number of chromosomes is reduced to the haploid number (one-half pairs) in the gametes, or sex cells. This reduction is necessary in order for the normal chromosomal number for the species to occur in the progeny. If the normal chromosomal number

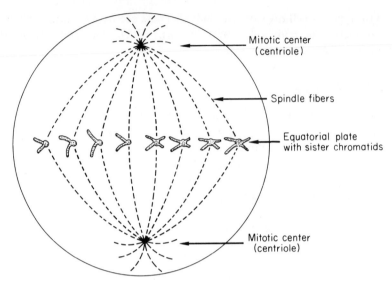

Mitotic center
(centriole)

Spindle fibers

Equatorial plate
with sister chromatids

Mitotic center
(centriole)

Fig. 2.4   The metaphase of cell division in mitosis.
Only four pairs of homologous chromosomes are shown
for purposes of illustration. Each chromosome at the
metaphase is doubled and then consists of sister
chromatids connected at the centromere.

were not maintained, the species would eventually become extinct. The halving of the chromosome pairs of body cells in the sex cells also maintains a certain amount of genetic variability because the chromosomes contributed by the two parents are seldom, if ever, exactly genetically alike. If they were, this would produce offspring that were homozygous for all pairs of genes.

The formation of gametes is known as *gametogenesis*. In the female it is called *oogenesis* (see Fig. 2.5) and in the male *spermatogenesis* (see Fig. 2.6).

In meiosis, the paired homologous chromosomes in the primitive sex cells *synapse* (pair up and come together) in the prophase of the first meiotic division. The chromosomes shorten and thicken, and each of them doubles forming a pair of sister chromosomes joined at the centromere. Since each original chromosome is doubled when they synapse, they appear as four chromosomes and are called *tetrads*.

Some time after the tetrads are formed, one chromatid may lie across the other at an angle. The point where they make contact is called a *chiasma*. At this time the chromosomes may exchange parts among themselves at the *chiasmata* in a phenomenon called "crossing over." The exchange of parts of the homologous chromosomes when crossing over occurs results in new chromosomes in the gametes that are genetically different from those of either parent chromosome in the original homologous pair.

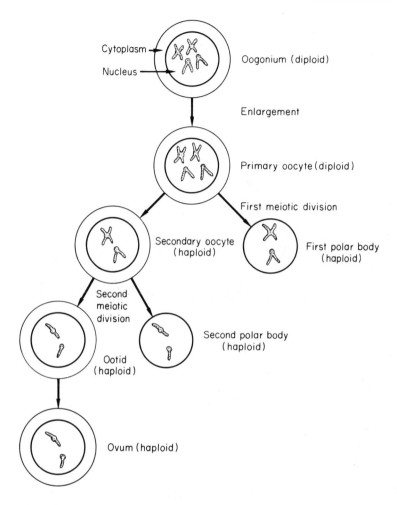

Fig. 2.5 Outline of oogenesis using only two pairs of homologous chromosomes. Note that the polar body has no cytoplasm.

In the metaphase of the first meiotic division, the homologous chromosomes line up at the equatorial plate of the cell much as they did in the metaphase of mitosis. The tetrads then separate into two parts called *dyads* (sister chromosomes connected at the centromere), and each dyad moves to the opposite pole. A division of the cell occurs at the equatorial plate, forming two different cells each with the haploid number (one-half pairs) of chromosomes. This is what is known as a reductional division. The new cells then go into a short period of inactivity or the interphase.

The second meiotic division begins after this short interphase. Cell divi-

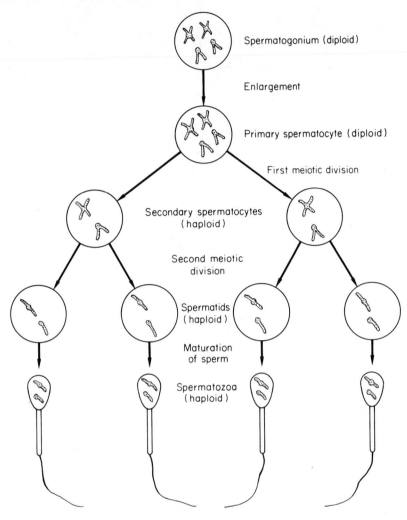

Fig. 2.6 Illustration of spermatogenesis using only two pairs of homologous chromosomes and using only the nucleus.

sion again goes through the prophase, metaphase, anaphase, and telophase similar to that described for mitosis. However, in the second meiotic division in the metaphase the two sister chromatids (dyads) split or separate into two identical chromosomes. In the anaphase they move to opposite poles (centrioles), and in the telophase the germ cells are formed that possess the haploid number (one-half pairs) of chromosomes.

2.4 SUMMARY

In this chapter we have emphasized that all life comes from preexisting life and that the cells are the building blocks of which the animal body is made. The various parts of the cell were discussed, but the major emphasis was placed upon the nucleus, which carries the genetic material. Each species has a characteristic number of chromosomes, which usually differs in number and shape from those of other species. Abnormal chromosome numbers (aberrations) sometimes occur, and when they do they cause the individual to be abnormal and may even cause its death.

Two types of cell division, mitosis and meiosis, take place in the animal body. Cell division by mitosis occurs in body cells, with each new daughter cell possessing the same number of chromosomes (diploid, or $2n$ number) as the original parent cell. Cell division by meiosis occurs in the ovary of the female and the testes of the male. Meiosis results in the reduction of chromosome numbers from homologous pairs to half pairs. Thus, when the sperm fertilizes the egg the homologous pairs found in body cells are restored. The maintenance of the normal number of chromosomes for the species is absolutely necessary for animal life to survive.

REFERENCES

1. Benirschke, K., N. Maloup, R. J. Low, and H. Heck. "Chromosome Complement: Differences Between *Equus caballus* and *Equus przewalskii*, Poliokoff," S, 148: 382, 1965.

2. Basrur, P. K., and Y. S. Moon. "Chromosomes of Cattle, Bison, and their Hybrid, the Cattalo," AJVR, 28: 1319, 1967.

3. Moorhead, P. S., P. C. Nowell, W. J. Mellman, D. M. Batteps, and D. A. Hungerford. "Chromosome Preparations of Leukocytes Cultured from Human Peripheral Blood," ECR, 20: 613, 1960.

4. Rory, J. M., V. G. Henry, G. H. Malsche, and R. L. Murphree. "The Cytogenetics of Swine in Tellico Wildlife Management Area Tennessee," JH, 59: 201, 1968.

5. Kieffer, N. M., and T. C. Cartwright. "Sex Chromosome Polymorphism in Domestic Cattle," JH, 59: 35, 1968.

6. Chu, E. H. Y. "Mammalian Chromosome Cytology," AZ, 3: 3, 1963.

7. Carr, D. H. "Chromosome Studies in Spontaneous Abortions," OG, 26: 308, 1965.

8. Thibault, C. "Analyse de la Fecondation de lóeuf de la triu Apres Accouplement au Insemination Artificielle," CORAI, INRA, Paris, 165, 1959.

9. McFeely, R. A. "Chromosomal Abnormalities in Early Embryos of the Pig," JRF, 13: 579, 1965.

10. Henricson, B., and L. Backstrom. "Translocation Heterozygosity in a Boar," H, 52: 166, 1964.

11. Ohno, S., J. M. Trujillo, C. Stenius, L. C. Christian, and R. L. Teplitz. "Possible Germ Cell Chimeras among Newborn Dizygotic Twin Calves (Bos taurus)," CG, 1: 258, 1962.

12. Henry, T. A., and T. H. Ingalls. "Teratogenesis of Craniofacial Malformations in Animals. IV. Chromosomal Anomalies Associated with Congenital Malformations of the Central Nervous System in Sheep," AEH, 13: 715, 1966.

STUDY QUESTIONS

1. What is the law of biogenesis?

2. Name the major parts of a body cell and give the function of each.

3. What is an "organelle"?

4. What is the main function of the Golgi apparatus? The mitochondria? The Ribosomes? The Lyosomes?

5. What are homologous chromosomes?

6. Distinguish between autosomes and sex chromosomes.

7. What is a karyotype?

8. Define the following: metacentric, submetacentric, telocentric, and acrocentric.

9. How does the chromosome complement of the wild horse differ from that of the domesticated horse? What is a possible explanation for this difference?

10. How do the chromosomes of the American bison (Bos bison) differ from those of domestic cattle (Bos taurus)?

11. What is a possible explanation for sterility in the mule and hinny?

12. What are the different kinds of numerical chromosomal abnormalities?

13. What are the different kinds of structural chromosomal abnormalities?

14. Define mosaicism and chimerism and distinguish between them.

15. What usually happens when a major part or an entire chromosome is missing from the karyotype? Why?

16. Distinguish between the following: Klinefelter's syndrome, Downe's syndrome, Turner's syndrome.

17. Explain why an individual of YO composition (possesses a Y but no X chromosome) has never been observed?

18. Why haven't as many chromosomal abormalities been reported for domestic animals as for humans?

19. It is possible that chromosomal abnormalities may be responsible for some embryonic death losses in domestic animals. Explain.

20. What is a hermaphrodite? Could this condition be due to a chromosomal abnormality? Explain.

21. What is a freemartin? What test could be conducted to determine if a freemartin heifer were sterile?

22. Can drugs and other compounds cause chromosome abnormalities when consumed by humans and animals? What is the possible importance of this?

23. What is mitosis? In what cells does it occur?

24. How does meiosis differ from mitosis?

25. What might happen if there were no provision in nature for keeping the number of chromosomes constant from generation to generation? Explain.

3

Genes—Their Function and Role in Animal Genetics

The use of refined techniques and more effective equipment has yielded excellent results in recent years in the study of the gene and has added a great deal to the knowledge of its nature and function. Much more is to be learned in the future. This phase of genetic research has reached the exciting state, and much effort is being spent in more intensive studies of the nature of the gene.

A knowledge of the gene and of some of its functions is necessary to obtain a good foundation in the principles of animal breeding. This chapter will discuss some of the fundamental concepts of the gene and how it functions.

3.1 NATURE OF THE GENE

The gene is the smallest biological unit of inheritance, and is a structural part of a chromosome. Hundreds and possibly thousands of genes are present in each chromosome, each in a fixed or special position called a *locus.* The existence of a particular gene can be determined genetically only because it exists in at least two alternative forms having different effects that can be observed readily in the organism. These forms are given the name *allelomorphs,* and they usually affect a trait in a contrasting or different manner.

Genes are so small that they cannot be seen with the ordinary type of microscope. However, special methods have been used recently to isolate a single gene in bacteria (*Escherichia coli*), and micrographs have been taken by means of an electron microscope.

The chemical composition of the gene has been studied indirectly, by the chemical analysis of chromosomes [2]. Chromosomes have also been studied by special staining techniques, by their absorption of ultraviolet light, and by digestion experiments with the enzyme, deoxyribonuclease, and with proteolytic enzymes. These studies have shown that chromosomes contain proteins and nucleic acids bound together in the form of nucleoproteins.

A gene may now be defined as a portion of a deoxyribonucleic acid (DNA) molecule. This molecule in animal cells is found in the nucleus and extends the length of the chromosome, more or less in its center. The DNA molecule is a long helical (spiral) structure [1] (Fig. 3.1), which resembles a long twisted ladder with the two sides [2] (strands) joined together by rungs, or steps. Each of the strands is called a polymer (*poly* meaning many; and *mer* meaning parts) because it is composed of many repeated units called *nucleotides*. A nucleotide consists of a nitrogenous base (either a purine or a pyrimidine) linked to a sugar. The sugars in the nucleotide sequences are linked together by a phosphoric acid molecule. The sugar in DNA is a 5-carbon sugar, deoxyribose. Thus, a nucleotide consists of a sugar, a base, and a phosphate (Fig. 3.1). The many nucleotides in the DNA molecule are joined by the linkage of the sugar in one nucleotide to the phosphate of another.

The two strands of the DNA molecule are joined by two bases linked together by hydrogen bonds. Only four different bases are found in the DNA molecule. These are adenine (A), thymine (T), guanine (G), and cytosine (C). Adenine always joins with thymine and guanine with cytosine. The average gene, sometimes called a *cistron*, probably consists of about 600 consecutive base pairs. Of course, some genes contain more and some fewer consecutive base pairs.

Another nucleic acid, ribonucleic acid (RNA), is found in both the nucleus and the cytoplasm of the cell. RNA differs from DNA by containing the sugar ribose instead of deoxyribose found in the DNA molecule. RNA also contains the base, uracil (U) instead of thymine (T) found in DNA. Thus, in RNA, adenine always joins with uracil, whereas in DNA it always joins with thymine.

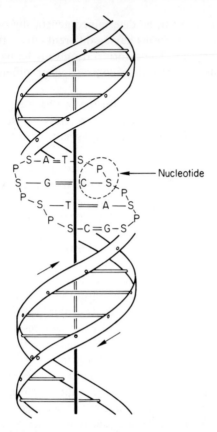

Nucleotide

Fig. 3.1 Diagrammatic representation of the
Watson-Crick DNA structure. P, phosphate; S, sugar;
A, adenine; T, thymine; G, guanine; C, cytosine. The
horizontal parallel lines symbolize hydrogen bonding
between complimentary bases. (From Missouri
Agricultural Experiment Station Research Bulletin 588.)

3.2 GENE FUNCTIONS

The gene has two major functions within the cell. One function is to reproduce
itself (duplicate chromosomes). A second is to produce RNA molecules,
which assist the ribosomes in the cytoplasm in the synthesis of proteins.

3.2.1 Gene Duplication

We have already noted that in cell division (both mitosis and meiosis)
chromosomes and genes are duplicated. This duplication is exact except

for new gene mutations and the "crossing over" among homologous chromosomes in the first meiotic division.

In duplication, the DNA molecule is thought to uncoil or straighten and the two strands to separate into single strands (Fig. 3.2). Each of the two

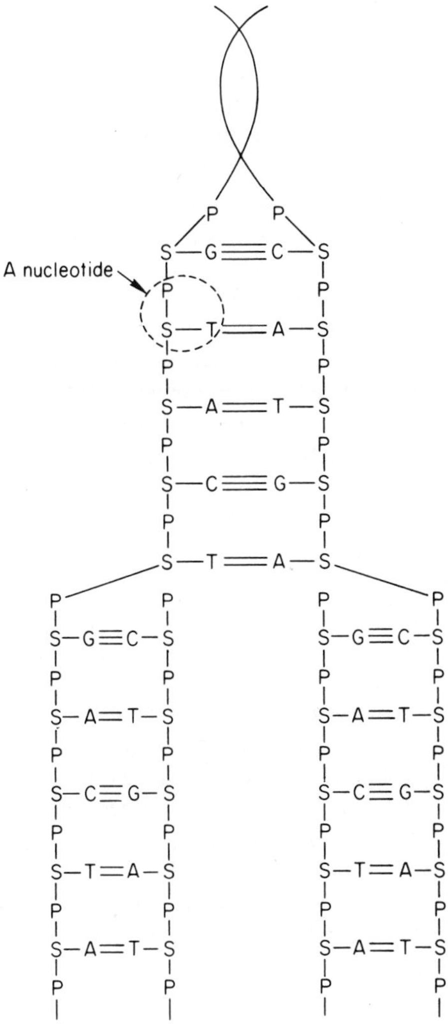

Fig. 3.2 Illustration of a double-stranded DNA molecule duplicating itself. The two strands split and assemble the various parts on each strand, resulting in the exact duplication of the molecule. P, phosphoric acid; S, sugar deoxyribose; A, base adenine; T, base thymine; G, base guanine; and C, base cytosine. Also note that a nucleotide containing P, S, and T is illustrated.

single strands then forms a new double-strand DNA molecule with the bases uniting in the A to T and the G to C combinations. This results in the production of two new double-stranded DNA molecules exactly like the original double-stranded molecule.

3.2.2 RNA Production

DNA also produces RNA molecules. In the production of RNA it is thought that the DNA molecule splits and a single strand of this molecule serves as a template (pattern, or mold) upon which a single strand of RNA is assembled with each base in the RNA strand being complementary to the bases on the DNA molecule strand. Thus, when the RNA strand is formed, each base on the RNA strand takes up its proper position on the DNA strand with the base combinations of A to U, U to A, C to G, and G to C (Fig. 3.3). The free ends of the bases in the RNA strand become connected in a new ribose, phosphate, base sequence. The single strand of RNA produced separates from the DNA template and passes into the cytoplasm.

Three forms of RNA are known. These are messenger RNA (mRNA), transfer RNA (tRNA), which is also sometimes referred to as soluble RNA (sRNA), and ribosomal RNA (rRNA). These three kinds of RNA molecules differ among themselves and perform separate functions in assembling proteins in the cytoplasm. It is likely that all three forms of RNA are produced in the nucleus by the method described.

3.2.3 Protein Production

The DNA molecule is responsible for determining the code sent by means of mRNA to the ribosomes to build proteins in most organisms. The only exception known is found in certain viruses, which use RNA directly as the genetic material. Proteins are made up of 20 different amino acids. Each protein differs from all others by the kind, number, and arrangement of amino acids in its molecule. Thus, thousands of different proteins can be built from 20 amino acids.

The single-stranded mRNA molecule transmits the code received from DNA in groups of three consecutive bases (Fig. 3.3) known as triplets, or codons, such as UUU, UUC, UUG, and so on. A total of 64 different codons is possible. Each amino acid is specified by a codon, but since there are many more codons (64) than amino acids (20) there is a potential surplus of codons. Therefore, the code contains some codons that specify the same amino acids (called degenerate codons) and others that do not code for any amino acid (called nonsense codons). Other codons may specify the beginning or ending of the chain of amino acids.

mRNA transmits the code received from DNA for a certain protein

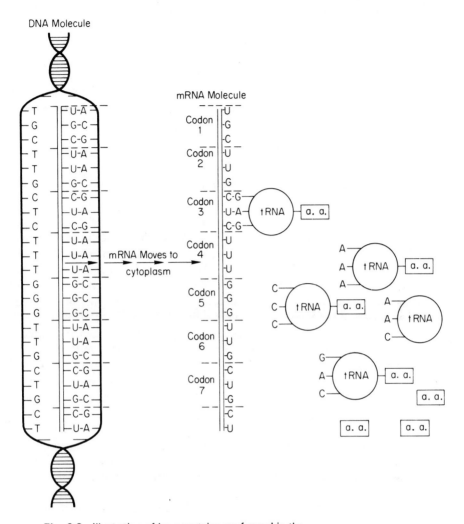

Fig. 3.3 Illustration of how proteins are formed in the cytoplasm of the cell. mRNA is built in the nucleus by DNA and carries the code for protein formation to the ribosomes in the cytoplasm. The code is sent by codons (three consecutive bases), each of which specifies a particular amino acid. tRNA identifies a particular amino acid by means of an anticodon corresponding to the codon of the mRNA. tRNA carries this amino acid to the ribosome, where proteins are built along the mRNA molecule. The various amino acids are bound together to form molecules in the ribosomes according to the code sent by means of mRNA. The kind of protein molecule built depends upon the kind, number, and arrangement of amino acids in the protein molecule.

to the ribosomes in the cytoplasm. tRNA identifies a particular amino acid by means of an anticodon corresponding to the codon in the mRNA molecule and transfers it to the ribosomes. Several kinds of tRNA are present in the cytoplasm and recognize certain amino acids, assembling them in the sequence specified by the gene, or DNA molecule. A change in the code sent by DNA causes a different protein to be formed in the ribosomes. Thus, it would be a new mutation.

3.3   CONTROL OF GENE FUNCTION

Each body cell that has a nucleus contains a sample of all of the genes the individual possesses. Thus, the genetic information in each cell of the many-celled organism is identical. It is apparent that not all of the many genes in the single cell are functional, many of them being in an inactive state at different times during embryonic development of the individual or later in life. Since the DNA molecule contains the primary genetic information and expresses it through the production of mRNA, it is thought that the regulation of gene function depends upon the ability to control the synthesis of mRNA. Most of the basic information has been obtained in studies with microorganisms and might not apply in strictly the same manner in the function of genetic material in higher forms of animal life. Further studies may show differences in gene function in microorganisms and many-celled organisms [11].

The most widely accepted theory of how genes function in microorganisms proposes that there are two principle kinds of genes [5]: the *structural* gene, which is responsible for the synthesis of various proteins, and *control* genes (regulator and operator genes), which regulate the activity of the structural genes but are not directly involved in protein synthesis. They control or regulate indirectly through the structural genes the amount of protein that is produced, however.

The operator gene is thought to be located on the same chromosome as the structural genes and is adjacent to them. Each operator gene may control the activity of several structural genes, having the ability to turn them on and off like a switch. The *regulator* gene is thought to be located on a different chromosome from the structural gene but in some cases may be located on the same chromosome. The regulator gene is thought to produce a substance called a repressor (protein) which blocks the action of the operator gene (Fig. 3.4), not allowing the structural genes to produce proteins because they cannot produce mRNA. An *inducer* substance (also sometimes called a depressor), usually a substrate or hormone, converts the repressor substance into an inactive compound, which in turn allows the operator gene to function and the structural genes to produce mRNA.

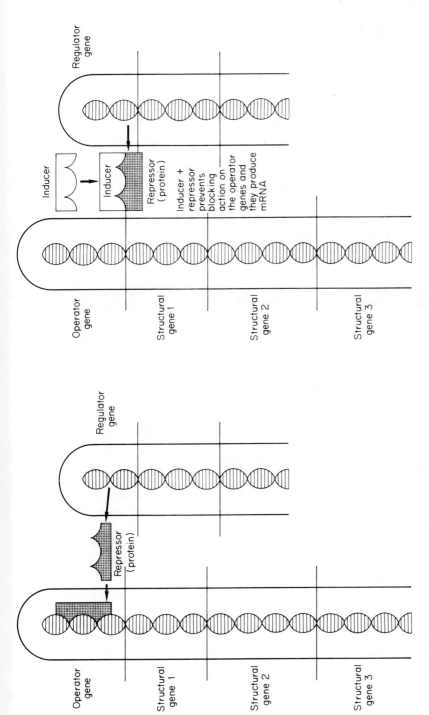

Fig. 3.4 The interaction of structural, operator, and regulator genes in the transcription of genetic information in a single cell (microorganisms).

Collections of adjacent nucleotides (portion of DNA molecule) that code for single mRNA molecules and are under the control of a single repressor substance are called *operons*. An operon may contain one, two, or even several genes.

This theoretical explanation of the control of gene function is of interest in higher animals because of the future potential of switching gene activity on and off when desired. It may be a long time before the ability to do this is realized, but it is already known that various chemical substances such as drugs and hormones and some viruses apparently block proper gene function and cell division in the developing embryo, resulting in defects in body parts and organs in the newborn.

3.4 EXAMPLES OF HOW GENES FUNCTION

Gene action can be determined only if it affects some trait in the individual. Most of the knowledge of gene action has come through studies of genetic defects, resulting from mutations. In some cases, the mutations have been caused by exposure of the individuals to X-rays, but in most cases they have been caused by some, as yet unidentified, factors in the external or internal environment. These studies have shown that the function of genes is chemical in nature.

One of the first known cases of the genetic control of a specific chemical reaction was found in the rare metabolic disease in man called alcaptonuria. This disease is characterized by the hardening and blackening of the cartilage of the bones and the blackening of the urine when exposed to the air. The black urine is caused by an accumulation of homogentisic acid. In the normal person, the enzyme is present that is responsible for the change of homogentisic acid to aceto-acetic acid, which is clear in the urine. The person with alcaptonuria lacks this enzyme, so homogentisic acid accumulates in abnormal amounts in the urine.

Phenylketonuria in humans is another example of insufficiency of a particular enzyme. In this disease, phenylalanine hydroxylase, necessary for the normal metabolism of phenylalanine, is lacking. Abnormal metabolites accumulate in the tissues, and most affected individuals are idiots or imbeciles unless phenylalanine is excluded from the diet.

Albinism is a condition caused by an enzyme deficiency of genetic origin. Albinos lack the pigment melanin in the hair, skin, and eyes. Melanin is probably formed from tyrosine through the action of tyrosinase. When tyrosinase is not active, no pigment is formed in the individual, and it is an albino.

Another group of genetic defects in humans appears to involve the production of abnormal forms or the failure to produce a certain protein. Agammaglobulinemia seems to fit into this group, and has been described in the human population [3]. This term refers to the failure of production of gamma globulin in the body and and its lack in the blood. A newborn baby receives a supply of gamma globulins from its mother before birth, but this supply gradually decreases to near zero at four months of age. Normally, the baby's own gamma globulin production begins at about three weeks of age and reaches a high level by five to eight months. The lack of gamma globulins results in increased susceptibility to bacterial infections because of the lack of resistance from antibodies in the blood.

Table 3.1 How genes function as illustrated by some reported genetic abnormalities in humans and animals

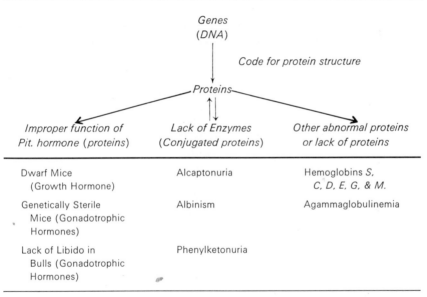

Improper function of Pit. hormone (*proteins*)	Lack of Enzymes (*Conjugated proteins*)	Other abnormal proteins or lack of proteins
Dwarf Mice (Growth Hormone)	Alcaptonuria	Hemoglobins *S, C, D, E, G, & M.*
Genetically Sterile Mice (Gonadotrophic Hormones)	Albinism	Agammaglobulinemia
Lack of Libido in Bulls (Gonadotrophic Hormones)	Phenylketonuria	

Note: Many other examples are known that could be added to this list, but these show how genes act.

A second example of genes in man that appear to function in the control of the production of a particular protein is that revealed by the gene defect causing sickle-cell anemia [9]. In the individual having this disease, the red blood cells have the shape of a sickle instead of the normal round shape. This disease is confined largely to populations in central Africa, southern India, and a region of Greece, or descendants of these peoples. In normal humans, only hemoglobin A is found in the red blood cells. The sickle-cell gene apparently results in the production of an abnormal form of hemoglobin called hemoglobin S.

Individuals of genotype (actual genetic makeup) H_aH_a produce hemoglobin A and no hemoglobin S. Those of genotype H_aH_s produce both hemoglobin A and hemoglobin S, and both are detectable in the blood with neither gene affecting the expression of the other. Individuals of this genetic makeup seldom have anemia, and when they do it is slight. When the cells are deprived of oxygen, however, they may assume the sickle rather than the normal shape. This genetic makeup is usually referred to as the sickle-cell trait. Individuals of genotype H_sH_s produce mostly hemoglobin S and usually die of anemia. Studies of the hemoglobin in the blood of normal and affected individuals show that the two hemoglobins differ slightly in their electrical charge and can be separated by electrophoresis and paper chromotography.

Individuals of the H_sH_s genetic makeup seldom live to reproduce, unless given frequent blood transfusions and special medical care. Thus, there is a strong natural selection against the H_s gene. However, in certain tribes in Africa it has been found that as high as forty percent of the individuals in the tribe were of genotype H_aH_s or showed the sickle-cell trait. This was very difficult to understand, since H_sH_s individuals seldom, if ever, lived to reproduce. Later work showed that the H_aH_s individuals were more resistant to malaria than the normal H_aH_a individuals, and thus natural selection favored the H_aH_s individuals, whereas both those which were normal or had sickle-cell anemia died at a more rapid rate.

More recently, sickle-cell anemia has been used as a very important means of studying the mode of action of genes [4]. By a very detailed separation and chemical analysis of the amino acids in the hemoglobin of normal and sickle-cell individuals, it was found that the two hemoglobins contained the same types of amino acids but differed in the amounts of one of the polypeptides. The peptide from the normal persons contained two glutamic acid units and one valine, whereas the abnormal peptide contained one glutamic acid unit and two valines. Thus, valine replaced one glutamic acid unit in the abnormal hemoglobin. The importance of this study was that a mutation caused an abnormal protein to be produced. This gives added support to the theory that a gene carries the code which controls the construction of the protein molecule for which it is responsible.

Evidence is also accumulating that hormone production by the pituitary gland, or the action of these hormones, may be under genetic control. If true, this would fit in very well with the theory that mutations result in the production of defective proteins. Hormones from the anterior pituitary are protein in nature.

Dwarfism in mice has been shown to be a genetic defect [8] resulting in a lack of growth hormone secreted by the anterior pituitary gland, which stimulates body growth.

Sterility in an obese strain of mice has been reported to be due to a genetic defect [10]. The ovaries of the females seem to have normal reproductive capacity but remain immature, either because of an inadequate release of gonadotrophic hormones from the pituitary or the inactivation of the hormones after their release. Ovulation, gestation, parturition, and lactation have been induced in these mice after a series of treatments with the various hormones. An inherited defect in bulls causing lack of libido (sex drive) has been attributed to the failure of the anterior pituitary gland to secrete adequate amounts of a particular hormone or to the inactivation of the hormone [7].

Further study of such inherited defects should be quite helpful in determining how genes function in growth and reproduction in all species, including farm animals, and could lead to improved methods of more efficient and economical livestock production.

3.5 SEGREGATION AND RECOMBINATION OF GENES

The survival of each mammalian species depends upon the maintenance of the normal diploid ($2n$) number of chromosomes in body cells, as shown in Chapter 2. Since genes are carried on the chromosomes, the maintenance of the normal complement of gene pairs is also necessary. Nature has provided for the maintenance of gene pairs and chromosome pairs in body cells through their reduction to one-half pairs in the gametes in the process of meiosis. The normal paired number of genes and chromosomes is restored when the sperm and egg unite in fertilization to initiate the development of a new individual.

3.5.1 Segregation of Genes in the Gametes

All of the inheritance an individual possesses comes from its two parents. Each parent contributes approximately one-half of this inheritance. An exception to this is in poultry, where it has been found that some turkeys develop from unfertilized eggs. They, therefore, have a mother but no father. In higher animals, fatherless progeny are a rare exception to the rule that each parent contributes approximately equally to the inheritance of each offspring. No cases of fatherless individuals in farm animals and humans have been verified.

Each individual possesses thousands of pairs of genes in its body cells.

Some of the genes in these pairs are *homozygous*, which means that within each pair the genes are identical. Other pairs of genes are *heterozygous*. This means that within each pair the genes are different (*hetero* means unequal). Because gene pairs in the individual's cells are often unlike, or heterozygous, this means that the many gametes produced by the same individual do not always carry identical genes. Even though they are not identical, genes of these unlike pairs do follow certain laws of probability in segregation in the gametes.

One pair of genes.   To illustrate how genes of each pair segregate in the gametes according to the laws of probability, we will first use one pair of genes and will assume that no new mutation occurs. We will use genes *P* for polled and *p* for horns. We are interested here in showing how they segregate in the gametes and are not interested in how they express themselves phenotypically. With two genes (alleles) we can have three different combinations of genes (genotypes): *PP*, *Pp*, and *pp*.

The probability that an individual of genotype *PP* will transmit the *P* gene in a gamete is one, because this is the only gene it possesses in that pair. The probability that it will transmit gene *p* in a gamete is zero, because it does not possess that gene. Similarly, the probability of an individual of genotype *pp*'s transmitting gene *p* in a gamete is one, and of transmitting gene *P* is zero. The probability that an individual of genotype *Pp* will transmit gene *P* in a gamete is one-half, and the probability of transmitting gene *p* is also one-half, because the gene pair contains both of these genes. The law of probability is followed in this manner for all gene pairs, just as we have illustrated with genes *P* and *p*.

Two pairs of genes. Let us now illustrate how the law of probabilities applies to the segregation of two pairs of genes in the gametes, using two different traits in cattle, horns and coat color. The black gene is *B* and the red gene is *b*; the polled gene is *P* and the horned gene is *p*, as in the previous example. The genes for coat color are carried on one pair of homologous chromosomes, and the genes for horn conditions are carried on another pair. First, let us use as an example a bull that is pure for both polled and black color or of genotype *PPBB*. If we consider each pair of genes separately, we see that the probability of a sperm from this bull carrying *P* is one and the probability of its carrying *p* is zero. The probability of that same sperm's carrying *B* is also one and for its carrying *b* is also zero. What is the probability that such a bull will produce a sperm carrying both *P* and *B*? The answer is one, because the probability of both of these genes occurring together in the same sperm (two independent events) is the product of the two independent probabilities, which is 1 times 1, or 1. The probability of a sperm from such a bull carrying both *p* and *b* genes, of course, is 0 times 0, or 0.

Probabilities of the various combinations of *P*, *p*, *B*, *b* occurring in the sperm of a bull that is heterozygous (*PpBb*) is handled similarly and would be

Gene	Probability of a sperm carrying this gene
P	$\frac{1}{2}$
p	$\frac{1}{2}$
B	$\frac{1}{2}$
b	$\frac{1}{2}$

The probability of various combinations of the two different pairs of alleles' occurring together would be

Possible combinations of genes in the sperm	Probability of a sperm carrying these two genes
PB	$\frac{1}{2} \times \frac{1}{2}$ or $\frac{1}{4}$
Pb	$\frac{1}{2} \times \frac{1}{2}$ or $\frac{1}{4}$
pB	$\frac{1}{2} \times \frac{1}{2}$ or $\frac{1}{4}$
pb	$\frac{1}{2} \times \frac{1}{2}$ or $\frac{1}{4}$

3.5.2 Recombination of Genes in the Zygotes

The concept of probability and the combining of probabilities can now be expanded to the union of genes in the zygote (fertilized egg). The examples used, of course, assume that the different pairs of alleles assort and recombine independently. First, let us use one pair of alleles for polled and horned cattle. The probability that gametes from parents of the three genotypes would carry one of each allele would be

	Genotype of parent		
	PP	*pp*	*Pp*
Probability of *P* in a gamete	1	0	$\frac{1}{2}$
Probability of *p* in a gamete	0	1	$\frac{1}{2}$

Now we can calculate the probability of various combinations of gametes occurring in the offspring of parents that are both heterozygous (*Pp*).

Genotype of offspring	Probability of this genotype
PP	$\frac{1}{2} \times \frac{1}{2}$ or $\frac{1}{4}$
Pp	$\frac{1}{2} \times \frac{1}{2}$ or $\frac{1}{4}$
pP	$\frac{1}{2} \times \frac{1}{2}$ or $\frac{1}{4}$
pp	$\frac{1}{2} \times \frac{1}{2}$ or $\frac{1}{4}$

$\left. \begin{array}{l} \\ \end{array} \right\} \frac{1}{2}$

This corresponds to the 1:2:1 genotypic ratio in the offspring of crosses of heterozygous individuals.

The following example illustrates the probability of various combinations of genes in the gametes when two separate pairs of genes are involved and the parents are heterozygous for both pairs of genes:

	Genotypes of:	
	Sire	Dam
	$PpBb$	$PpBb$
Probability of PB in gamete	$\frac{1}{4}$	$\frac{1}{4}$
Probability of Pb in gamete	$\frac{1}{4}$	$\frac{1}{4}$
Probability of pB in gamete	$\frac{1}{4}$	$\frac{1}{4}$
Probability of pb in gamete	$\frac{1}{4}$	$\frac{1}{4}$

The probabilities of different combinations of the genes in the offspring of such parents would be

Genotype of offspring	Gamete from: Sire	Gamete from: Dam	Probability of gamete from: Sire		Dam	Probability of genotype
$PPBB$	PB	PB	$\frac{1}{4}$	$\times$	$\frac{1}{4}$	$\frac{1}{16}$
$PPBb$	PB	Pb	$\frac{1}{4}$	$\times$	$\frac{1}{4}$	$\left.\begin{array}{c}\frac{1}{16}\\ \frac{1}{16}\end{array}\right\}\frac{2}{16}$
	Pb	PB	$\frac{1}{4}$	$\times$	$\frac{1}{4}$	
$PPbb$	Pb	Pb	$\frac{1}{4}$	$\times$	$\frac{1}{4}$	$\frac{1}{16}$
$PpBB$	PB	pB	$\frac{1}{4}$	$\times$	$\frac{1}{4}$	$\left.\begin{array}{c}\frac{1}{16}\\ \frac{1}{16}\end{array}\right\}\frac{2}{16}$
	pB	PB	$\frac{1}{4}$	$\times$	$\frac{1}{4}$	
$PpBb$	PB	pb	$\frac{1}{4}$	$\times$	$\frac{1}{4}$	$\left.\begin{array}{c}\frac{1}{16}\\ \frac{1}{16}\\ \frac{1}{16}\\ \frac{1}{16}\end{array}\right\}\frac{4}{16}$
	Pb	pB	$\frac{1}{4}$	$\times$	$\frac{1}{4}$	
	pB	Pb	$\frac{1}{4}$	$\times$	$\frac{1}{4}$	
	pb	PB	$\frac{1}{4}$	$\times$	$\frac{1}{4}$	
$Ppbb$	Pb	pb	$\frac{1}{4}$	$\times$	$\frac{1}{4}$	$\left.\begin{array}{c}\frac{1}{16}\\ \frac{1}{16}\end{array}\right\}\frac{2}{16}$
	pb	Pb	$\frac{1}{4}$	$\times$	$\frac{1}{4}$	
$ppBB$	pB	pB	$\frac{1}{4}$	$\times$	$\frac{1}{4}$	$\frac{1}{16}$
$ppBb$	pB	pb	$\frac{1}{4}$	$\times$	$\frac{1}{4}$	$\left.\begin{array}{c}\frac{1}{16}\\ \frac{1}{16}\end{array}\right\}\frac{2}{16}$
	pb	pB	$\frac{1}{4}$	$\times$	$\frac{1}{4}$	
$ppbb$	pb	pb	$\frac{1}{4}$	$\times$	$\frac{1}{4}$	$\frac{1}{16}$

3.6 SUMMARY

In this chapter it has been shown that the gene is a portion of a DNA molecule, with a gene consisting on the average of about 600 consecutive base pairs. The gene has the ability to exactly reproduce itself when new cells are formed, and it sends the code to the ribosomes in the cytoplasm for the production of specific proteins. The code message is transmitted by means of mRNA. A change in the code sent by the gene is known as a mutation.

The normal function of genes in the expression of various traits is determined when mutations occur many of which may cause genetic defects. Many genetic defects in humans and animals are known which show that genes are involved in the production of certain hormones and blood antigens, as well as other protein substances in the body.

Although each body cell with a nucleus contains a sample of all the genes that the organism possesses, the expression of individual genes is altered to permit the specific functions of that tissue. Some genes function in the production and activity of some cellular tissues, whereas others function in another. The present theory of the control of gene activity in microorganisms that seems most acceptable involves structural genes and control genes (operator and regulator) which are responsible for the activity or inactivity of structural genes. The control of the activity of structural genes depends upon the production of substances that block or allow the production of mRNA. It may be more complex than this in multicellular animals.

A thorough understanding of the control of gene activity would be of great value in the future in improving the health and welfare of mankind and the efficient production of livestock.

Members of a gene pair in body cells segregate (separate) into gametes independently of each other and members of other gene pairs according to laws of probability, provided that the different pairs of genes are carried on different pairs of homologous chromosomes. These same genes carried by the gametes also recombine in the zygote according to laws of probability when fertilization occurs.

REFERENCES

1. Beadle, G. W. "Gene Structure and Gene Function," MoAESRP, 588, 1952.
2. Crick, F. H. C. "Nucleic Acids," SA, 197: 188, 1957.
3. Gitlin, D., and C. E. Janeway. "Agammaglobulinemia," SA, 197: 93, 1957.
4. Ingram, V. M. "How Do Genes Act?" SA, 198: 68, 1958.

5. Jacob, F. and J. Monod. "Genetic Regulatory Mechanism in the Synthesis of Proteins," JMB, 3: 318, 1961.

6. Knox, E. "Hereditary Molecular Diseases (Inborn Errors)," BRNR, 20: 23, 1959.

7. Lagerlof, N. "Hereditary Forms of Sterility in Swedish Cattle Breeds," FS, 2: 230, 1951.

8. Mirand, E. A., and C. M. Osborn, "Insulin Sensitivity in the Hereditary Hypopituitary Dwarf Mouse," PSEBM, 82: 746, 1953.

9. Pauling, L., H. A. Itano, S. J. Singer, and I. C. Wells. "Sickle-cell Anemia, a Molecular Disease," S, 110: 543, 1949.

10. Smithberg, M., and M. N. Runner. "Pregnancy Induced in Genetically Sterile Mice," JH, 48: 97, 1957.

11. Tompkins, G. M., et al. "Control of Specific Gene Expression in Higher Organisms," S, 166: 1474, 1969.

STUDY QUESTIONS

1. What is a gene? A locus? What are allelomorphs?

2. Describe a DNA molecule. What is a nucleotide?

3. What four bases are found in the DNA molecule, and which ones always are found together in pairs?

4. What is RNA and how does it differ from DNA?

5. What are the main functions of genes?

6. Explain the details involved in the duplication of genes or chromosomes.

7. What kinds of RNA are found in cells? Where are they produced?

8. How can only 20 different amino acids form thousands of different kinds of proteins?

9. What are codons?

10. Explain the basic principles as to how genes function.

11. Define: regulator gene, operator gene, structural gene, operon, inducer substance, and a repressor substance.

12. Give examples of genetic defects that illustrate what genes do when functioning normally.

13. Why is it important to know how genes function?

14. Describe how genes segregate in the gametes according to the law of chance.

15. Do two different pairs of genes carried on the same homologous chromosomes segregate independently? Explain.

16. What conditions are necessary for two or more pairs of genes to recombine in the zygote according to the laws of probability?

17. Assume that an individual is heterozygous for four pairs of genes with each pair of genes carried on different pairs of chromosomes. If the individual is of genotype *AaBbCcDd*, what is the probability that it will produce gametes carrying *ABCD*?

18. What is the probability that the individual in question 17 will produce a gamete carrying *AaBCD*? Why?

19. Assume that two individuals are of genotype *AaBbCcDd* with each pair of genes (alleles) carried on a different chromosome. What is the probability that they will produce an *AABBCCDD* offspring? An *aabbccdd* offspring?

20. Which individual is more likely to breed true, one of genotype *AABBCCDD, AaBbCcDd*, or *aabbccdd*? Will the *aabbccdd* individual always breed true? Why?

4

Phenotypic Expression
of Genes

In Chapter 3 we discussed the segregation of genes in the gametes and their recombination in the zygote without reference to how they express themselves phenotypically, alone or in different combinations. All pairs of genes segregate and recombine in the way described as long as they are carried on different pairs of homologous chromosomes. We shall now discuss the different ways genes can express themselves phenotypically. By phenotype is meant the expression of genes in such a way that we can measure this expression by our senses.

The phenotypic expression of genes may be divided into two general types: nonadditive and additive. We shall discuss each type separately. A knowledge of the different ways in which genes can express themselves phenotypically is important in order to devise effective selection and mating systems for livestock breeding.

4.1 NONADDITIVE GENE EXPRESSION

Nonadditive gene action refers to the nonlinear expression of genes. This means that the addition of a gene to the genotype does not add an equal amount to the phenotype. For example, where dominance is complete, the heterozygote or *Aa* is not midway between genotypes *AA* and *aa*, but it is approximately the same as *AA*. This would be a nonlinear phenotypic expression. If genotype *Aa* were midway between *AA* and *aa*, this would be linear for this one pair of genes.

Several different kinds of nonadditive gene action are known. These will be discussed separately.

4.1.1 Dominance and Recessiveness

We shall again use the condition of horns in cattle to describe this type of nonadditive gene action. The symbol *P* will be used to designate the gene for absence of horns and *p* to represent the gene for presence of horns. If a pure polled bull (*PP*) is mated to a group of horned cows (*pp*), what kinds of calves will be produced? The calves receive one-half of their genes from the bull and one-half from the cow. It can be seen from the following diagram, which shows a single mating, that the calves will receive the *P* gene from the sire and the *p* gene from the dam, for the sire carries only genes for polledness, and the dam carries only genes for horns.

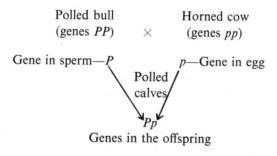

Polled bull   Horned cow
(genes *PP*)   ×   (genes *pp*)

Gene in sperm—*P*   *p*—Gene in egg

Polled calves

Pp

Genes in the offspring

The polled gene *P* is said to be *dominant*. Therefore, the calves will all be polled even though they also carry the gene for horns. The ability of the polled gene to block out or cover the expression of the horned gene is called *dominance*. The gene for horns, then, is said to be *recessive* to the polled gene, because it takes a back seat, so to speak, in its presence. When two genes affecting a character in different manners occur together in an individual, and occupy identical locations on each member of a pair of homologous chromosomes, this individual is said to be *heterozygous* (in this example, the calves are *Pp*). For this reason, all of the calves from the mating of a pure polled bull to pure horned cows will be heterozygous; that is, each will carry one gene for horns and one for polled. These two genes in the heterozygous animal are called *alleles* (allelomorphs) of one another, because they both affect the same characteristic in the individual but in different manners. When the two genes affect the character in the same manner, as is the case in the horned cows (*pp*) and the polled bull (*PP*), the animal is said to be *homozygous*. This genetic construction is called the *genotype*. The expression of the genotype is called the *phenotype*.

What would the offspring be like if two of these heterozygous polled (*Pp*) individuals were mated? If we remember that genes occur in pairs in the animal's body cells and that one or the other of these genes of each pair is present in the sperm or egg, we can see that each of these polled individuals, which are heterozygous, can produce two kinds of sex cells. One sperm or egg will carry the gene *P* and the other the gene *p*. By a diagram that combines the genes from the eggs and the sperm in all possible ways, it can be seen that one-half of the offspring have the same genotype as that of the parents, one-fourth have the same genotype as that of the polled grandparent, and the other one-fourth have the same genotype as that of the horned grandparent.

Polled bull		Polled cow
(genes *Pp*)	×	(genes *Pp*)
Genes in sperm		Genes in egg
P or *p*		*P* or *p*

Possible combinations of sperm and eggs:

1 *PP*   Pure for polled (homozygous)
2 *Pp*   Polled (heterozygous)
1 *pp*   Pure for horns (homozygous)

The same type of diagram can be used to demonstrate gene distribution for all traits where only one pair of genes is involved and where dominance is complete. This also assumes that the penetrance of the gene for polledness is 100 percent and that its expression is constant.

It will be observed that three genotypes appear, in a definite ratio: 1 *PP*: 2 *Pp*: 1*pp*. This, then, is called a *genotypic ratio*. Two of the genotypes give rise to the same phenotype, however. And the distribution of phenotypes in the offspring is called the *phenotypic ratio*. All the calves carrying *Pp* and *PP* are polled, while only the one carrying *pp* has horns. From the above diagram, it can be seen that the phenotypic ratio in the offspring of this mating would be 3 polled: 1 horned.

The *PP* and the *pp* individuals are said to breed true, or to be pure, since they are homozygous and can pass only one kind of gene to their offspring. The *Pp* individuals, however, never breed true; that is, they do not

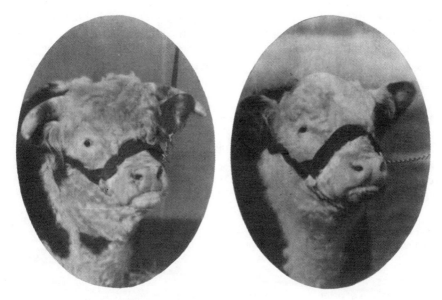

Fig. 4.1 The presence or absence of horns in cattle is determined by genes with the absence of horns dominant to the horned condition. (Courtesy of the American Hereford Association.)

have 100 percent *Pp* offspring, because they are heterozygous and pass two kinds of genes through their sex cells to their offspring.

From the foregoing discussion, we can state some facts that must be remembered clearly for a better understanding of the inheritance of a trait where only one pair of genes is involved. These are: (1) Genes occur in pairs in the body cells of the individual. (2) One of each pair came from the father and one from the mother. (3) The genes of a pair separate during formation of sex cells. (4) On fertilization, the genes are restored to the paired condition.

The law of chance determines which sperm and which egg will combine at the time of fertilization to form the new individual. The 1: 2: 1 genotypic ratio for the offspring of two heterozygous (*Pp*) individuals works well on paper, but under practical conditions the actual ratio may be far from the expected. Actually, when a small number of offspring are produced from such a mating, all might be of one genotype of either *PP*, *Pp*, or *pp*. The larger the number of offspring produced, however, the more likely it is that the expected ratio will occur. A similar condition is observed in the number of boys and girls that may occur in any one family. The chances of any one baby's being a boy is one out of two, for it has an equal chance of being a boy or girl. Yet, many of us have seen even large families where all the children are of the same sex. This happens because of the law of chance.

Six different kinds of matings, with reference to parental genotype, are possible when one pair of genes is involved. These are listed in Table 4.1 together with other comments about such matings.

4.1.2 Incomplete Dominance

Occasionally, we find modifications of this simple two-gene inheritance, as for example, the case of coat color in Shorthorn cattle. Three different coat colors appear in the Shorthorn breed: red (RR), roan (Rr), and white (rr). When a roan bull (Rr) is mated to a group of roan cows (Rr), the off-

Table 4.1   Various kinds of matings which can be made and the results of such matings where one pair of genes affect a given trait and dominance is complete

Genotype of parents	Genotypic ratio of offspring	Breeding ability of offspring*
$PP \times Pp$	⎰1 PP	Breeds true
	⎱1 Pp	Doesn't breed true
$PP \times PP$	All PP	All breed true
$PP \times pp$	All Pp	None breed true
$Pp \times Pp$	⎧1 PP	Breeds true
	⎨2 Pp	None breed true
	⎩1 pp	Breeds true
$Pp \times pp$	⎰1Pp	None breed true
	⎱1 pp	All breed true
$pp \times pp$	All pp	All breed true

*The term *breeds true* means that individuals of this genotype transmit only one kind of gene to their offspring. Heterozygous genotypes do not breed true because they may transmit either of two genes to their offspring.

spring will be colored in the ratio of one red (RR) to two roan (Rr) to one white (rr). It should be noted that this is the same genotypic ratio that is obtained when dominance is complete, as in the example where polled was dominant to horned. But in this instance, the heterozygote expresses itself phenotypically in a different manner from either homozygote, and the genotypic and phenotypic ratios are the same, 1:2:1. This type of gene action is called *incomplete dominance*, or a blending type of inheritance. One practical point here is that, even though it is not possible to develop a pure breeding roan herd (Rr) because the roan individuals are all heterozygous and will not breed true, it is always possible to produce roan cattle (Rr) by mating a red parent (RR) to a white parent (rr).

Many cases of inheritance are known where dominance is incomplete and the genotypic and phenotypic ratios of the offspring of a mating between heterozygous parents are the same. For example, palomino horses will not breed true, because they are heterozygous for genes for color, but they can always be produced by mating chestnut sorrels with pseudo albinos [4]. Another example is in Comprest Herefords, which are heterozygous for the comprest gene. When mated, comprests produce offspring in the ratio of 1 normal (*cc*) to two comprest (*Cc*) to one dwarf (*CC*). Of course, in this instance it is impossible to mate the normals (*cc*) with the dwarfs (*CC*), to produce all comprest (*Cc*), because the mortality rate of the comprest dwarfs is so high that they do not survive to maturity and cannot be used for breeding purposes.

Fig. 4.2 A comprest Hereford bull on the left and a conventional Hereford bull on the right. (Courtesy H. H. Stonaker, Colorado State University.)

An important fact concerning selection when dominance is incomplete is that the genotype can be determined by inspection of the phenotype. Therefore, an undesirable gene can be eliminated from the herd by discarding all heterozygotes.

4.1.3 Overdominance

Overdominance is another kind of gene action that may be of importance in animal breeding. It has been proposed as a theory, but in the past few years some examples have been found that give support to the proposal.

Overdominance is the interaction between genes that are alleles and results in the heterozygous individual's being superior to either of the homozygotes. For example, three different genotypes such as A_1A_1, A_1A_2, and A_2A_2 may be used. According to this theory, alleles A_1 and A_2 together produce a reaction that they do not produce separately.

The principle of overdominance may be illustrated by the inheritance of a particular blood type in rabbits [5]. Two alleles seem to be involved. One allele is responsible for the production of one antigen, and the other is responsible for the production of a second, different antigen. The heterozygous individual produces still a third, separate antigen not found in either homozygote. Thus, the two alleles interact to cause the production of an antigen that they do not cause in the homozygous state. A similar condition seems to exist in human haptoglobins [2], which are proteins with the specific property of binding hemoglobins. It has been proposed that two autosomal genes are involved. One homozygous individual causes the production of one haptoglobin, the other homozygote causes the production of a second, and the heterozygous individual causes the production of these two haptoglobins plus a third not found in either of the two homozygous individuals.

Overdominance probably involves many pairs of genes all affecting the same trait. The effect of any one pair of genes, however, could be rather small, but the combined effect of many pairs could result in a considerable advantage to the heterozygote. This kind of gene action, together with epistasis and dominance, could be responsible for the expression of hybrid vigor when certain lines, strains, or breeds are crossed.

4.1.4 Dihybrid Inheritance

Examples of dihybrid inheritance may be used to illustrate how two different traits may segregate and recombine in various combinations from one generation to another. Dihybrid inheritance should not be confused with polygenic; in dihybrid inheritance, two factors are involved, each affected by a single pair of genes. Although the same manner of segregation and recombination of genes is found in both types of inheritance, the manner of gene expression may be quite different.

As an example of dihybrid inheritance, we shall use two traits that are familiar to most breeders: horns and coat color. In cattle, polledness (P) is dominant to horns (p), and black (B) is dominant to red (b). What kind of offspring can be expected when a homozygous black polled bull is mated to homozygous red horned cows? The genotype of the bull would be $PPBB$, and that of the cows $ppbb$. The genotypes and phenotypes of the different generations may be illustrated as follows:

Generation

P_1 black Polled bull   red Horned cows

     PPBB   ×   *ppbb*

F_1 *PpBb*

All black Polled (mated *inter se*)

		Eggs			
		PB	*Pb*	*pB*	*pb*
F_2 Sperm	*PB*	*PPBB* Polled black	*PPBb* Polled black	*PpBB* Polled black	*PpBb* Polled black
	Pb	*PPBb* Polled black	*PPbb* Polled red	*PpBb* Polled black	*Ppbb* Polled red
	pB	*PpBB* Polled black	*PpBb* Polled black	*ppBB* Horned black	*ppBb* Horned black
	pb	*PpBb* Polled black	*Ppbb* Polled red	*ppBb* Horned black	*ppbb* Horned red

Some new genetic terms are introduced in this example. The P_1 generation is that of the parents. The F_1 is the first filial generation, or the first generation of a given mating. The F_2 generation is produced by crossing the F_1 individuals. The term *inter se* means the mating of the F_1 generation among themselves.

All the sperm of the bull of genotype *PPBB* contain one *P* and one *B* gene. All the eggs of the cows of genotype *ppbb* contain only *p* and *b* genes. When the sperm and egg cells unite in fertilization, the genotype of all F_1 individuals is *PpBb*. Their phenotype is polled and black.

Individuals of the F_1 generation have the genotype *PpBb* whether they are male or female, and they produce four different kinds of sex cells, *PB*, *Pb*, *pB*, and *pb*, in approximately equal numbers. Thus, any egg produced by the cows has an equal chance of carrying one of these four combinations of genes, and each of the eggs has an equal chance of being fertilized by a sperm of one of the four different kinds. From the checkerboard, we see that among the 16 F_2 individuals there are 9 different genotypes and 4 different phenotypes. The phenotypic ratio of the F_2 generation is 9 polled black to 3 polled red to 3 horned black to 1 horned red.

An interesting point of such a cross is that the traits which were possessed by the grandparents have now been combined in different combinations in the F_2 generation, that is, the grandchildren. This illustrates one of the laws of Mendel known as the Law of Independent Assortment of Characters and means that in this case genes affecting horns and coat color were inherited independently and therefore were on different pairs of homologous chromosomes.

The principle of independent assortment of characters is often used in livestock breeding. For example, the Santa Gertrudis breed of cattle was developed from a cross of the Brahman and the Shorthorn breeds. In the formation of this new breed, selections were made to combine the heat and disease-resistance of the Brahman with the beefiness of the Shorthorns. The principle has also been used in many other livestock breeding programs.

4.1.5 Epistasis

Epistasis is another type of variation in gene expression that may be of importance in livestock breeding. Epistasis is the name given to nonlinear interactions (where the effect of one gene does not add to the effect of another to influence the phenotype) of various kinds between any nonallelic genes. Epistasis may also be defined as the interaction of two or more pairs of genes that are not alleles. These may be interactions between genes on the same or on different chromosomes. In contrast, dominance is the nonlinear interaction between genes that are alleles.

Epistasis may be illustrated by coat color in horses [4]. The genetic bases of all the colors have not been established definitely, but they will serve to illustrate the point. In most animals, there is a color gene C, which is necessary for the expression of color. When it appears in the homozygous recessive form cc the coat color is white, a condition called albinism. Another gene that affects coat color is also found in some animals. This is the gene D for the dilution of color, and its effects are manifest in the palomino color in horses.

When a chestnut stallion ($AAbbCCdd$) is mated to a type A albino mare ($AAbbCCDD$), a palomino foal ($AAbbCCDd$) is produced. When two palominos are mated, it is possible to get foals of which one-half are palomino, one-fourth chestnut, and one-fourth type A albino. This is illustrated as follows:

P_1 Chestnut stallion   Type A albino mare
             $AAbbCCdd$ $\times$ $AAbbCCDD$

F_1 Palomino foals (mated *inter se*)
                        $AAbbCCDd$

F_2
$$
\begin{array}{lll}
1 & AAbbCCdd & \text{Chestnut} \\
2 & AAbbCCDd & \text{Palomino} \\
1 & AAbbCCDD & \text{Type A albino}
\end{array}
$$

In this instance, the gene D^* is the controlling factor in coat-color expression. When present in the homozygous recessive state (dd), color is present in the full amount (chestnut). In the palomino there is a dilution of color, and in the homozygous dominant state (DD), there is an absence of color, and the animal is white.

Many examples of epistatic gene action are known in mammals, and the genes act or interact in many different ways. In the preceding example, the gene D was dominant, or at least partially dominant, to its own allele,

Fig. 4.3 Grand Champion Palomino mare at a show in 1952. (Courtesy of the Palomino Horse Breeders of America.)

*Some geneticists consider this gene an allele of the color gene (C) and refer to it as (Ccr).

so this is referred to as dominant epistasis. Instances of recessive epistasis are known in which the epistatic gene is recessive to its own allele and has to be present in the homozygous recessive state in order to have its effect. One form of albinism in rats is due to recessive epistasis. Still other instances are known in which genes that are not alleles complement each other to produce an effect that neither pair of genes produces alone.

It is possible that epistasis has an important influence on the expression of genes that affect many of the economic traits in farm animals. Since these probably would deal with fundamental types of gene action, they might be difficult to identify. In spite of this difficulty, however, we can devise breeding and selection procedures to make the greatest use of these types of gene expression.

4.2 ADDITIVE GENE ACTION

This is still another kind of gene expression in animals. In this type of inheritance, there is no sharp distinction between genotypes, but many gradations between the two extremes. Davenport's theory of skin-color inheritance in humans is a good example of additive gene action. Two different pairs of genes are thought to affect the production of pigment in the skin, although this may be an oversimplification of the actual mode of inheritance. In general, however, the theory seems to fit the mode of inheritance fairly well, although there are probably other genes involved. The following example illustrates the principle of additive gene action:

P_1	Black skin		White skin
	AABB	$\times$	*aabb*
F_1	Medium (mulatto) (*inter se*)		
	AaBb		
F_2	1 *AABB*	Black	
	2 *AABb*	Dark	
	1 *AAbb*	Medium	
	2 *AaBB*	Dark	
	4 *AaBb*	Medium	
	2 *Aabb*	Light	
	1 *aaBB*	Medium	
	2 *aaBb*	Light	
	1 *aabb*	White	

Five different phenotypes are observed, with a continuous gradation between white and black and no sharp distinction between any two classes. The proportion of offspring of different skin colors in the F_2 generation would be

1   black,
4   dark,
6   medium (mulatto),
4   light,
1   white.

These are shown in the form of a frequency-distribution curve in Fig. 4.4, and it may be observed that a normal, bell-shaped curve is formed.

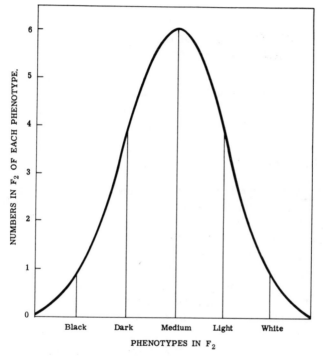

Fig. 4.4 Illustration of how the different phenotypes in the F_2 of a mating of F_1 medium individuals (mulattoes) assume a frequency distribution curve.

Genes A and B are called contributing genes, because they make a contribution to the darkening of the skin. Genes a and b are called neutral genes, because they contribute nothing, or at least very little, to skin color. An individual of the genetic composition of *aabb* is said to be of the residual genotype, with no contributing genes present.

An important aspect of additive gene action is that none of the genes is dominant or recessive. Each contributing gene adds something that makes the skin darker in color, and the effects of each gene accumulate. Hence, the term, additive gene action.

Additive gene action is thought to affect most traits in farm animals that are of economic importance, although it has a much greater influence on some traits than on others. Growth rate, milk production, conformation, carcass quality, as well as other traits are affected by this type of inheritance. Because many different pairs of genes may affect such traits, such traits are polygenic (*Poly* means many).

4.3 SOME CAUSES OF VARIATIONS IN PHENOTYPIC RATIOS

The existence of multiple alleles, linkage of genes, and sex linked genes may be responsible for some variations in the expression of the phenotypes in the progeny of certain parents.

4.3.1 Multiple Alleles

Previously we defined alleles as those genes which occupied the identical loci on homologous chromosomes but affected the same trait in a different manner. Research work in genetics and animal breeding has demonstrated that in a population there may be more than two alternative genes that can occupy the loci. Such genes are called *multiple alleles*.

The best-known series of multiple alleles is that which affects blood types in humans. Three different genes are known to be involved, and these are called genes *A*, *B*, and *O*. These three genes may be combined to form several genotypes, as follows: individuals belonging to blood group *A* may be of the genotype *Aa* or *AA*; those of blood group *B*, of genotype *Ba* or *BB*; those of blood group *O* of genotype *aa*, and of blood group *AB*, of genotype *AB*. Only two of the three alleles occur in any one individual, since there are only two locations on homologous chromosomes that are available for these genes.

A large series of multiple alleles and several blood groups is known to exist in farm animals, and a knowledge of these may be used to identify the parents of offspring in particular instances.

4.3.2 Linkage of Genes

Linkage is the presence in the same chromosome of two non-allelic genes affecting two traits. Instead of segregating and recombining in a random manner in subsequent generations, the genes travel together in the process of oögenesis and spermatogenesis and thus are inherited together.

It has been pointed out earlier that each chromosome carries hundreds of genes affecting different traits. The different traits determined by genes on the same chromosome tend to be inherited as a single group, and this is called a *linkage group*. The number of linkage groups corresponds to the number of pairs of chromosomes in a particular species. Not much is known about traits belonging to a single linkage group in farm animals, but our knowledge in this field should increase as research continues.

How to test for linkage groups. The method used to test for the linkage of genes is to backcross a double heterozygous individual to one that is homozygous recessive for both pairs of genes. If no linkage is involved, four phenotypes will appear in the offspring, in approximately equal numbers. If the genes are linked, only two phenotypes will appear, and in approximately equal numbers.

To illustrate this method, let us use characters that are not linked and thus are inherited independently: polledness (P), which is dominant to horns (p), and black (B), which is dominant to red (b). When an individual of geno-type $PPBB$ is crossed with one of genotype $ppbb$, all of the F_1 offspring will be of genotype $PpBb$ and of phenotype polled and black. When these F_1 individuals of the $PpBb$ genotype are backcrossed (or mated) to the homozygous recessive $ppbb$, the offspring produced will be

4 $PpBb$— polled, black
4 $Ppbb$— polled, red
4 $ppBb$— horned, black
4 $ppbb$— horned, red

Thus, the four different phenotypes will occur in approximately equal numbers, showing that the two pairs of genes are on two different pairs of homologous chromosomes.

Now, to show what would happen if these genes were linked, let us assume that P and B are together on one member of a pair of chromosomes and p and b are together on the other member. If linkage were complete and close, we would expect the following results from crosses:

P_1	Polled black		Horned red
	$PPBB$	$\times$	$ppbb$

F_1		Polled black
		$PpBb$

F_2	1 $PPBB$	Polled black
	2 $PpBb$	Polled black
	1 $ppbb$	Horned red

In other words, polledness and black coat color would be inherited together, as would horns and red coat color. No horned black or polled red individuals would be observed in offspring, as is the case when there is an independent assortment of the characters.

Crossing-over. However, in almost all instances where linkage is involved and the test cross is made, all four phenotypes do occur in the next generation, but two appear more often than the other two. For instance, suppose that a back cross were made by mating F_1 individuals in the preceding example to the homozygous recessive genotype, and the phenotypes were the following:

45 $PpBb$— polled, black
5 $Ppbb$— polled, red
45 $ppbb$— horned, red
5 $ppBb$— horned, black

This is not a 1:1:1:1 ratio expected in independent assortment, and we would have to conclude that linkage must be involved. But, during meiosis in the heterozygote, an "accident" has occurred to the members of a chromosome pair in some of the sex cells. When they came together in the intimate process of synapsis, they broke at about the same points, and exchanged parts. The result is that on some chromosomes P and b are together and on some p and B are together. This phenomenon is called *crossing-over*. In the hypothetical example, the percentage of crossing-over is 10 (phenotypes where crossing-over occurred) divided by 100 (total number of phenotypes), times 100, or 10 percent.

An example of linkage in poultry. Linkage of comb shape and leg length in chickens has been reported [7]. Rose comb (R) is dominant to single comb (r), and the creeper (C) is dominant to normal leg length (c). Creeper chickens have very short legs that cause them to take very short steps, so that they appear to creep when they walk. Individuals homozygous for the creeper gene die early in life, and thus this gene is a semilethal gene. When the test cross was made between a male heterozygous for both rose comb and creeper legs and females homozygous for both single comb and normal legs, the progeny produced were as follows:

Rose comb creeper   Single comb normal
RrCc   ×   *rrcc*

		Offspring
22	*RrCc*	Rose comb, creepers
1	*Rrcc*	Rose comb, normal
33	*rrcc*	Single comb, normal
4	*rrCc*	Single comb, creepers

Thus, the genes for rose comb and creeper legs were carried on one member of a chromosome pair, and the genes for single comb and normal legs on the other member. Some crossing-over occurred, however, as shown by the appearance of some rose comb normal and some single comb creepers in the progeny. The amount of crossing-over was about eight percent.

In the double heterozygous offspring, where R and C remained together on the one member of chromosome pair, and r and c remained together on the other, the linkage is called *coupling*; where crossing-over occurred and genes R and c became linked, and r and C became linked, the disruption of linkage is called *repulsion*.

Genes located further apart on the same chromosome cross over more often than those that are closer together. In fact, the further apart they are, the more difficult it becomes to distinguish crossing-over from the independent assortment of characters; thus, some linkages remain undetected. For this reason, there may be more instances of linkage in the traits of farm animals than we have been able to discover.

The degree of crossing-over has been used to determine approximately the loci of certain genes in some species. Maps of chromosomes have been made, showing approximate locations. Mapping has progressed quite rapidly for the fruit fly (Drosophila) and to a lesser extent for mice and humans. Very little chromosome mapping has been done for the various species of farm animals.

Practical significance of linkage in farm animals.   Very little is known about the linkage of genes in farm animals. Detailed and extensive studies have been made of the blood groups in cattle, where it has been found that 10, and possibly 12, different pairs of the 30 pairs of chromosomes carry genes for blood groups. These genes must be linked with genes that determine other characters. The possibility of detecting such linkage associations depends on how closely the genes are linked and how frequently the linked genes are manifested. If the linkage between blood group genes and genes for some other trait is strong, the blood group genes might be useful as chromosome markers. It seems doubtful that attempts to discover linkage associations between blood group genes and genes affecting economic traits would be very fruitful. The main reason for this is that so many different genes are involved and the expression of any single pair is weak. In addition, linkage associations would have to be determined for each animal, for linkage relationship would vary between individuals.

Furthermore, it is often difficult to distinguish between effects of two closely linked genes and the pleiotrophic effects (where the same gene may affect more than one trait) of the same gene. A good example of this is found in chickens, where there is a series of alleles affecting blood groups at the B locus on the chromosomes. By test crosses, it has been found that hens homozygous for a certain B allele such as BB or $B'B'$ produced eggs with decreased hatchability as compared to those heterozygous for BB'. Viability to nine weeks of age was slightly greater in the heterozygous offspring.

In other instances in farm animals, there does seem to be a correlation between certain quantitative traits, such as rate of gain and efficiency of gain, but it is not assured that this correlation is due to linkage. These two traits are associated, probably because the genes affecting them do so through a common physiological pathway. Linkage could be involved, but it is less likely, because of crossing-over and consequent equilibrium, which is dependent upon gene frequency and not on the degree of linkage.

4.4 SEX-LINKED INHERITANCE

Farm animals are known to possess from 19 to 32 pairs of chromosomes in the body cells, depending on the particular species involved. Each pair of homologous chromosomes segregate independently of the other pairs when the sex cells are formed. One pair of homologous chromosomes is called the *sex chromosomes*. One of the pair is called the X- and the other the Y-chromosome. All other pairs of chromosomes are referred to as *autosomes* (Fig. 4.5).

In mammals the female possesses two X-chromosomes in the body cells, and the male possesses an X- and a Y-chromosome. Many different genes are carried on each of the sex chromosomes, as is true of the autosomes.

The following example illustrates how the X- and Y-chromosomes are transmitted by the parents to the offspring:

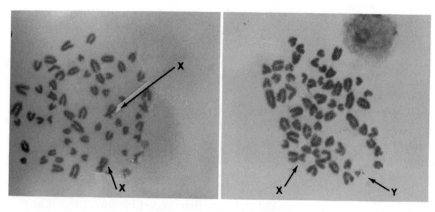

Fig. 4.5 Microphotographs of chromosomes of male
(right) and female (left) cattle. These were taken at the
metaphase of cell division when each chromosome is
doubled with both being connected at the centromere.
All autosomes are acrocentric but the sex chromosomes
are metacentric or sub-metacentric. Note how much
larger the *X* chromosomes are than the *Y* chromosome.
(Courtesy of Dr. C. L. Chrisman, University of
Missouri.)

Female   Male

XX   *XY*

Kind of sex chromosome in the sex cells:

Eggs (*X* only)   Sperm (*X* and *Y*)

Sex of offspring:

XX or Female

XY or Male

Thus, the male offspring receives its *Y*-chromosome from the father and its
X-chromosome from the mother, and the female receives one *X*-chromosome from the mother and one from the father.

When we speak of sex-linked traits, we refer to those traits determined
by genes carried (linked) on the *X*-chromosome. Very little is known about the
importance of sex linkage in farm animals, probably because it has not been
studied as much in animals as it has in humans. Undoubtedly, future studies
in genetics will reveal several sex-linked traits in farm animals.

The Y-chromosome is shorter than the X-chromosome and so cannot carry all the genes corresponding to those on the X-chromosome. Therefore, in the male, it is possible for a single recessive gene on the "extra" portion of the X-chromosome to express itself. In the female, there must be two recessive genes on the X-chromosomes, just as on autosomes, for a recessive trait to be expressed.

A further difference between sex chromosomes and autosomes is that a portion of the Y-chromosome is not homologous to the X-chromosome, and in a few instances known in humans, a gene is carried on the Y-chromosome and is responsible for the appearance of a trait in the male. This is referred to as *holandric inheritance.* An example in humans is webbed toes, and it is passed from father to son for generation after generation. In general, however, sex-linked traits are determined by genes carried on the X-chromosomes.

Agammaglobulinemia in humans, mentioned in Chapter 3, is a sex-linked recessive trait, and the gene for it is carried on the X-chromosome. Let us use the letter A for the normal gene and the letter a for the gene that causes agammaglobulinemia. Let us also assume, for an example, that a woman who is a carrier of this gene (genotype $X_A X_a$) marries a man who is not affected by this disease ($X_A Y$). The following diagram gives the genotypes of the boys and girls one would expect, with respect to this trait:

Carrier woman   Normal man

$X_A X_a$ $\times$ $X_A Y$

Genotypes expected in offspring:

1 $X_A X_A$ Normal girl
1 $X_A X_a$ Carrier girl
1 $X_A Y$ Normal boy
1 $X_a Y$ Boy with agammaglobulinemia

Daughters from such a marriage would have equal chances of being carriers of the defective gene, but none would contract the disease. The sons, on the other hand, would have equal chances of receiving a normal or a defective gene from their mother, and since only one gene carried on the X-chromosome is necessary for the expression of this trait in the male, those receiving the gene for agammaglobulinemia from the mother would be afflicted by this disease.

The importance of sex-linked genes for type and performance in farm animals is not definitely known; but they are probably not very significant, for type and performance in animals are influenced by many pairs of genes, most of which are carried on the autosomes.

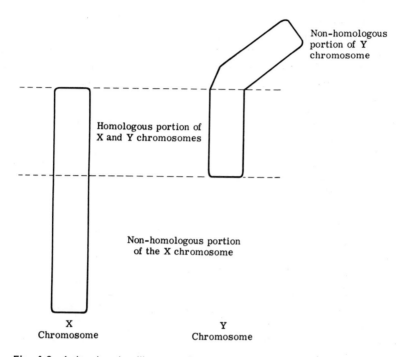

Fig. 4.6 A drawing that illustrates that some parts of the *X*- and *Y*-chromosomes are not homologous. Thus, only one recessive gene on the nonhomologous portion of either sex chromosome is all that is needed for the gene to express itself.

4.5 SEX-INFLUENCED INHERITANCE

This type of inheritance is often confused with the sex-linked type. The genes for sex-influenced inheritance are carried on the autosomes, and their expression is influenced by the sex of the individual. In the heterozygote, the genes usually are expressed as dominant in the male and as recessive in the female. The inheritance of horns in sheep is a good example of a sex-influenced trait. Table 4.2 gives the genotypes and phenotypes for this trait.

Since the sex of the individual, or more probably the male hormone production, can cause a definite difference in the expression of genes for horns in sheep, it is interesting to speculate how the same kind of situation could exist for other traits in farm animals. For instance, it is known that boars possess from .25 to .35 inch less backfat than litter-mate barrows, but there is considerable variation among litters. Possibly a boar could have very thin backfat at market weight because of a high level of male hormone production.

This condition might not be transmitted to his castrate male offspring, because their male hormone production would practically cease at the time of castration. Another way of stating this problem is to ask whether we actually know what characteristics we would look for in the intact breeding males in order to select those that will produce the most desirable castrate male offspring.

Table 4.2 Inheritance of horns in sheep—A sex-influenced trait

Genotype	Phenotype of rams	Phenotype of ewes
HH	Horned	Horned
Hh	Horned	Hornless
hh	Hornless	Hornless

4.6 WHEN GENES EXPRESS THEMSELVES

Many people are of the opinion that traits determined by genes are always present or visible at birth. This is not true, because the time at which genes express themselves is variable. In sheep, an inherited condition has been found in which the lambs are born with short tails. The genes for this trait express themselves early in embryonic life when the bones are being formed. Genes for eye color in humans usually begin to show their effects a few weeks after birth, and not at birth. A form of muscular dystrophy in humans is not expressed until the age of seven to fifteen years. Hereditary baldness affects most individuals only after maturity, at 25 to 30 years of age. The gene for Huntington's Chorea, a nervous disorder in humans, may not affect the individual until he is past 50 years of age.

The fact that genes do not always express themselves early in life may also be important in animal breeding. It is probably unwise to make a practice of selecting breeding stock at the time of birth. It is more advisable to make selections at market age, for by that time the genes will have expressed themselves in a favorable or unfavorable manner.

4.7 GENES VARY IN THEIR EXPRESSION

Some genes are constant in their expression, whereas others are quite variable. For instance, "snorter" dwarfism in beef cattle is characterized by a great variation in size, from the very small animals that die soon after birth to those that approach normal size and may live to be several years of age. Short-tailedness in sheep, discussed earlier, is inherited, and the length of the tail varies from almost normal to a condition where a portion of the spinal column may be missing. This variability in the expression of genes is called "variable expressivity."

Fig. 4.7 Variation in the expression of genes for tail length in sheep.

"Penetrance" of a gene is the frequency of the actual expression of a trait as compared with the frequency at which it is expected to be expressed. The degree of penetrance of a gene is usually referred to in terms of percentages, and can vary from almost zero to 100 percent. A gene that expresses itself only 30 percent of the time is said to possess 30 percent penetrance.

Genes may vary in both penetrance and expressivity. Little is known of the importance of this in selection for more efficient production of livestock except that environment can mask or change the way genes express themselves. Often environmental effects are mistaken for gene effects, and selection for traits so influenced is not effective. This will be discussed in more detail in a later chapter.

4.8 SOME CAUSES OF VARIATIONS IN GENE EXPRESSION

Many factors may be responsible for variations in gene expression. In general, variations may be divided into two groups, those due to external environmental conditions and those due to internal environment.

4.8.1 External Factors

Temperature can play a very important part in the expression of genes. One of the best-known examples of this effect in animals is in the Himalayan breed of rabbits [1]. The gene responsible for coat pattern causes the production of an enzyme which is necessary for the formation of a black pigment. However, only the extremities of the body, such as the nose, ears, feet, and tail are pigmented, whereas the main parts of the body are white. The explanation of this difference is that the enzyme does not form at normal body temperature but does form at slightly lower temperatures; hence, pigment appears in the body extremities, where the temperature is presumably lower. A similar explanation has been made for the coat pattern in the Siamese cat.

Studies of bread mold, or Neurospora, give a clue as to the possible explanation of the temperature sensitivity of the pigment-forming system in Himalayan rabbits and Siamese cats [1]. Melanin is formed in this organism, and in one strain the pigment formation shows a strong temperature effect. Cell-free tyrosine-containing preparations of a strain of Neurospora in which temperature does not affect pigment formation and a similar preparation of the temperature-sensitive strain show quite clearly that their differences in response to temperature are due to different temperature stabilities of their tyrosinases. The tyrosinase of the temperature-sensitive strain is much less stable in vitro at 35°C.

Sunlight is also known to affect the ability of genes to express themselves. One example is found in certain Southdown sheep, where hypersensitivity

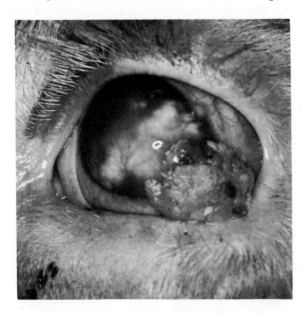

Fig. 4.8 Cancer eye in cattle. This condition is quite prevalent in animals in the southwestern United States. It is heritable and seems to be increased in incidence in areas of intense sunlight. (Courtesy of the University of Texas, M.D. Anderson Hospital & Tumor Institute.)

to sunlight is a heritable trait. In affected lambs, the liver does not function properly, and there is a failure to excrete phylloerythrin, an end product of chlorophyll metabolism. This product accumulates in the bloodstream and in certain areas of the skin, where it is activated by sunlight. Eczema develops over the face and ears, and the animals may even die if left outdoors; but if they are kept indoors and allowed to graze at night, they do not develop symptoms. Cancer eye in Hereford cattle is also thought to be inherited [3], and many ranchers believe that the condition is aggravated in regions of intense sunlight, such as the southwestern portion of the United States.

4.8.2 Internal Factors

Of the various internal environmental factors that may affect gene expression, the hormones are probably among the most important. Earlier, we pointed out that a gene for dwarfism in mice apparently has its influence through the failure of the production of the growth hormone by the anterior pituitary gland. Other genes for growth are present but cannot express themselves because of the lack of this hormone.

In humans, there is considerable evidence that the tendency to contract diabetes is inherited, but this tendency often is not manifested unless the pancreas is overloaded by the individual's consumption of large amounts of carbohydrates over a long period of time.

The sex hormones are also known to influence gene expression. Baldness in humans is often hereditary, but is expressed more often in the male. The same is true of the mahogany and white color in Ayrshire cattle. Boars are known to possess from .25 to .35 inch less backfat at market weight than barrows from the same litters on the same kind of feed. Bull calves, on the average, almost always outweigh heifer calves at weaning time, and the difference seems to widen as the calves grow older. Dairy bulls carry genes for milk production, and roosters carry genes for egg production, but neither of the traits is expressed in the males of the species.

Genes on one pair of chromosomes may modify the expression of another pair of genes on a different pair of chromosomes. These are called modifying genes and epistasis is involved. Spotting of the coat in Holsteins is due to a single pair of genes. Yet coats vary from almost white to almost black, and have black spots of various sizes. Selection in either direction for size of spots is successful, indicating that other genes (modifying genes) are responsible for the expression of the genes for spotting. The possibility of the existence of other modifying genes in farm animals could be important from the standpoint of avoiding errors in selective breeding for important traits.

Fig. 4.9 Spotting in some breeds of cattle is due to a single pair of recessive genes, but modifying genes cause a variation in spots from individuals which are almost white to those which are almost black. These three sets of identical twins show the uniformity of spotting within twin sets but great variation between different sets. (Courtesy of Moorman Manufacturing Company, Quincy, Illinois.)

REFERENCES

1. Beadle, G. W. "Gene Structure and Gene Function," MoAESRP, 588, 1952.
2. Bearn, J. G., and E. C. Franklin. "Some Genetic Implications of Physical Studies of Human Haptoglobins," S, 128: 596, 1958.
3. Blackwell, A. L., D. E. Anderson, and J. H. Knox. "Age Incidence, and Heritability of Cancer Eye in Hereford Cattle," JAS, 14: 43, 1956.
4. Castle, W. E. "The ABC of Color Inheritance in Horses," C. 35: 122, 1948.
5. Cohen, C. "Occurrence of Three Red Blood Cell Antigens in Rabbits as the Result of the Interaction of Two Genes," S, 122: 35, 1956.
6. Rendel, J. "Blood Groups of Farm Animals," ABA, 25: 223, 1957.
7. Serebrousky, A. S., and S. G. Petrov. "A Case of Close Autosomal Linkage in the Fowl," JH, 19: 305, 1928.

STUDY QUESTIONS

1. What are the two general types of phenotypic expression of genes?

2. Why is it important to know how genes express themselves phenotypically?

3. Define: dominance, recessiveness, homozygote, heterozygote, genotype, phenotype.

4. Which would be the easiest to establish, a pure herd of polled cattle or a herd of horned cattle? Why?

5. How does incomplete dominance differ from complete dominance?

6. Explain how you would develop a pure breeding strain of red Shorthorns; of white Shorthorns; of roan Shorthorns.

7. How many different matings can be made if two genes, a dominant and a recessive allele, are present in a population?

8. Outline a method of eliminating a partially dominant gene from a population. Use the comprest gene in Herefords as an example.

9. Explain what is meant by Mendel's Law of Independent Assortment of Characters. How could this principle be used to advantage in livestock breeding?

10. Explain the basic difference between epistasis and overdominance.

11. Explain how additive gene action differs from nonadditive.

12. What is meant by multiple alleles? Are examples known in farm animals?

13. Explain what is meant by the "linkage" of genes; by a linkage group.

14. What is the accepted procedure for testing for the linkage of two pairs of genes?

15. Explain the following: crossing-over, coupling phase, repulsion phase, mapping of chromosomes.

16. What is the practical significance of linkage in farm animals?

17. Distinguish between sex chromosomes and autosomes. How do the females differ from the male in sex chromosome composition in farm mammals? In poultry?

18. What is holandric inheritance? How is it transmitted in the human?

19. Why do many more males than females express a trait phenotypically when the gene for it is carried on the nonhomologous portion of the *X*-chromosome?

20. Agammaglobulinemia in humans is a sex-linked recessive trait. How could a female have such a trait?

21. How does sex-influenced inheritance differ from sex-linked inheritance?

22. What is incorrect about this statement, "A gene-determined trait is always expressed at birth?"

23. In what way could the fact that some genes do not express themselves phenotypically early in life be of practical value in animal breeding?

24. Explain the difference between "variable expressivity" and "penetrance."

25. What external environmental factors may be responsible for variations in the phenotypic expression of genes?

26. What internal environmental factors may be responsible for variations in the phenotypic expression of genes?

27. Explain what is meant by "modifying genes."

28. Of what practical value is the knowledge that certain factors in the environment may be responsible for the phenotypic expression of genes?

29. Do you agree with the following statement? "Over a period of several generations flies developed an immunity to DDT."

30. Can you explain why snowshoe rabbits (hares) in the Arctic turn white in winter and dark in summer, but nothing similar is found in rabbits in the central midwest of the United States?

5

Mutations

A mutation may be defined as a change in the self-duplication process of a gene so that a new allele is produced that differs from the original gene in its effect upon the expression of a particular trait. Mutations may also be due to a change, or changes, in chromosomes (chromosome aberrations), which may change the phenotype. In a gene mutation the new gene duplicates itself exactly for succeeding generations or until another mutation occurs. This is the explanation for the occurrence of several alleles in a multiple allelic series.

Mutations may occur in genes either in the somatic cells or in the germinal cells. Somatic cell mutations are not transmitted from parents to offspring, so they are not of great importance genetically. An example of a somatic cell mutation is the occurrence of a black hair spot on the red portion of the coat of Hereford cattle. If a mutation should occur in the somatic genes for coat color in this breed, it would be dominant, since black is dominant to red.

Mutations are responsible for genetic variations in our farm animals. If genes could duplicate themselves perfectly for generation after generation over a period of thousands of years without a single mistake, members of a given species would probably have the same color, would be alike for the qualitative traits, and would not be divided into distinct types and breeds. All variations that exist would be superficial environmental effects and would not be transmitted from parents to their offspring.

Familiar mutations in farm animals are due to changes in the composition of the genes themselves. In plants and animals, however, it is known that mutations also are often due to changes in the chromosomes, and these are called *chromosome aberrations*. Similar mutations also occur in farm animals, but not many have been identified. Since mutations involving changes in specific genes seem to be of the greatest importance in animal breeding, we shall discuss these first.

Fig. 5.1   Albino Hereford dwarf illustrating the occurrence of a rare mutation in cattle. (Courtesy of M. E. Ensminger, Washington State University.)

5.1   OCCURRENCE OF MUTATIONS

Many mutations in farm animals have occurred in the past and will continue to occur in the future. References in the Holy Bible show that mutations occurred in humans many years before the birth of Christ, although the reason for their occurrence was not then known. In 2 Samuel 21:20, it is written, "And there was yet a battle in Gath, where was a man of great stature, that had on every hand six fingers, and on every foot six toes, four and twenty in number; and he also was born to the giant." This particular characteristic is now known as polydactyly and is known to be inherited as a dominant trait. It had its origin from a mutation affecting finger number.

One of the first mutations affecting body form in farm animals appearing in American farm records is the case of the Ancon sheep that was mentioned earlier [3]. Polled cattle developed from an original mutation of the horned to the polled gene; and the different colors in all of our farm animals probably are the result of mutations from some original color in animals before domestication. Many inherited defects have been reported [6, 7] for the different species of farm animals, showing that a tremendous number of mutations have occurred in the past and are still present in the animal populations in the recessive state.

Research in genetics shows that some genes are more stable than others and that mutations occur in them less often than in others. In humans, it has been estimated [2] that the rate of new mutations per 100,000 gametes ranges from 1.4 to 8. Other estimates [5] suggest that the total rate of mutations for the 5000 or more loci assumed to exist in the chromosome entity of humans results in an average of about one new mutation in 10 or fewer gametes. It is also believed that the hidden recessive and semidominant load of mutations in the human species may average eight or more per person. If this is true in humans, undoubtedly the same is true in farm animals.

The discovery or the appearance of new mutations depends to a certain extent upon whether they are dominant or recessive in their expression. If they are dominant, they are recognized immediately; but if they are recessive, they may be carried in a population for generation after generation, and may not be discovered unless two individuals are mated that carry the same recessive gene. Even then, for any one mating, the chances are only one out of four that a homozygous recessive individual will result. From this we can say that the discovery of a new mutation is dependent entirely upon our ability to measure or observe the expression of a gene that is different from other alleles in that particular series.

5.2 IMPORTANCE OF MUTATIONS

Observations made by many geneticists indicate that an overwhelming majority of mutations are recessive, and detrimental to the organism. A few, such as the polled trait in cattle, may be beneficial, but we have to observe a large number of mutations before we find one that is beneficial. In the past, the animal breeder has spent very little time in eliminating harmful mutations from his herd, because he has not recognized them as being very important. The appearance of the condition of "snorter dwarfism," in which there must have been some selective advantage for the heterozygous individuals, is an exception; breeders directed much attention to the elimination of this defect. Mutations resulting in defective development of farm animals in which the heterozygote does not have a selective advantage will not become very important in the future unless inbreeding is practiced, but our ideas on this could change as more information on the subject becomes available.

5.3 HOW MUTATIONS BECOME ESTABLISHED IN A POPULATION

If most mutations are detrimental and recessive, how do they become established in a population? Mutations have originated initially from a mutation in a single individual, and there are several possible reasons for a gene increasing in frequency at the expense of its alleles. One is the rate of recurrent mutation at that particular locus on the chromosome. It was pointed out earlier that genes differ in their rates of mutation, some mutating much more often than others. Thus, a gene that undergoes mutation very frequently would have a better chance of gaining a foothold in a population than would one that mutated much less frequently. Another reason is that the heterozygous individual may be favored by either natural or artificial selection in a particular environment. In such a case, the frequency of the recessive gene would be increased, even though the homozygous recessive individual might not be capable of reproduction. The selective advantage of the heterozygous individuals need not be exceedingly great if the gene mutates frequently and if there is a long period of time for the selection forces to operate. A third reason is that the segregation ratio may be abnormal, with the recessive gene being carried in the gametes more often than normal. As was pointed out by Crow [1] the short-tailed condition in mice occurs more frequently than expected, and this may be due to some process whereby the gene for this trait segregates in the gametes at a higher rate than is normal or expected.

5.4 CHROMOSOME ABERRATIONS

Chromosomes do not always behave in a normal manner in the process of mitosis and meiosis [8]. They may fail to separate at meiosis, so that the gamete either has an extra chromosome or lacks one. In some instances, whole sets of chromosomes may be duplicated in the gamete. A chromosome may rearrange itself so that the genes may be located in a different order, and this may have a decided effect on the expression of a particular gene. Instead of an entire chromosome being duplicated, only a portion may be involved. In addition, there may be an exchange of parts between two chromosomes that are not homologous. These abnormalities are known as chromosomal aberrations (deviations from normal) and may cause decided changes in the phenotype of an individual.

Many chromosome changes have been reported in plants and lower forms of animals, and, as was pointed out in Chapter 2, defects such as mongolism and Klinefelter's syndrome in humans are due to such abnormalities. The importance of chromosome aberrations in farm animals has been given little attention, but this is a promising field of research in the future. Possibly many defects in farm animals are due to some form of chromosome abnormality.

Many different agents are known to cause chromosome aberrations in plants and lower forms of animals. The duplication of entire sets of chromosomes (polyploidy) in some plants, such as tobacco, yeast, and chrysanthemums, has been useful in some instances in producing larger and more vigorous plants. The principle has not been applied to farm animals, and we do not know for certain the possible application to animal breeding. Indications are, however, that the induction of polyploidy in farm animals may not be possible because of accompanying sterility or failure of embryo development.

5.5 INDUCTION OF MUTATIONS

Research work with plants [9] and with insects [4] has shown that mutations may be caused by exposure to certain chemicals and X-rays. These artificially induced mutations are similar to those which occur under natural conditions. From the study of induced mutations, one important fact has been discovered—that even though the mutation rate can be greatly increased by artificial means, the mutation of a particular gene cannot be controlled at will. That is, one cannot attempt to induce desirable mutations artificially. For the livestock breeder, this means that it is not practical to use X-rays and chemicals on breeding stock to induce mutations; for, even if this could be done, many defective genes would appear before a favorable mutation was found and defective genes would occur with the favorable ones.

For the present, at least, the livestock breeder can watch for the rare favorable mutations that might occur to cause an improvement in the present-day breeds of livestock. This, however, is going to be much harder than looking for the proverbial needle in a haystack and conceivably is of little or no value in constructive livestock breeding. The more desirable practice is to weed out the detrimental mutations that now exist in our breeds of livestock and to improve existing stocks through proper mating and selection systems, for these have proved to be reliable methods.

REFERENCES

1. Crow, J. F. "Ionizing Radiation and Evolution," SA, 201: 138, 1959.

2. Haldane, J. B. S. "Mutations in Man," P81CG, Hereditas, Supplement 267, 1949.

3. Landauer, W., and T. K. Chang. "The Ancon or Otter Sheep, History and Genetics," JH, 40: 103, 1949.

4. Muller, H. J. "The Gene," PSB, 134: 1, 1947.

5. Muller, H. J. "Our Load of Mutations," AJHG, 2: 111, 1950.

6. Roubicek, C. B., R. T. Clark, and O. F. Pahnish. "Dwarfism in Beef Cattle— a Literature Review," compiled by research workers in the Western Region cooperating under Regional Research Project W-1, 1955.

7 Roubicek, C. B., R. T. Clark, and O. F. Pahnish. "Range Cattle Production, 7, Genetics of Cattle—A Literature Review," ArizAESR 149, 1957.

8. Srb, A. M., and R. D. Owen. *General Genetics* (San Francisco: W. H. Freeman and Co., 1953), Chap. 10.

9. Stadler, L. J. "Genetic Effects of X-rays in Maize," PNAS, 14: 69, 1928.

QUESTIONS AND PROBLEMS

1. Explain the meaning of the term *mutation*.

2. What is the mode of inheritance of most mutations? What is the importance of a knowledge of this to the animal breeder?

3. Explain the difference between gene mutations and chromosome aberrations.

4. What is meant by a somatic cell mutation?

5. What causes a new mutation to become established in a population?

6. What is the importance of mutations to the animal breeders?

7. Explain why the artificial induction of mutations in farm animals is of little value at the present time.

6

Detrimental and Lethal
Genes in Farm Animals

The literature contains numerous reports of detrimental genes in farm animals. Some of these genes have such a drastic effect that they cause the death of the young during pregnancy or at the time of birth. Such genes are referred to as *lethal* (deadly) genes. Other genes, which are called *sublethal* or *semilethal*, cause the death of the young after birth or some time later in life. Still other genes do not cause death, but definitely reduce viability or vigor. These will be referred to as *nonlethal* or *detrimental* genes.

6.1 WHEN LETHAL OR DETRIMENTAL
EFFECTS OCCUR

A lethal gene may have its effect any time from the formation of the gamete until birth or shortly afterward. In a strain of horses, a sex-linked recessive lethal gene has been reported that kills approximately one-half of the male offspring of carrier females, so there are approximately twice as many females as males at birth. It is possible that such a genetic defect may also be present in other species of farm animals. Frequently a cow or mare is mated, and apparently conceives, because she does not show signs of estrus at the time of the next regular heat period, but returns to estrus at a later date. Possibly conception takes place, but the zygote or embryo dies because of lethal gene effects; it is resorbed, and the female resumes the normal estrous cycle. There is good evidence that lethal genes may cause losses in swine during pregnancy, because inbreeding increases embryonic death losses, whereas crossbreeding decreases them. This suggests that genes with nonadditive effects are involved.

The largest proportion of death losses in young farm animals occurs at the time of birth or within a few hours thereafter. Very often, when the animals also have some obvious external defect, studies have been made to determine whether lethal genes are involved. In many cases they have been found to be responsible, and the mode of inheritance has been determined. When the dead young show no obvious external defects, a possible genetic cause of death is less likely to be investigated.

Many times, lethal genes have their effects on the internal organs. Although such inherited traits have not been fully investigated, we know they must exist, for increased inbreeding is followed by higher death losses in the young from conception to weaning, whereas lower death losses during this period accompany crossbreeding. In fact, in most animals inbreeding and crossbreeding seem to have their greatest effects during this period of life. As was true with embryonic death losses, genes with nonadditive effects must be involved.

Sublethal or semilethal genes are responsible for some death losses in farm animals. Dwarfism in Herefords, resulting from the mating of Comprest with Comprest, is such an example. The dwarfs are born alive, as a general rule, but almost invariably they die before they are one year of age.

Other genes with no obvious visible effects may be detrimental; they express themselves in the reduction of life span or of vigor. Undoubtedly, many of these genes have escaped detection, even though they may be of as great economic importance as those genes with lethal or semilethal effects.

6.2 MODE OF EXPRESSION OF LETHAL OR DETRIMENTAL GENES

Most detrimental and lethal genes are either recessive or partially dominant and must be present in the homozygous state to have their full effect. In some instances, the partially dominant genes affect the heterozygous individuals so that they are intermediate in phenotype between the normal and the homozygous recessives.

Detrimental recessive genes are generally present at low frequencies in a population, and, in many cases, only inbreeding or linebreeding will cause their occurrence in the homozygous state. Knowing that homozygous recessives of detrimental genes will appear at greater rates with the practice of inbreeding, breeders have avoided and still avoid it. Because of the low frequency of these genes, breeders direct little effort toward eliminating or controlling their appearance in their herds. Sometimes, however, the frequency of a lethal or detrimental gene may become relatively high, as was the case with snorter dwarfism, which caused considerable economic loss to some breeders of purebred cattle in the 1950's. The frequency of the dwarf gene was so high in many herds that numerous individuals of the homozygous recessive genotype were produced even without inbreeding.

6.3 COAT COLOR AND DETRIMENTAL GENES

Experiments, mostly with small laboratory animals, indicate that in some instances genes affecting coat color also affect the vigor of the individual. In mice, it is known that coat color is affected by genes located at 24 or more loci on the chromosomes and that many of these genes have highly specific effects on other characters, such as skeletal growth and the development of certain body tissues [25, 30]. One of the first of these effects was found in a certain strain of yellow mice many years ago. When yellow mice were mated, they produced approximately two yellow to one nonyellow offspring rather than the 3:1 ratio expected if yellow were dominant and nonyellow were recessive and yellow mice were heterozygous. It was found that homozygous yellow individuals died at an early stage of gestation, and the surviving yellow animals were heterozygous. Thus, a lethal gene was related to the homozygous yellow color. Platinum foxes are also known to be heterozygous, because they produce two platinum to one silver offspring when mated. The homozygous platinum individuals apparently die before birth as a general rule.

Some lethal coat colors have also been reported in farm animals. In sheep of certain gray breeds, the mating of gray with gray individuals results in progeny of which one-fourth are black and three-fourths are gray. This indicates that black is recessive. A large proportion of the gray lambs possess an abnormal abomasum, as well as other defects of the digestive tract, that causes death within a few months after birth. In horses, there is also some evidence that the true albino of the homozygous genotype may die before birth [12].

Many other examples could be given that suggest a relationship between coat color and some other traits. It is not known for certain whether one gene is having the multiple effect or two or more genes are so closely linked on the same chromosome that only one seems to be responsible. In nearly all instances, the color genes having a detrimental effect are either dominant or partially dominant in their phenotypic expression.

6.4 EXAMPLES OF DETRIMENTAL AND LETHAL GENES IN FARM ANIMALS

Numerous reports in the literature have described the occurrence of detrimental and lethal genes in farm animals, and new reports appear each year. The mode of the inheritance of some of the traits has not been definitely established, because in some instances only a few affected animals were observed and mating tests to prove the mode of inheritance have not been made. This should be kept in mind when statements are made about the mode of inheritance of any one trait. Sometimes, traits are determined by more than one pair of genes, or by genes that vary in their expression or by genes showing incomplete penetrance. This serves to complicate the situation.

Some traits that are undesirable or lethal will be discussed in the remainder of the chapter. No attempt has been made to make this list complete. The list of references should emphasize the fact that many genetic defects have occurred and that breeders should recognize them when they occur in their herds and flocks. Methods of controlling or eliminating such genetic defects are discussed in Chapter 11.

6.4.1 Examples in Cattle

Achondroplasia 1

Affected calves have short vertebral columns, inguinal hernia, rounded and bulging forehead, cleft palates, and very short legs. The homozygous dominants are bulldog. the heterozygotes are Dexters, and the homozygous normals are Kerrys. About one-fourth of the bulldog calves are aborted after their death in the sixth to eighth month of pregnancy, following a pronounced accumulation of amniotic fluid in the dam. Similar genes have been observed in the Jersey, Hereford, and British Friesian breeds. The mode of inheritance is partially dominant, requiring two genes to have the lethal effect [14, 42].

Achondroplasia 2

This condition has been described in the Telemark cattle of Norway and is similar to the bulldog calf. Affected calves are carried to full term but die within a few days after birth due to respiratory obstruction. A similar condition has also been reported in Jerseys, Guernseys, and Ayrshires. The mode of inheritance appears to be recessive [57].

Achondroplasia 3

Described in the Jersey breed, the defect is quite variable in expression and is usually, but not always, lethal. Both the axial and appendicular skeleton may be affected. The head is deformed, being short and broad, and the legs are slightly reduced in length. In extreme cases, the calves are stillborn or die soon after birth. One affected calf lived for 14 months, when it was slaughtered. A recessive gene seems to be involved [22].

Agnathia

This lethal condition has been reported in Angus and Jersey cattle. The lower jaw is several inches shorter than the upper, and it has been observed only in male calves, so it may be a sex-linked recessive [18].

Amputated

The forelimbs end with the arm or humerus, and the hind legs are sometimes present and sometimes absent from the hock down. Cleft palate and hydrocephalus have also been observed. Calves are stillborn or die soon after birth. The defect was observed in Swedish Friesian cattle, and is a recessive [56].

Bulldog head (prognathism)

Observed in a grade Jersey herd. The skull is broad, the eye sockets large, the nasal bones short and broad, and the forehead broader than normal. The condition is associated with impaired vision in partial or full daylight. Recessive [7].

Cerebral hernia

Described in Holstein-Friesian calves. The affected calves have an opening in the skull because of a failure of ossification of the frontal bones. The brain tissue protrudes and is easily seen. Affected calves are stillborn or die soon after birth. Probably a recessive [50].

Comprest Herefords

An extreme form of compactness in body conformation involving a partially dominant gene. One gene produces a comprest individual, and two a dwarf, the comprest, or heterozygote, being more or less intermediate in phenotype between the dwarf and normal. The homozygous condition is usually lethal [11].

Congenital lethal spasms

Affected calves show a continual intermittent spasmodic movement of the head and neck, usually in the vertical plane. Lethal recessive [24].

Congenital cataract

The lens of the eyes of the affected calves shows an opaque body beneath the cornea. The cornea usually becomes enlarged as the animal grows older and becomes distorted in shape. Vision is somewhat reduced. Nonlethal recessive [23].

Curved limbs

A lethal character reported in Guernsey cattle. The hind legs are grossly deformed, with the hocks held close to the body and scarcely flexed forward. Probably a recessive [21].

Doddler cattle

Affected calves suffer extreme muscle spasms, convulsions, nystagmation, and dilation of the eyes. Respiratory movements are uncoordinated and difficult. A lethal recessive [28].

Ducklegged cattle

Observed in grade Herefords. The body is of normal size, but the legs are greatly shortened. Probably dominant and nonlethal [37].

Epilepsy

Symptoms are the lowering of the head, chewing of the tongue, foaming at the mouth, and finally a collapse into a coma. Attacks occur at irregular intervals and are usually brought on by undue excitement. Dominant [2].

Flexed pasterns

A semilethal condition in Jersey cattle in which the toes of the front feet in some cases are completely turned under. The toes of the hind feet are not affected. Affected calves are as vigorous as their normal sibs, but severely afflicted calves cannot nurse without aid. The abnormality is present at birth and persists for a few days or a week but gradually disappears. Recessive [3, 41].

Harelip

Described in Shorthorn cattle. Affected calves are unilaterally harelipped, and the dental pad on that side is missing, but the hard palate is formed. Young calves so affected experience difficulty nursing. Epistasis may be involved [55].

Hairlessness

This condition has been described in several breeds, but it is not known if it is determined by a recessive gene at the same locus. There is variation from partial to almost complete lack of hair. Most reports indicate that it is due to a recessive gene [32].

Hydrocephalus

Affected calves have a bulging forehead and enlargement of the cranial vault. The limbs and other bones are sometimes involved. It has been described in several breeds, and more recently in the Hereford. The trait is lethal in most cases and is recessive [13].

Hypoplasia of the ovary

An underdeveloped condition of the gonads in both sexes. When both gonads are involved, the animal is sterile; when one is involved, the animal is less fertile than normal. A recessive gene with reduced penetrance seems to be the cause [19].

Impacted molars

Described in milking Shorthorns. Impaction of the premolar teeth in the mandible, which is greatly reduced in length and width, giving a parrot-mouth appearance. The calves die within the first week after birth. Recessive [27].

Long-headed dwarf

Body proportions are similar to those of the snorter dwarf, but with advancing age the head becomes longer and narrower. Seems not to be due to the gene that causes snorter dwarfism. Recessive [6].

Multiple Lipomatosis

A large growth, consisting of adipose tissue in the peritineal area, appears at about 3.5 years of age and becomes progressively larger. In some cases, the fat deposition invades the udder and prevents the mammary system from functioning normally. It occurs in both males and females and appears to be a dominant gene with complete penetrance [4].

Muscle contracture

The limbs are bent, and the joints are rigid and ankylosed. The head is stiff and drawn up toward the back. Recessive lethal [29].

Muscular hypertrophy

First described in a crossbred Africander-Aberdeen Angus line, but present in other breeds also. The thighs are extremely thick and full, with a deep groove between the vestus lateralis and semimembranus muscles. Affected animals often assume an unusual stance, with the fore and rear legs extended anteriorly and posteriorly. Appears to be due to a recessive with variable expressivity. The heterozygotes appear to be favored in selection [33] in some present-day cattle because of the emphasis on muscular development in slaughter and breeding cattle. This trait is associated with lower fertility and increased difficulties at calving.

Polydactylism

Individuals with extra toes on one or all feet have been reported in several breeds. The trait is sometimes accompanied by lameness and therefore is undesirable. The mode of inheritance is not clear, but a dominant gene may be involved [47].

Prolonged gestation

Gestation is prolonged to 310 to 315 days, with calves weighing from 110 to 168 pounds at birth. Calves are thought to be homozygous for a lethal recessive gene [43].

Semihairlessness

Hair is absent from the margin of the ears and along the underline from the brisket to the udder, on the inside of the legs, the side of the neck, shoulder vein, sides, and thighs. Recessive [15].

Screwtail

This trait is caused by a fusion of one or more pairs of the coccygeal vertebrae at the end of the tail. Some calves show a double and some a single kink. Nonlethal recessive [35].

Short spine

The vertebral column is shortened to about one-half the normal length. Calves are stillborn or die shortly after birth. Recessive [44].

Snorter dwarfism

Dwarf calves are usually thick and blocky at birth, and the difference between dwarfs and normals becomes more noticeable with increasing age. Dwarfs have difficulty in breathing, hence the name "snorter." The same gene is present in both the Angus and Hereford and possibly in the Shorthorn. It may also be present at a low frequency in other breeds. Semilethal recessive [31].

Strabismus

The eyes of affected animals are crossed, and protrude abnormally. The trait is not evident at birth, but develops at 6 to 12 months of age. Recessive [46].

Stumpy

Affected individuals have curly hair coats, the tail switch is smaller than normal in amount and length, and achondroplastic conditions are more apparent in the forelegs than in the rear legs. Nonlethal recessive [5].

Syndactylism

Individuals have one rather than two toes on one or more of their feet. Probably recessive [17].

Umbilical hernia

Described in Holstein-Friesian cattle. The hernia appears at the age of 8 to 20 days and persists until the calves are 7 months of age. At that time the hernial sac seems to contract, permitting the hernial ring to close. Appears to be limited to males and is dominant [53].

White heifer disease

Observed in white dairy Shorthorn heifers. The hymen is constricted, the anterior vagina and cervix are missing, and the uterine body is rudimentary. Sex-limited recessive gene seems to be involved [51].

Wrytail

A malformation resulting in the distortion of the tail head, with the base of the tail being set at an angle to the backbone instead of in line with it. Found in several breeds. Recessive [1].

6.4.2 Examples in Horses

Abracia

This term refers to the absence of the fore limbs, and was observed in an inbred line. Probably a recessive lethal [39].

Aniridia

Characterized by the absence of the iris, with secondary cataracts. Observed in the Belgian breed. Dominant [20].

Atresia coli

A condition resulting from the closure or partial closure of the ascending colon in the region of the pelvic flexure. Observed in inbred Percherons. Lethal recessive [57].

Bleeding

The presence of fragile blood vessels in the nasal mucosa, which have a tendency to burst. Has been observed in the English Thoroughbred. It is a semilethal condition and seems to be recessive [57].

Epithelio-genesis imperfecta

Foals are born alive, but the hair coat is lacking in some areas of the body. Sometimes a hoof is missing. All foals die within a few days of birth. Recessive [36].

Frederiksborg lethal

Attempts to propagate a small group of rare white horses in Denmark led to inbreeding and apparently the dissolution and disappearance of the breed. It was suggested that a lethal recessive factor caused the death of the fetuses [57].

Hereditary foal ataxia

This has been reported in a German breed, the Oldenburg. Symptoms appear at 3 to 8 weeks of age. Affected animals first show periodical failure of muscular coordination or irregular muscular action, then collapse. Death occurs in 8 to 14 days after the symptoms become evident. Recessive [34].

Scrotal hernia

This trait appears to be due to an incompletely dominant gene with low penetrance [52].

Umbilical hernia

Reported as a simple recessive trait [26].

Wryneck

A contraction of the cervical muscles results in a twisted neck and unnatural position of the head. One form may not be inherited, but a congenital condition in foals is inherited as a lethal recessive [38].

6.4.3 Examples in Sheep

Amputated

The legs are missing at the fetlock joints in newborn lambs. Mode of inheritance not established [36].

Dwarfism

Parrot-mouth dwarfs have been observed in a strain of Southdown sheep. All lambs affected die within a month of birth. Semilethal recessive [9].

Lethal gray

The homozygous gray individuals appear to die either during embryonic life or early in postnatal life. The gray color is apparently due to a partially dominant gene [16].

Muscle contracture

The limbs are rigidly fixed in many abnormal positions at birth, with only a small amount of movement possible in the joints. This often makes parturition difficult. The lambs are nearly always dead at birth. Lethal recessive [48].

6.4.4 Examples in Swine

Atresia ani

A congenital condition resulting in the lack of an anus in both sow and boar pigs. Male pigs die within 2 or 3 days after birth, whereas sow pigs sometimes live and reproduce. An opening of the colon into the vagina in sow pigs allows defecation to take place through the vulva. The condition has been observed in several different breeds. It has been suggested that two pairs of dominant genes (epistasis) is involved, but other modes of gene expression and other genes, as well as environment, may be involved [8].

Hair whorls

Whorls of hair appear on different parts of the body. This condition is undesirable but is not lethal. Two pairs of dominant genes (epistasis) seem to be involved [45].

Hemophilia

The failure of the blood to clot has been observed in some animals within an inbred Poland line at about 2 months of age. The abnormality appears to increase in severity as the pigs grow older. Some boar pigs that were castrated late in life bled to death. Semilethal recessive [10].

Hydrocephalus

Affected pigs are born dead or die within a day or two. Lethal recessive [54].

Mule foot

A condition in swine where the hoof is solid, as in the mule. Nonlethal dominant [40].

Paralysis

The hind legs are affected, and the pigs crawl only by means of the fore-legs, which are less affected. All pigs die within a few days. Recessive [8].

Red eyes

Observed in Hampshire swine which also had a light-brown or sepia hair coat. Not lethal, but undesirable, because Hampshires are black with a white belt. Probably a recessive [49].

REFERENCES

1. Atkeson, F. W. "Prevalence of Wrytail in Cattle," JH, 35: 11, 1944.

2. Atkeson, F. W. "Inheritance of an Epileptic Type Character in Brown Swiss Cattle," JH, 35: 45, 1944.

3. Atkeson, F. W., F. Eldridge, and H. L. Ibsen. "Bowed Pasterns in Jersey Cattle," JH, 34: 25, 1943.

4. Albright, J. L. "Multiple Lipomatosis in Dairy Cattle," JH, 51: 231, 1960.

5. Baker, M. L., C. T. Blunn, and M. M. Oloufa. "Stumpy, a Recessive Achondroplasia in Shorthorn Cattle," JH, 41: 243, 1950.

6. Baker, M. L., C. T. Blunn, and M. Plum. "Dwarfism in Aberdeen-Angus Cattle," JH, 42: 141, 1951.

7. Becker, R. B., and P. T. Arnold. "Bulldog Head Cattle," JH, 40: 282 1949.

8. Berge, S. "The Inheritance of Paralyzed Hind Legs, Scrotal Hernia and Atresia Ani in Pigs," JH, 32: 271, 1941.

9. Bogart, R., and A. J. Dyer. "The Inheritance of Dwarfism in Sheep" (Abstract), JAS, 1: 87, 1942.

10. Bogart, R., and M. E. Muhrer. "The Inheritance of a Hemophilia-like Condition in Swine," JH, 33: 59, 1942.

11. Chambers, D., J. A. Whatley, Jr., and D. F. Stephens. "The Inheritance of Dwarfism in a Comprest Hereford Herd" (Abstract), JAS, 13: 956, 1954.

12. Castle, W. E. "The ABC of Color Inheritance in Horses," C, 35: 122, 1948.

13. Cole, C. L., and L. A. Moore. "Hydrocephalus, a Lethal in Cattle," JAR, 65: 483, 1942.

14. Craft, W. A., and H. W. Orr. "Thyroid Influence in Cattle," JH, 15: 255, 1924.

15. Craft, W. A., and W. L. Blizzard. "The Inheritance of Semihairlessness in Cattle," JH, 25: 385, 1934.

16. Eaton, O. N. "A Summary of Lethal Characters in Animals and Man," JH, 28: 320, 1937.

17. Eldridge, F. E., W. H. Smith, and W. M. McLeod. "Syndactylism in Holstein-Friesian Cattle," JH, 42: 241, 1951.

18. Ely, F., F. F. Hull and H. B. Morrison. "Agnathia, a New Bovine Lethal," JH, 30: 105, 1939.

19. Eriksson, K. "Hereditary Forms of Sterility in Cattle," in *Biological and Genetical Investigations* (I. Lund: Häkan Ohlssons "Boktryckeri," 1943).

20. Eriksson, K. "Hereditary Aniridia with Secondary Cataract in Horses," NV, 7: 773, 1955.

21. Freeman, A. E. "Curved Limb, a Lethal in Dairy Cattle, "JH, 5: 229, 1958.

22. Gregory, P. W., S. W. Mead, and W. M. Regan. "A New Type of Achondroplasia in Cattle," JH, 33: 317, 1942.

23. Gregory, P. W. "A Congenital Hereditary Eye Defect in Cattle," JH, 34: 125, 1943.

24. Gregory, P. W., S. W. Mead, and W. M. Regan. "Congenital Lethal Spasms in Jersey Cattle," JH, 35: 195, 1944.

25. Haldane, J. B. S. *The Biochemistry of Genetics* (London: George Allen and Unwin, Ltd., 1954).

26. Hamori, D. "Inheritance of the Tendency to Hernia in Horses" (t.t.), AL, 63: 136, 1940.

27. Heizer, E. E., and M. C. Hervey. "Impacted Molars—a New Lethal in Cattle," JH, 28: 123, 1937.

28. High, J. W. "Doddler Cattle, an Inherited Congenital Disorder in Hereford Cattle," JH, 49: 250, 1958.

29. Hutt, F. B. "A Hereditary Lethal Muscle Contracture in Cattle," JH, 25: 41, 1934.

30. Hutt, F. B. *Genetic Resistance to Disease in Domestic Animals* (Ithaca: Cornell University Press), Chap. 2, 1958.

31. Johnson, L. E., G. S. Harshfield, and W. McCone. "Dwarfism, a Hereditary Defect in Beef Cattle, JH, 41: 177, 1950.

32. Kidwell, J. F., and H. R. Guilbert. "A Recurrence of the Semi-hairless Gene in Cattle," JH, 41: 190, 1950.

33. Kidwell, J. F., and E. H. Vernon. "Muscular Hypertrophy in Cattle," JH, 43: 62, 1952.

34. Koch, P., and H. Fisher. "Oldenburg Foal Ataxia as a Hereditary Disease" (t.t.), TU, 7: 244, 1952.

35. Knapp, B. Jr., M. W. Emmel, and W. F. Ward. "The Inheritance of Screw Tail in Cattle," JH, 27: 269, 1936.

36. Lerner, I. M. "Lethal and Sub-lethal Characters in Farm Animals," JH, 35: 219, 1944.

37. Lush, J. L. "Duck-legged Cattle on Texas Ranches," JH, 21: 85, 1927.

38. Mauderer, H. "Hereditary Defects in the Horse" (t.t.), DTW, 46: 649, 1938.

39. Mauderer, H. "Abrachia and Torticollis; Lethal Factors in Horse Breeding" (t.t.), Dissertation, Hanover, Abstract in ZTZ, 51: 216, 1940.

40. McPhee, H. C., and O. G. Hankins. "Swine—Some Current Breeding Problems," USYA, p. 887, 1936.

41. Mead, S. W., P. W. Gregory, and W. M. Regan. "Hereditary Congenital Flexed Pasterns in Jersey Cattle," JH, 34: 367, 1943.

42. Mead, S. W., P. W. Gregory, and W. M. Regan. "A Recurrent Mutation of Dominant Achondroplasia in Cattle," JH, 37: 183, 1946.

43. Mead, S. W., P. W. Gregory, and W. M. Regan. "Prolonged Gestation of Genetic Origin in Cattle," JDS, 32: 705, 1949.

44. Mohr, O. L., and C. Wriedt. "Short Spine, a New Recessive Lethal in Cattle, with a Comparison of the Skeletal Deformities in Amputated Calves," JG, 22: 279, 1930.

45. Nordby, J. E. "Inheritance of Whorls in Hair of Swine," JH, 23: 397, 1932.

46. Regan, W. M., P. W. Gregory, and S. W. Mead. "Hereditary Strabismus in Jersey Cattle," JH, 35: 233, 1944.

47. Roberts, E. "Polydactylism in Cattle," JH, 12: 84, 1921.

48. Roberts, J. A. F. "The Inheritance of a Lethal Muscle Contracture in the Sheep," JG, 21: 57, 1929.

49. Roberts, E. "Occurrence of Red Eye in Swine," JH, 36: 207, 1945.

50. Shaw, A. O. "A Skull Defect in Cattle," JH, 29: 319, 1938.

51. Spriggs, D. N. "White Heifer Disease," VR, 58: 405, 1946.

52. Tuff, P. "Inheritance of Inguinal Hernia in Domestic Animals," NV-T, 57: 332, 1945.

53. Warren, T. R., and C. W. Atkeson. "Inheritance of Hernia in a Family of Holstein-Freisian Cattle," JH, 22: 347, 1931.

54. Warwick, E. J., A. B. Chapman, and B. Ross. "Some Anomalies in Pigs," JH, 34: 349, 1943.

55. Wheat, J. D. "Harelip in Shorthorn Cattle," JH, 51: 99, 1960.

56. Wriedt, C., and L. Mohr. "Amputated, a Recessive Lethal in Cattle; with a Discussion of Lethal Factors on the Principles of Livestock Breeding," JG, 20: 187, 1928.

57. Wriedt, C. *Heredity in Livestock* (London: Macmillan and Company, Ltd., 1930).

STUDY QUESTIONS

1. What is the difference between lethal, sublethal and semilethal genes?

2. At what time in the life of the individual may lethal genes have their effect?

3. On what part of the body may lethal genes have their effects?

4. May death losses at birth and shortly afterward be due to detrimental or lethal genes? Explain.

5. How do most lethal genes express themselves genotypically?

6. In what ways can coat color be related to lethal genes in animals?

7. A cattle producer complains that he has had some difficulty with extremely large calves at birth. These calves even cause difficulties in large, mature cows. Is it possible that his trouble is genetic? Explain.

8. A swine breeder reports that he has some pigs that are born with solid hooves, sometimes called the mule-footed condition. How could he eliminate this condition from his herd? How could he develop a pure strain for the mule-foot condition? (The mule-foot condition is dominant.)

9. Why is the mating of related individuals often associated with the occurrence of lethal or detrimental traits in farm animals?

10. What is indicated about the frequency of an undesirable gene when it is often expressed when nonrelated individuals are mated? Explain.

11. How important are lethal and undesirable genes in animal breeding? Why?

7

The Concept
of Gene Frequencies

In previous chapters the discussion of genetic principles was directed largely to individuals. In animal breeding, however, we also deal with the genetics of populations of individuals, and this is called *population genetics*. Certain principles are involved in population genetics that do not necessarily apply to single individuals. These will be discussed in this and subsequent chapters.

A population may be defined as the total of all individuals in a breed, species, or other such groupings, or as those individuals that inhabit a particular area. When we speak of population genetics, we refer to the genetic composition of one such particular population.

Controlled matings of individuals of known genotypes were used as examples in previous chapters to illustrate certain fundamental principles of Mendelian inheritance, and we learned that one can predict the genotypic ratios that will occur in the F_1 and F_2 generations with a certain degree of accuracy if large numbers are involved. In populations of animals where the genotypes of the breeding animals for a particular trait are seldom known and matings are not controlled, it is more difficult to predict with reasonable accuracy the genotypic ratios of the offspring. Research work in genetics of populations, however, has shown that the Mendelian laws of inheritance that operate in individual matings and crosses also apply to randomly mating populations.

7.1 GENE FREQUENCIES

The term *gene frequency* refers to the relative abundance or the relative rarity of a particular gene in a population as compared to its own alleles in that particular population. In the example of single-factor inheritance in Chapter 4, where polled bulls (*PP*) were mated to horned cows (*pp*), the alleles *P* and *p* were introduced into the F_1 generation in equal numbers. In actual practice, however, this may not occur, and these two genes may be present in the population in ratios quite different from those in the example.

For illustration, let us assume that we own a herd of horned Hereford cows and have used nothing but horned Hereford bulls for many, many years. The calf crop during this time has never included a naturally polled calf. Under such conditions, the frequency of the horned gene *p* would be one and the frequency of the polled gene *P* would be zero. Notice that we assume that the frequency of a gene varies between zero and one. When the frequency of a gene is one, as in this example for the horned gene, the population or herd is completely homozygous for that gene. On the other hand, if the frequency of the polled gene were one, which is very unlikely, the frequency of the horned gene would be zero, and we could say that the herd or population was completely homozygous for the polled gene.

For further illustration of the meaning of gene frequencies, let us make a mating of a homozygous polled bull with homozygous horned cows as follows:

P_1	Polled bull		Horned cows
	PP	$\times$	*pp*
F_1	All *Pp* Polled (*inter se*)		
F_2	1 *PP* Polled		
	2 *Pp* Polled		
	1 *pp* Horned		

We may now ask what the frequencies of the genes *P* and *p* would be in the F_1. Obviously, all individuals in this generation would possess one gene of each kind, so that frequency of each gene would still be 0.5, since the total of the frequencies of both genes must equal one.

We might also ask what the frequencies of the two genes are in the F_2. By calculation, we see that in the F_2 there are 4*P* and 4*p* genes, showing that the frequency of each gene is still 0.5. If we discarded all of the horned animals, however, there would be a total of 4 *P* genes and 2 *p* genes left, and the frequency of the two genes would be 0.67 and 0.33, respectively.

7.2 CALCULATING GENE FREQUENCIES IN A POPULATION WHEN NO DOMINANCE EXISTS

To illustrate how gene frequencies in a population may be calculated, it is best to use one pair of alleles in which no dominance exists. For such a trait, the genotype can be determined from the phenotype. An example is coat color inheritance in Shorthorn cattle, where the red gene (*R*) and the white gene (*r*) are involved. Red Shorthorns would be of genotype *RR*, roan Shorthorns of genotype *Rr*, and white, *rr*.

Let us assume that in a herd of 100 Shorthorn cattle 50 are red, 40 are roan, and 10 are white. What are the frequencies of the red and white genes in this herd? The calculations may be made as shown in Table 7.1. Note that in Table 7.1 there are 100 individuals and therefore a total of 200 red and white genes, because genes occur in pairs in body cells.

Table 7.1 Illustration of the calculation of gene frequencies in a population where there is no dominance and where the genotype may be determined from the phenotype

Coat color (phenotype)	Genotype	Number of individuals	Number of	
			(R) genes	(r) genes
Red	RR	50	100	0
Roan	Rr	40	40	40
White	rr	10	0	20
Totals		100	140	60

To calculate the frequencies of the red and white genes, note that 140 out of the 200 genes are red (*R*), and thus $\frac{140}{200}$ gives 0.70, the frequency of the red gene. The frequency of the white gene (*r*) would be $\frac{60}{200}$, or 0.30. Thus, the frequency of the red gene (0.70) plus the frequency of the white gene (0.30) is one. In instances where the frequency of one of the two alleles is known, the frequency of the other allele may be calculated by subtracting the known frequency of the allele from 1.00, which gives the frequency of the other allele. For example, if the frequency of the red gene in our example is known to be 0.70, the frequency of the white gene would be 1.00 minus

0.70, or 0.30. In a situation where more than two alleles are present in a population (multiple alleles) one must determine the frequency of all alleles but one. Subtracting the sum of the frequencies of these alleles from one will give the frequency of the remaining allele. The sum of the frequency of all alleles is one.

In working with gene frequencies in populations, it may be helpful to think of genes in terms of pools rather than in terms of pairs. This is illustrated by the following circle, which represents the confines of a particular population containing two alleles with different frequencies. If we use the red gene (*R*) and the white gene (*r*) with the frequencies calculated previously, there will be 7 red genes for every 3 white genes in that population. If we are able to reach into this circle and pull out a single gene without knowing

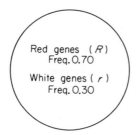

which we are getting, what would be the probability that we would get a red gene? Obviously, if 0.70 of the genes are red, the probability of picking a red gene is 0.70. The probability of picking a white gene is 0.30. The same probabilities hold true for any one gamete (sperm or egg) that would be produced in this population as long as the genes are thoroughly mixed and we select a single gamete at random. In other words, *the probability of any one gamete carrying a particular gene is equal to the frequency of that particular gene in the population.*

We may ask what is the probability that a sperm from this population carrying the red gene will fertilize an egg carrying the red gene? To answer this question we must use the law of probability, which states that the probability of two or more independent events, occurring together is equal to the product of the probabilities for each event separately. Thus, the probability of a sperm and egg from this population getting together at fertilization to pair two red genes in the zygote is 0.70 times 0.70, or 0.49, the latter figure being an estimate of the proportion of the individuals produced in this population that would be homozygous red (*RR*). The probability of two gametes (sperm and egg) carrying the white gene getting together at fertilization would be 0.30 times 0.30 or 0.09, which is the proportion of homozygous white (*rr*) individuals we would expect from such a population.

The proportion of roan individuals (*Rr*) can also be calculated in a similar manner. In this same population the probability of a sperm carrying the red gene fertilizing an egg carrying the white gene is 0.70 times 0.30, or 0.21. However, a sperm carrying the white gene can also fertilize an egg carrying the red gene, giving a combined probability of 0.30 times 0.70, or 0.21. Since the (*Rr*) and (*rR*) individuals are exactly alike for these two alleles, the proportion of roan individuals produced by such a population is 0.21 plus 0.21, or 0.42. The proportion of heterozygous individuals in a population, may be calculated by multiplying the product of the two gene frequencies by two.

It is very unlikely that in a population the frequency of one gene would be 1.00 and that of its allele 0 unless inbreeding were almost 100 percent and there had been continuous and vigorous selection against one of the two alleles. For example, selection against a dominant allele would reduce its frequency to 0, providing a new mutation to the recessive did not occur. The polled gene in Angus cattle appears to have a frequency close to 1.00, since purebred horned Angus probably are never seen. However, it is not uncommon for Angus cattle, especially males, to have scurs or rudimentary horns. Scurs are controlled by a different pair of genes at a locus different from that of horns.

As a general rule, two or more genes may exist at a definite ratio in a large population and may remain at or near this same ratio more or less constantly for many generations. A number of forces may be responsible for maintaining this balance of alleles in a large population, one of which is mutation equilibrium, to be discussed later. The presence of a balance of two or more alleles in a population for a long period of time is known as polymorphism (*poly* means many; *morphism* means forms). A good example is the *A*, *B*, and *O* blood group in humans.

7.3   CALCULATING GENE FREQUENCIES WHEN DOMINANCE IS COMPLETE

A different procedure must be used to calculate the frequency of two alleles in a population when dominance is complete. This is because only two phenotypes are expressed for the three genotypes, with the homozygous and heterozygous dominant genotypes being indistinguishable phenotypically. The Hardy-Weinberg Law must be used in this case.

7.3.1   The Hardy-Weinberg Law

In 1908, Hardy, a British mathematician, and Weinberg, a German physician, working independently, published certain fundamental ideas of gene distributions in populations. These are known as the Hardy-Weinberg Law. This law states that in large populations, where (1) the frequency of one of two alleles is equal to a, (2) the frequency of the other is equal to b, (3) the sum of the frequencies, $a + b$ equals one, and (4) matings are at random, the offspring of the three genotypes will occur in a definite ratio, or will be in equilibrium in the next generation, at the frequencies of a^2, $2ab$, and b^2. From this definition, then, after one generation of random mating, individuals of the genotypes involving the two alleles will occur at the following frequencies:

> Number of homozygous dominant individuals will equal a^2
> Number of heterozygous individuals will equal $2ab$
> Number of homozygous recessive individuals will equal b^2

The use of this law in calculating gene frequencies will be illustrated later.

One important requirement for the Hardy-Weinberg Law to be correct is that the matings must be at random. Random mating in a population for a certain character means that matings of individuals are made without attention to their genotype for that trait. Probably one of the best examples of the results of random mating is the frequencies of human blood types involving the (M) and (N) alleles. Men and women pay no attention to the blood types of their prospective spouses. In fact, very few people know what their genotype is for these blood types, because little physiological importance has been shown for this trait.

Some traits in animals are due to homozygous recessive genes. For example, snorter dwarfism in beef cattle is recessive, whereas the normal size is dominant, so we cannot tell by visual observation which animals are heterozygous for the recessive dwarf gene and which are homozygous normal. We may, in some instances, be able to determine the proportion of heterozygous carriers of the dwarf gene in the population. The accuracy of these calculations depends upon (1) the size of the population, (2) the correct determination of the proportion of homozygous recessive individuals, and (3) the random mating of the individuals in the population for at least one generation.

To illustrate the use of a gene frequency analysis, let us assume that, in purebred cattle of a certain breed, four out of every 100 calves dropped are dwarfs. Our problem is to determine the probable frequency of the dwarf gene in the progeny and the probable frequency of the heterozygous individuals. For the calculations, we shall refer to the normal allele as D and the dwarf allele as d. Substituting D for a and d for b in the binomial, we obtain the formula $D^2 + 2Dd + d^2$. The frequency, or the proportion of dwarfs in the progeny, would be equal to d^2 or 0.04, and the frequency of the dwarf gene in the progeny would be the square root of 0.04 or 0.20. The frequency of the normal gene D would be $1 - 0.20$, or 0.80. The proportion of heterozygous individuals in the progeny would be 2(0.80 × 0.20), or 0.32. In other words, 32 out of each 100 of the calves should be carriers of the dwarf gene.

Records indicate that approximately one black Angus calf out of every 200 is red instead of black. Using the same procedure as for the dwarf gene, we find that the frequency of the red gene is the square root of 0.005, or 0.07, and the frequency of the black gene is 1 − 0.07, or 0.930 in the progeny. The expected frequency of the heterozygous black progeny from parents mated at random for the red and black color would be 2(0.07 × 0.93), or about 13 out of each 100. If only black parents were used for breeding, the frequency of the heterozygous black individual would be about 13 percent.

The calculation of gene frequencies for a sex-linked recessive trait is rather simple, although the same conditions necessary for the application of the Hardy-Weinberg Law to recessive traits also apply. By sex-linked is meant those genes carried on the portion of the X chromosome that is not homologous with the Y chromosome (Chapter 4). For a sex-linked recessive trait the frequency of the gene is equal to the proportion of men in the population that possess and express the trait phenotypically. The frequency of the dominant allele of a sex-linked recessive gene is calculated as usual by subtracting the frequency of the sex-linked recessive gene from 1.00. For example, if 8 percent of the men in a population show a sex-linked recessive trait, the frequency of this gene is 0.08. The frequency of its dominant allele would be 1 minus 0.08, or 0.92.

Knowledge gained from a gene frequency analysis may be used to determine the possible mode of inheritance of a particular trait in a population. The basic concept of gene frequencies and the factors responsible for their variation within a population are important in understanding genetic variation within and between populations as well as the genotypic effects of selection, inbreeding, linebreeding, outbreeding, and crossbreeding. In fact, most, if not all, of the genetic variation among breeds within a species is due to differences in gene frequencies.

7.4 FACTORS MODIFYING GENE FREQUENCIES

The frequency of the same allele may be quite different in separate populations within a single species. For example, the frequency of the M blood type gene determined from samples of the human population show that it is 0.913 in Eskimos, 0.719 in Russians, 0.524 in the English, and 0.176 in Australian aborigines. The frequency of the black coat color gene is 0 in Hereford cattle but very high in Angus. The frequency of many other genes, some of which do not have a major phenotypic effect, also varies in different populations.

This raises the question of why they vary. The following are some factors responsible for such variations.

7.4.1 Selection*

This is a very important factor that may be responsible for changes in gene frequencies. With respect to farm animals, selection means that individuals possessing certain desirable traits are caused or allowed to produce the next generation. Thus, individuals of a certain genotype may be retained in larger numbers for breeding purposes than others. This causes an increase in the frequency of some genes and a decrease in the frequencies of others.

To illustrate how selection may change gene frequencies, assume that we have a herd of 100 Shorthorn cows consisting of 25 red, 50 roan, and 25 white. In this herd, the frequency of the red and white genes would be equal (0.50 each). Let us assume that for some reason we decide to cull and sell all of the white individuals. Selling these would leave 25 red and 50 roan and would increase the frequency of the red gene to 0.667 and decrease the frequency of the white to 0.333. By culling the white and roan individuals in this herd, the frequency of the white gene would be reduced to 0 and that of the red gene would be increased to 1.00.

Selection can be of two general types, artificial and natural. Artificial selection is that practiced by man, whereas natural selection is that done by nature in causing the death of the less viable individuals. Both types of selection have been very effective in the past when considered over a long period of time, as shown by the presence of different breeds of sheep, horses, hogs, dairy cattle, and beef cattle. One of the main differences between breeds is that they differ in the frequencies of genes for certain traits.

The genetic principles involved in selection will be discussed in more detail in a later chapter.

7.4.2 Gene Mutations†

It was pointed out in Chapter 5 that a point, or gene mutation, may be due to a change in the code sent by the DNA molecule by means of mRNA to the ribosomes in the cytoplasm, where amino acids are assembled into proteins. Much of the wide genetic variations among individuals within a

---

*Selection has important effects on gene frequencies. It is the only tool available to man that he can use to make permanent changes in the productivity of populations by changing the frequency of desired genes.

†The effects of mutations on gene frequencies are extremely slow and usually are important only over a period of many years.

species originated within the genetic material. It appears that certain individuals possess the preferred genetic composition that makes them better adapted to a particular environment, and therefore they are more likely to survive than are those individuals that have a less preferred genetic composition for that environment. This is called *natural selection.*

Even detrimental genes may occur at a high frequency in certain populations and may be completely absent or at a low frequency in another. Since all such genes began with a new mutation, it is interesting to discuss some of the factors that may be responsible for a newly mutated gene obtaining a foothold in a population. These include a frequent mutation rate, selection favoring the heterozygote, and possibly a change in the usual segregation ratio of genes in the gametes so that the chromosome carrying the new mutation segregates at a much higher rate than normal, or linkage to another gene which has a decided advantage in selection.

Genes differ in their rate of mutation. Some mutate more often than others. For example, the mutation rate must be very high for certain recessive traits in humans, such as thalassemia and sickle-cell anemia, in which a large proportion of the individuals in a population may be homozygous recessive. Other traits, such as albinism and pseudohypertrophic muscular dystrophy, appear to have a much lower mutation rate, ranging between 25 and 60 per million per generation [5]. No doubt a high recurrent mutation rate alone is not responsible for the high frequency of a newly mutated recessive gene in a population, but it is one of the predisposing factors involved in a gene obtaining a foothold in a population.

Selection favoring the heterozygote is another factor that may be responsible for a gene gaining a foothold and rising to a relatively high frequency in a population. Many examples in which selection favors the heterozygote are known. The sickle-cell trait is one of the best-known examples in humans. Sickle-cell anemia is a disease most common among African Negroes and results when the individual is homozygous for the sickle-cell hemoglobin gene (Hb_sHb_s). Affected individuals suffer from a very severe anemia and often die in childhood, although a few may survive to adult life. In some parts of Africa up to 4 percent of the population (and even more in some areas) suffer from this disease. This means that about 32 percent of the individuals in the population are carriers of the sickle-cell gene (genotype Hb_sHb_a). Since individuals suffering from anemia seldom live to reproduce, it is obvious that the sickle-cell gene must be propagated by heterozygous individuals. Research [1] has shown that at least in some areas of Africa individuals heterozygous for the sickle-cell gene are preferred by natural selection, because a larger proportion of the homozygous normal individuals than heterozygous individuals die of malaria, and those homozygous for the sickle-cell gene die of anemia. The heterozygote does not seem to have a selective advantage in the areas free of malaria.

Another example in which the heterozygote appears to have been preferred to either homozygote in selection is the snorter dwarf trait in beef cattle. This recessive trait became a problem in purebred beef cattle in the late 1950's and the early 1960's. In spite of the fact that few, if any, homozygous recessive dwarfs were used for breeding, the incidence of dwarfism increased to the point where it was an important economic problem. The reason for an increased incidence of dwarfism in certain breeds of beef cattle was thought to be due to a certain degree of heterozygous advantage over the nondwarf gene carrier individuals (homozygous normal genotype), probably in the selection of herd bulls. At that particular time the animals of medium to small size with a thick, compact body, a beautiful head, and showing a great deal of masculinity were favored in selection. These characteristics were probably found more often in the heterozygous carriers of the dwarf gene, and may have been the main cause of increased frequency of dwarfs in purebred beef cattle herds. A frequent mutation rate of the normal to the dwarf gene may also have been a minor factor.

In selection practiced by man the homozygous recessive individual may be preferred in selection. A good example in cattle is the deliberate establishment of a Red Angus breed by mating Red Angus with Red Angus. The red gene in Angus is recessive, so it should breed true.

Favoring the heterozygous individual in natural selection depends upon whether or not it is more adaptable to a certain environment and has a higher reproductive and survival rate than either homozygote. In such an environment even a slight heterozygote advantage may increase the frequency of a mutant gene although the greater the advantage the heterozygote enjoys, the more rapid the increase in the frequency of the mutant gene. In areas where there is no adaptive advantage to the heterozygous individual, no increase in the frequency of a gene should occur. In fact, the frequency of a detrimental gene would decrease under such conditions if selection were against the homozygous recessive individual.

Selection practiced by man often results in the preference of individuals of a certain genotype over those of another. The trait preferred may, or may not, have an advantage in natural selection. For example, selection for the palomino color in horses automatically results in selection for the heterozygous individual, which probably has no advantage in natural selection. Man sometimes selects for homozygotes that may not have an advantage in natural selection. Man may even select for a particular genotype in direct opposition to natural selection. For example, man has selected for the albino rabbit, but nature selects against this color because individuals that are albinos may be less vigorous and certainly more susceptible to death from predators than those of a more drab color that would blend into the surrounding landscape.

Any factor that distorts the usual segregation ratio of gene pairs into the haploid number (half pairs) in the gametes may cause a change in the frequency of alleles in a population. For example, the probability that an individual of the heterozygous genotype *Aa* will produce a gamete-carrying gene *A* is 0.50. The probability of producing a gamete-carrying gene *a* is also 0.50. If something would cause the destruction of some of the gametes (or chromosomes) carrying the *A* gene, the proportion of gametes carrying the *a* gene would be much higher than normal and would favor the increased frequency of gene *a* in a population. A segregation distorter such as this has been reported in insects, but has not been reported in farm animals.

When new mutations gain a foothold in a population, this can cause a change in the frequency of the other allele. For example, suppose that a population is completely homozygous for the *A* gene (frequency of 1.00). It is conceivable that, with new generations of individuals being produced, this gene may mutate to *a*. As more generations are produced, it is possible for gene *a* to increase in frequency at the expense of gene *A*. Actually, for mutations to have much effect in changing gene frequencies, they must occur quite often, and there must be a definite selective advantage for the new gene. Under such conditions, it is conceivable that eventually the new allele would largely replace the original one. A point should be reached, however, before the original gene is eliminated, where selection in one direction to eliminate one gene will finally equal the mutation rate in the other, so that the gene frequency in the population would become stable, that is, in equilibrium. This point, where the elimination of a gene becomes equal to its replacement by mutation, is called a *mutation equilibrium*.

Mutation equilibrium may be reached, theoretically, for another reason. It is well known that mutations are reversible, in that there is not only a mutation from gene *A* to *a*, but one from *a* back to *A* again. Furthermore, these reversible mutations are not always equal in their occurrence. If we suppose that before a mutation occurs, the frequency of the *A* gene is 1.00, and that the mutations from *A* to *a* occur twice as rapidly as those in the opposite direction, a mutation equilibrium will eventually be reached. At first, the shift in the gene frequency would be toward a higher level of the *a* allele, Eventually, however, if the frequency of the *a* allele became twice as great as that of the *A* allele, which mutated at twice the rate, there would be a status quo, or equilibrium of the alleles in the population.

7.4.3 A Mixture of Populations

This may be another factor responsible for changes in gene frequencies in a population.

Suppose we have a population of humans in which the frequency of the *M* blood type gene is 0.915. In another population some miles away the frequency of this same gene is 0.215. If individuals from the first group migrated to the second and they intermarried, the frequency of the M gene in the children (descendants) would average somewhere between 0.215 and 0.915, depending upon how much intermarriage occurred.

Another example of a mixture of populations is similar to what often occurs in herds of beef cattle and other farm animals. Assume that a farmer has a herd of 100 horned Hereford cows, but he decides to use a polled bull to produce polled calves (polled is dominant to horns in beef cattle). If he purchases a homozygous polled bull (genotype PP) and each cow produces a calf, the frequency of the polled gene among the 100 calves would be 0.50, since all calves would receive the polled gene (P) from their sire and the horned gene (p) from their dam. Thus, the mixture of populations in this example increased the frequency of the polled gene in the calves to 0.50 in one generation as compared to a frequency of 0 in their dams. A mixture of populations could cause a change in the frequency of any or all genes in a population in the same manner. Undoubtedly, people in the United States who have ancestors of many nationalities in their background have different gene frequencies, as a population, than the pure stock.

7.4.4 Genetic Drift or Sampling Nature of Inheritance

This is another factor that may be responsible for changes in gene frequencies and may result in a type of genetic variation within a population that is not due to natural selection. It is a random statistical fluctuation independent of natural selection that sometimes is also referred to as the sampling nature of inheritance. Wright [6] has called this mechanism affecting frequencies *random genetic drift*.

Theoretically, genetic drift might happen in the following way. Suppose that in a large population of humans the frequency of the M gene were 0.60 and that of the N gene were 0.40. Let us also assume that a small group of people (say a dozen) left the larger group and migrated to a new country to settle in a remote area where they were isolated from other humans. It is possible that the frequency of the M gene in this small group might be very low and that of the N gene very high, or vice versa, just because of chance, due to a small population (sample) coming from a larger one. In actual practice, the genetic makeup of each individual in the small group leaving the larger one might be almost the same, because often they could be related. Relatives are more likely to be carrying the same genes than nonrelatives. They probably would be relatives, because those couples leaving the larger population would be expected to take their own children and possibly other relatives with them. Undoubtedly, many tribes of American Indians and many different nationalities of people originated in this way from a relatively small group. Under favorable circumstances, a small population could greatly expand in numbers in a few generations.

Genetic drift may also be important in the development of inbred lines of livestock, where a parent-offspring or a brother-sister mating system is followed for several generations. This could result in several sublines being developed within the original inbred population, and genetic drift could be responsible for pulling the sublines apart because the original base for establishing each subline was very narrow or small.

An example illustrating genetic drift when inbreeding was practiced was reported in a study of blood groups in poultry [2]. Up to eight different alleles were present at the B-locus in a group of individuals before inbreeding was practiced. After inbreeding reached 66 percent or more, the number of alleles at this locus had been reduced to two. Other factors such as selection may have also been involved in the loss of certain of the B alleles, but undoubtedly genetic drift due to inbreeding in a small population was involved.

Population size [3] may be responsible for the importance of genetic drift in changing the frequencies of genes in a population. In a small population a gene might either disappear or become fixed in a few generations, because the population is so small that even a slight change in the number of people carrying a gene might cause a large change in the percentage of the total population showing the gene, or not showing it. In a large population a change in the frequency of a gene would affect a smaller percentage of the population. In a small isolated population inbreeding increases from necessity, whereas in a larger one it is often decreased.

In a small population, the chance segregation of genes in the gametes and recombination in the zygote may be the main cause of genetic drift. In fact, this may be a more important cause of genetic drift than population sampling in some instances.

7.5 SUMMARY

The concept of gene frequencies may be applied to the genetics of populations. By gene frequencies is meant the relative abundance or relative rarity of a particular gene as compared to other alleles in that population.

Methods for calculating gene frequencies in different populations are known. With two alleles considered, when the frequency of one allele is 1.0, the frequency of the other allele is 0. With a frequency of 1.00 for one allele, this means that all individuals in that particular population are homozygous for that allele. Differences in gene frequencies in a population may vary between 0 and 1, and the frequencies of the two alleles give some indication of the amount of genetic variation that exists among individuals in that population for that particular locus on the chromosome.

The frequencies of the same alleles may be quite different in separate populations within the same species. Four factors are responsible for these differences. They include mutations, selection, genetic drift, and mixtures of populations.

REFERENCES

1. Allison, A. C. "Protection Afforded by the Sickle-cell Trait against Subtertian Malarial Infection," BMJ, No. 4857, Pt. 1, p. 290, 1954.
2. Briles, W. E., C. P. Allen, and T. W. Millen. "The Blood Group System in Chickens; I. Heterozygosity in Closed Populations," G. 42: 631, 1957.
3. Cavalli-Sfarza, L. L. "Genetic Drift in an Italian Population." S. Z. 221: 30–37, 1969.
4. Hardy, G. H. "Mendelian Proportions in Mixed Populations," S, 28: 49, 1908.
5. Penrose, L. S. and H. L. Brown. "Recent Advances in Human Genetics," Little, Brown and Co., Boston, 1961.
6. Wright, S. "Fisher's Theory of Dominance," A. N., 63: 274, 1929.

STUDY QUESTIONS

1. What is a population? Are matings controlled in nature? Explain.

2. Define gene frequencies. What would be the frequency of the recessive allele in the progeny when two parents heterozygous *Aa* were crossed?

3. In a herd of 100 Shorthorns, 60 were red, 30 were roan and 10 were white. What is the frequency of the red gene in this herd? Calculate the frequency of the white gene in two different ways.

4. If the frequency of the red gene in a large herd of Shorthorns were 1.00, how many would you expect to be roan? How many would be white?

5. Assume that the frequency of the red gene in a large herd of Shorthorns is 0.80. How many of the individuals should be red? White? Roan?

6. Define polymorphism.

7. What are the fundamental points necessary for the Hardy-Weinberg Law to be accurate in calculating gene frequencies?

8. Assume that one calf in each 10,000 born in the United States has a recessive condition known as hydrocephalus. What is the frequency of this recessive gene? What is the frequency of its dominant allele? How many are heterozygous (percent) for the recessive gene?

9. How could a gene such as the one mentioned in question 8 have gained a foothold in the cattle population in the United States?

10. Assume that 4 percent of the males in a swine population have a recessive defect that is sex linked. What is the frequency of this recessive gene? What is the frequency of its dominant allele? What proportion of females should show this trait, providing the males live to reproduce? What percentage of females should be carriers of this sex-linked recessive?

11. Estimates indicate that 12 to 13 percent of the Black Angus are carriers of the red gene (*b*). What factor, or factors, may be responsible for this gene remaining at this relatively high frequency in Black Angus even though the Red Angus are not saved for breeding?

12. Explain how selection may modify gene frequencies in a population. What types of selection exist?

13. Basically, what is a mutation? Do all genes mutate at the same rate? Explain.

14. If the heterozygous individual is favored in a particular environment, will the same thing be true in another environment? Why?

15. Does man ever favor the heterozygous individual in selection? Explain.

16. What is meant by mutation equilibrium? Explain how this condition is attained.

17. Explain what is meant by a mixture of populations and how this might change gene frequencies.

18. What is meant by genetic drift? How can this affect gene frequencies?

19. How can the size of the population be responsible for inbreeding?

20. Inbreeding as such does not change gene frequencies, but when it is practiced the frequency of recessive genes is likely to decrease. Explain how this might be true.

8

Quantitative Inheritance and Its Measurement

Economic traits are those which have monetary value in the production of livestock. Some of these traits in purebred animals, such as coat color, are of importance from the standpoint of breed type and are determined by only one pair of genes, or only a few pairs, with a sharp distinction between phenotypes. For example, there is a sharp distinction between black and red coat color in Angus cattle. Traits such as these are known as *qualitative traits.* Many other traits in farm animals are affected by many pairs of genes and there is no sharp distinction between the phenotypes, but there is a more or less continuous range from one phenotypic extreme to another. These are known as *quantitative traits* and include such economic traits as gestation length, birth weight, weaning weight, rate and efficiency of gain in the feed lot, carcass quality and quantity, milk production, and egg production, as well as many others.

Most quantitative traits are greatly influenced by environmental conditions, whereas qualitative traits are influenced but little, as a general rule. The fact that environment has a great influence on the expression of most quantitative traits and they are influenced by many pairs of genes makes it necessary to consider certain special principles and procedures in improving such traits through breeding that are not necessary when only qualitative traits are involved. We shall emphasize these in the remainder of the text, although we shall consider the improvement of qualitative traits through breeding from time to time.

8.1 HYPOTHETICAL EXAMPLE OF QUANTITATIVE INHERITANCE IN FARM ANIMALS

For an example of quantitative inheritance, let us use backfat thickness in swine. Actually, many pairs of genes may be involved in the expression of this trait, and their action may be nonadditive as well as additive in nature. In addition, each gene may not contribute equally to the control of production of backfat. Environment can also play a very important part in the expression of this trait. However, for the sake of simplicity, we shall use only two different pairs of alleles, and we shall assume that each contributes equally toward backfat thickness.

Let us assume that the residual genotype is 0.80 inch of backfat thickness, since it has been suggested [3] that this probably is the least amount of backfat, from the genetic standpoint, that we can expect to obtain in pigs at the usual market weight of 200 to 225 pounds.

The following information will be used in this example:

B and *F* are contributing genes that add 0.20 inch of backfat thickness in swine.

b and *f* are neutral genes that add nothing.

bbff is the residual genotype, with 0.80 inch of backfat.

P_1	BBFF	×	bbff
	1.60 inches		0.80 inch
F_1		BbFf (*inter se*)	
		1.20 inches	
F_2	Genotypes	Phenotypes	
	1 BBFF	1.60 inches	
	2 BBFf	1.40 ″	
	1 BBff	1.20 ″	
	2 BbFF	1.40 ″	
	4 BbFf	1.20 ″	
	2 Bbff	1.00 inch	
	1 bbFF	1.20 inches	
	2 bbFf	1.00 inch	
	1 bbff	0.80 ″	

The phenotypic ratio in the F_2 would be

<pre>
1   1.60 inches
4   1.40   "   Mean = 1.20 inches
6   1.20   "   Range = 0.80 to 1.60
4   1.00 inch
1   0.80   "
</pre>

As shown in Fig. 8.1, the F_2 individuals form a normal frequency-distribution curve.

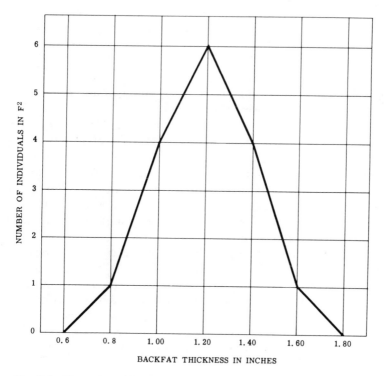

BACKFAT THICKNESS IN INCHES

Fig. 8.1 Illustration of how the phenotypic ratio of the dihybrid cross involving only additive gene action and no environmental effects forms a normal frequency distribution curve.

Let us look at the results of these crosses. Notice that the mean of the F_1 offspring was 1.20 inches in backfat, which coincides exactly with the mean or average of the two parents, and that the variation of individuals from this mean is zero. Considerable variation is noted, however, in the F_2 individuals, with a range from 0.80 inch to 1.60 inches. But when we calculate the mean for the F_2 individuals, we find that it also is 1.20 inches, the same as in the F_1. This is an excellent example of how two means may be the same but the individuals within the population may vary widely in one instance and not at all in another.

In actual practice, the instances will be rare in which the results are as perfect as those in the example above. One reason is that, in selecting individuals for the P_1 mating, it would be difficult, if not impossible, to find those that were homozygous for contributing genes or neutral genes. The trait will have been affected by environment, sex, and amount and kinds of feed, as well as other factors. These other effects would be confusing in the other generations as well. Another important factor, in practice, is that often the mean of the F_1 individuals does not coincide exactly with that of the parents, but may be closer to the mean of one or the other parental group. Many times, of course, this could be due to chance fluctuations, but at other times it may actually be a real observation.

In practice, the fact that the mean of the F_1 does not coincide with the mean of the parents would indicate that the genes involved contribute their effects through certain kinds of interactions rather than in an additive manner. Instead of adding or subtracting constant amounts, they seem to multiply or divide the deviation from the residual genotype by some constant amount. As a result, the mean of the F_1 population is closer to the geometric mean than to the arithmetic mean of parents. The geometric mean is the square root of the product of the parental means. If genes act in a multiplicative manner in affecting a quantitative trait, any deviation of the F_1 from the geometric mean of the two parental groups would indicate that dominance and epistasis are involved.

Evidence is now accumulating that additive gene action alone is not responsible for the inheritance of quantitative characters, but that other types of gene action such as overdominance, dominance, and epistasis may also be important. Furthermore, environment is an important cause of variation in most traits. The determination of the relative influence of each of these different kinds of gene action and of environmental factors on economic traits in livestock would be helpful in devising the most effective mating and selection systems.

8.2 ACTUAL EXAMPLE OF QUANTITATIVE INHERITANCE IN FARM ANIMALS

Backfat thicknesses of 270 pigs, measured at a market weight of 200 pounds, are presented graphically in Fig. 8.2. Note that these figures assume the bell-shaped frequency-distribution curve that is characteristic of quantitative traits in a population. In this actual example, environment, as well as heredity, has caused variations. For backfat thickness as well as for other traits, the environmental effects may be very large, and the breeder must control them as much as possible. For instance, animals to be compared should be given the same kind and amount of food. For factors that cannot be controlled, such as age and sex, adjustments should be made in the figures. The methods for making such adjustments will be discussed in each chapter on each of the species.

8.3 STATISTICAL METHODS FOR MEASURING QUANTITATIVE TRAITS

Since, in quantitative inheritance, the phenotypes are not distinct and separate but exhibit a series of variations between the extremes, mathematical methods have been devised for measuring and describing populations. Some of these methods will now be discussed, using the backfat thickness of 10 pigs picked at random in the population of 270 pigs recorded in Fig. 8.2

8.3.1 The Mean

Everyone has calculated averages, or means, so little time will be spent on this particular statistical measurement. Nevertheless, to help the student to become familiar with the use of symbols and formulas, the mean may be stated as follows:

$$\bar{X} = \frac{\sum X}{n}$$

Many statisticians refer to the mean for a group of individual observations as bar X or $\bar{X}$. The symbol X refers to each individual item or observation, and the Greek symbol $\sum$ means to add all items in the group. The letter n refers to the number of X items in the group of data to be summed. Using data tabulated in Table 8.1, we substitute in the formula and calculate the mean backfat thickness of the 10 pigs:

$$\bar{X} = \frac{385}{10} = 38.5 \text{ millimeters}$$

The mean summarizes all values into a single figure that is typical of the entire set of figures in that it is intermediate among the individual values. If successive samples of backfat thickness in groups of 10 pigs were taken from the data in Fig. 8.2, we would find that the means of the different samples of 10 would vary less than do the individual figures. This is one of the reasons why means are used to describe groups of individuals in a population.

8.3.2 The Range

The range is a very rough measure of the variation within a population. It is determined by finding the lowest and the highest values within a series or group of figures. The range of the items in Table 8.1 is from 31 to 47 millimeters, and this is the most extreme variation possible in this group of data. The chief disadvantages of the range as a measure of variation are that it is subject to chance fluctuations and that it becomes larger as the size of the sample increases. For instance, the range in backfat thickness in the entire 270 pigs included in Fig. 8.2 was from 25 to 51 millimeters, which is considerably larger than the range of the small sample of 10 pigs given in Table 8.1.

Table 8.1 Backfat thickness in live pigs measured at a market weight of approximately 200 pounds (A sample from those in Fig. 8.2)

Backfat thickness (mm)	Deviations from the mean	Squares of the deviations
47	8.50	72.25
38	−0.50	0.25
39	0.50	0.25
32	−6.50	42.25
34	−4.50	20.25
37	−1.50	2.25
31	−7.50	56.25
43	4.50	20.25
41	2.50	6.25
43	4.50	20.25
$\sum X$ 385 $\sum X^2$ 15063 $\overline{X}$ 38.50	0.00	240.50

8.3.3 The Variance

The variance is usually denoted by the symbol σ^2 and is defined as the average of the squared deviations from the mean. In Table 8.1 the deviation of each observation from the mean was determined (column 2); each deviation was then squared (column 3), and all the squared deviations were totaled. The sum of the squared deviations from the mean is 240.50.

A frequent question is: Why don't we simply use the average of the deviations from the mean as a measure of variation? One obvious answer is that, when we add all of the deviations and pay strict attention to the sign

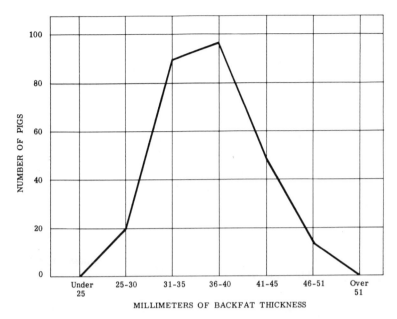

Fig. 8.2 Illustration showing the distribution of backfat thickness at 200 pounds live weight in 270 pigs.

of each deviation, the sum is equal to zero. Thus, the average of the deviations from the means would be zero also. The difficulty could be eliminated, however, by disregarding signs, and if this were done, the average of the deviations in our example would be 4.10 millimeters. Another, and probably more important, reason for squaring the deviations and averaging them is to magnify, or give more weight to, extreme values in the group of observations. Squaring the deviations and averaging them also makes it possible to perform other statistical measurements in a correct mathematical manner.

The variance σ^2, once the sum of the squared deviations from the mean is determined, may be calculated easily by dividing by the number one less than the total number of observations in the sample. Here, another departure from the usual is noted. Instead of dividing by the number of individuals in a sample n, we divide by the total minus one $(n - 1)$.

The calculation of the variance σ^2 from the data in Table 8.1 would be

$$\sigma^2 = \frac{240.5}{9} = 26.72$$

If we were to calculate the variance for a group of 200 to 300 pigs as we have done for the small sample of 10 pigs, it would be very time-consuming, and the chances for error would be greatly increased. Shortcuts have been devised by statisticians in which the variance may be determined by using a calculator. By using the following equation, the variance could be calculated:

$$\sigma^2 = \frac{\sum X^2 - (\sum X)^2/n}{n-1}$$

Where $\sum X$ is the sum of all items, $\sum X^2$ is the sum of all items squared, and n is the number of observations in the sample.

Substituting, we get:

$$\sigma^2 = \frac{15063 - (385)^2/10}{9} = \frac{240.5}{9} = 26.72$$

By using the calculator, we could, of course, obtain the answer more quickly.

One other important point to mention here is that the numerator of the formula for the variance,

$$\sum X^2 - \frac{(\sum X)^2}{n}$$

actually is the sum of the squared deviations from the mean and is often written as $\sum x^2$. The sum of little x squared divided by the number of observations minus one is the variance.

One of the most useful properties of the variance is that it can be separated by a special analysis into its various component parts. Special adaptations of the analysis of variance can be used to determine the percentage of the variation in a population that is due to inheritance and that is due to environment. Many other uses can be made of the analysis of variance, but it is not the purpose here to discuss all of them.

8.3.4   The Standard Deviation

The standard deviation is a much more accurate measure of variation in a population than is the range, and can be used very effectively, together with the mean, to describe a population. Statisticians use various symbols to denote the standard deviation, but the one used here will be S.D. The standard deviation is the square root of the variance. The following formula may be used for machine calculation:

$$\text{S.D.} = \sqrt{\frac{\sum X^2 - (\sum X)^2/n}{n-1}} = \sqrt{\frac{\sum x^2}{n-1}} = \sqrt{26.72}, \text{ or } 5.17$$

Figure 8.3 demonstrates how the mean and the standard deviation may be used to describe the variation in a population. The mean plus or minus one S.D. should include approximately 68 percent of the individuals in the population. The mean plus or minus two S.D. should include about 95 percent of the individuals in a population. In other words, we might expect only about five percent of the individuals in a population to fall outside the

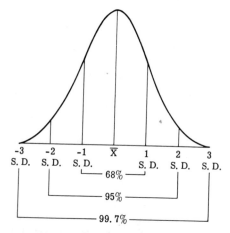

-3	-2	-1	X̄	1	2	3
S. D.	S. D.	S. D.		S. D.	S. D.	S. D.

Fig. 8.3 Normal frequency distribution curve showing how the mean and standard deviations may be used to describe the variation in a population.

mean plus or minus two S.D. This applies only to a normal distribution, however.

8.3.5 The Coefficient of Variation

The coefficient of variation is another method of expressing the amount of variation within a particular population. The formula is $S.D./\bar{X}$. When this coefficient is multiplied by 100, it is expressed as a percentage. The coefficient of variation is the fraction or percentage that the standard deviation is of the mean. One important use of this statistic is that it can be used to compare the variations of two unrelated groups. For instance, if the coefficient of variation for rate of gain in beef cattle is 25 percent and in hogs is 15 percent, we can say that there was a greater variation in daily gains in cattle than in hogs. Many other groups may be compared in a similar manner.

8.3.6 Standard Deviation of the Mean

In experimental work with livestock, we do not use an unlimited or indefinite number of animals. Actually, we use a very small sample of the entire population. This is true even if our sample includes the hundreds of animals for a given experiment. The question then arises as to how closely the mean of the sample we measure represents the true mean of the entire population. If we assume that the 270 pigs shown in Fig. 8.2 were the entire population, and we take 27 different samples of 10 pigs each at random and determine their mean, each group would probably have a different mean. But the means of the 27 different groups of pigs would have a definite characteristic, they would tend to fluctuate around a mean of their own. If we plotted them as a frequency distribution, we would find that they also would fall into a normal frequency-distribution curve. This very fact leads us to a method of calculating approximately the true mean of the population, and this statistic is called the standard deviation of the mean (S.E.). The standard deviation of the mean may be determined by dividing the standard deviation of the distribution (S.D.) by the square root of the number of items in the population. The formula can thus be written

$$\text{S.E.} = \frac{\text{S.D.}}{\sqrt{n}}$$

We can use the standard deviation of the mean together with the mean of the distribution to describe the true mean of an infinite number of means drawn from a population. The mean of the distribution plus or minus one S.E. should include about 68 percent of the means. The mean of the distribution plus or minus two S.E. should include approximately 95 percent of the means. In other words, we can say that there are only five chances out of 100 that the true mean of an infinite number of means drawn from a population would fall outside the mean of a sample plus or minus two S. E. Quite often in scientific reports, the mean of a sample is reported together with plus or minus the S.E.

If the means of two large samples have been derived independently, the information can be used to determine the standard deviation of a difference of means [1]. The formula for this is

$$\sqrt{(\text{S.E.}_1)^2 + (\text{S.E.}_2)^2}$$

If a difference between the means of two samples is at least twice as large as the standard deviation of the difference, we can accept this as a true difference at the five percent level of probability.

8.3.7 Coefficient of Correlation

This statistic and the expansion of the idea are often used in animal breeding and livestock production research. The coefficient of correlation is referred to as *r* and gives a measure of how two variables tend to move together. They are said to be positively correlated if they tend to move together in the same direction; that is, when one increases the other increases, or when one decreases the other decreases. They are said to be negatively correlated if they tend to move in opposite directions; that is, when one increases the other decreases. Thus, the coefficient of correlation for two variables lies somewhere between zero and ± 1.

Even though the coefficient of correlation tells us how two variables tend to move together in like or in opposite directions, it does not necessarily mean that the movement of one is the cause or the effect of the movement of the other. The cause and effect relationship must be determined, if possible, from other known facts concerning these two variables.

A particular coefficient of correlation is usually said to be significant, highly significant, or nonsignificant, the degree of significance depending upon the size of the coefficient of correlation and the number of individual items used to calculate it.

The formula for calculating the simple coefficient of correlation between two variables is

$$r = \frac{\sum XY - \frac{(\sum X) \cdot (\sum Y)}{n}}{\sqrt{\sum X^2 - \frac{(\sum X)^2}{n}} \cdot \sqrt{\sum Y^2 - \frac{(\sum Y)^2}{n}}}$$

where X is each individual observation for variable X, Y is each individual observation for variable Y, n is number of observations for each variable, and the Greek symbol $(\sum)$ means the summation of all items for each variable or pair of variables.

The coefficient of correlation will be calculated for the data presented in Table 8.2 as an example. Backfat thickness at 200 pounds (variable X) is the same as that given in Table 8.1; the rate of gain from weaning to 200 pounds for these same pigs has been included as a second variable (variable Y). Looking at the data in this table, we may ask if these two variables tend to move in like or in opposite directions and if the correlation coefficient is large enough to be statistically significant. All data necessary to calculate the coefficient of correlation are given in Table 8.2, and we merely substitute the appropriate figures for the symbols in the formula and proceed with the calculations as follows:

$$r = \frac{667.30 - \frac{(385) \cdot (17.3)}{10}}{\sqrt{15063 - \frac{(385)^2}{10}} \cdot \sqrt{30.15 - \frac{(17.3)^2}{10}}} = \frac{667.30 - 666.05}{\sqrt{240.5} \cdot \sqrt{0.22}}$$

$$= \frac{1.25}{15.51 \cdot 0.469} = \frac{1.25}{7.27} = 0.172$$

According to tables for the levels of significance of coefficients of correlation given in Snedecor's book (fifth edition) [2] on page 174, a coefficient of correlation with 8 degrees of freedom $(n - 2)$ should be 0.632 to be significant at the 5 percent level and 0.765 to be significant at the 1 percent level of probability. Thus, this coefficient of correlation of 0.172 is so small that it is very likely a chance correlation.

Table 8.2 Backfat thickness at 200 pounds live weight and rate of gain from weaning to 200 pounds in 10 market hogs

Number of pig	Thickness of backfat (X)	Rate of gain (Y)	Cross products (X times Y)
1	47	2.0	94.0
2	38	1.7	64.6
3	39	1.8	70.2
4	32	1.7	54.4
5	34	1.8	61.2
6	37	1.8	66.6
7	31	1.7	52.7
8	43	1.6	68.8
9	41	1.4	57.4
10	43	1.8	77.4
	$\sum X$ 385	$\sum Y$ 17.3	$\sum XY$ 667.3
	$\sum X^2$ 15063	$\sum Y^2$ 30.15	
	$\overline{X}$ 38.50	$\overline{Y}$ 1.73	

8.3.8 The Regression Line

Individual paired observations, such as backfat thickness and rate of gain for each pig in Table 8.2, may vary quite widely from others within a group of observations. In spite of this, however, a line can be calculated that will show the average relationship between two variables. This is called the regression line and is represented by the equation $Y = a + bX$.

The value b in this equation may be calculated as follows:

$$b = \frac{\sum XY - \frac{(\sum X) \cdot (\sum Y)}{n}}{\sum X^2 - \frac{(\sum X)^2}{n}}$$

The regression coefficient b from the data in Table 8.2 would be

$$b = \frac{667.3 - \frac{(385) \cdot (17.3)}{10}}{15063 - \frac{(385)^2}{10}} = \frac{1.25}{240.50} = 0.0052$$

The value b also refers to the slope of the regression line or the number of units change in Y with each unit change in X. Thus, for each change of 1 millimeter of backfat at 200 pounds, there was an average change of 0.0052 pounds in average daily gain.

The value a in the regression equation is called the Y intercept, because the regression line will cross the Y axis at this point when X is equal to zero. The Y intercept for the data in Table 8.2 may be calculated by using the following equation:

$$a = \bar{Y} - b\bar{X} = 1.73 - (0.0052 \times 38.5) = 1.53$$

The regression equation, then, becomes $Y = 1.53 + 0.0052X$. Two regression lines are possible when two variables are concerned, but the choice of the one to calculate and use depends upon which one seems to be dependent upon the other.

The regression equation as shown above may be used to estimate the value of Y when any value of X is known and substituted in the equation.

The above methods of deriving statistical ways to describe the distribution of individuals within a population are simple and can be calculated by anyone. There are many other aspects of statistics, however, that can be used in biology. These include methods for the determination of probable differences between means, methods of determination of both environmental and genetic correlations between two or more traits, and methods of separation of the variance of a population into its genetic and environmental portions. These methods necessarily require a wider knowledge of mathematics and statistics, and for that reason are not presented here. An understanding of these methods of statistical analysis is indispensable for the student who wishes to pursue further advanced studies and research in genetics and animal breeding. Some references for advanced study in the field are given at the end of this chapter.

REFERENCES

1. Crampton, E. W. "Estimating Statistically the Significance of Differences in Comparative Feeding Trials," SAg, 13: 16, 1932.

2. Snedecor, G. W. *Statistical Methods* (5th ed.), (Ames, Iowa: Iowa State College Press, 1956).

3. Thomsen, N. R. "Pig Breeding and Progeny Testing in Denmark," in *Report of the Meeting on Pig Progeny Testing in Denmark*, FAOUN, EAAP, pp. 9–12, 1957.

REFERENCES FOR FURTHER STUDY

Dickerson, G. E. "Techniques for Research in Quantitative Genetics," ASAP (monograph), p. 57, 1959.

Falconer, D. S. *Introduction to Quantitative Genetics* (New York: The Ronald Press Company, 1960).

Fisher, R. A. *Statistical Methods for Research Workers* (Edinburgh: Oliver and Boyd, 1950).

QUESTIONS AND PROBLEMS

1. List the important differences between qualitative and quantitative inheritance.

2. In a feeding trial, the rates of gain per day of 10 different steers in one lot were 1.50, 3.60, 3.00, 3.80, 1.55, 2.60, 2.10, 3.20, 1.80, and 2.70. Calculate the mean, standard deviation, coefficient of variation, and the standard deviation of the mean for this group.

3. Why is the range a poor way to express the variation of individual observations within a population?

4. What is the variance for the group of individuals in question 2?

5. Why do we use the value $n - 1$ instead of n in calculating the variance?

6. Explain in detail how the mean of the population together with a standard deviation may be used to describe a population.

7. Why do we need different methods of describing a quantitative than a qualitative trait?

8. Explain the meaning of "the coefficient of correlation."

9. Explain the meaning of "the coefficient of regression."

10. How do the coefficient of correlation and the coefficient of regression differ?

9

Variations
in Economic Traits
in Farm Animals

Phenotypic variation refers to the observable or measurable differences in individuals for a particular trait. This is the raw material with which the animal breeder must work. If there were no variations among individuals, there would be no need to select or cull animals for breeding purposes, because they would all look and perform alike, or at least there would be little difference between them.

The phenotypic variations that we see in animals in a herd are seldom, if ever, due completely to differences in genes. This chapter deals with the causes of variations and how we can estimate more effectively the portion of the total phenotypic variance that is due to genes, or heredity.

9.1 CAUSES OF PHENOTYPIC VARIATION
IN FARM ANIMALS

The total phenotypic variation for traits in farm animals is due to heredity, environment, or the interaction of both. The importance and influence of each of these factors will be discussed separately.

9.1.1 Heredity

The genotype of an individual is fixed at conception and, barring a mutation, remains the same for the remainder of its life. Its genetic makeup is determined by the genes that it receives from both parents.

The individual as well as its parents possesses thousands of genes; the exact number has never been determined. Probably the only time that two individuals are exactly alike genetically is when they are identical twins produced from a single fertilized egg. Members of an inbred line are more likely to be alike genetically than noninbred individuals. The degree of genetic similarity among individuals within an inbred line increases as the amount of inbreeding increases, especially if there is directional selection. Theoretically, in an inbred line in which the inbreeding is 100 percent, all members of the line should be genetically alike.

Since parents are not homozygous for all of the genes they possess and since they may differ genetically, no two of their offspring, with the exception of identical twins, are genetically alike. Personal experience tells us, however, that within some families the children have a closer resemblance to each other than in other families. This must be because the children in some families are more alike genetically than in others. How such a genetic difference could exist is illustrated in Table 9.1.

Table 9.1 Illustrating how heterozygosity of the parents is important in increasing the number of genotypes possible in their offspring

Family number	Genotype of the parents		No. of different genotypes possible in the offspring*
	Father	Mother	
1	AABBCCDD	AABBCCDD	1
2	AABBCCDD	aabbccdd	1
3	AaBBCCDD	AaBBCCDD	3
4	AaBbCCDD	AaBbCCDD	9
5	AaBbCcDD	AaBbCcDD	27
6	AaBbCcDd	AaBbCcDd	81

*Using four pairs of genes. The number of different genotypes may be calculated from the formula 3^n, where n is the number of pairs of genes heterozygous in both parents and each pair of genes is carried on a different pair of homologous chromosomes or if crossing over occurs.

Parents homozygous for many pairs of genes will have offspring that are more alike genetically than parents that are heterozygous for several pairs of genes. In fact, genetic variation within a species is almost unlimited.

For example, if two parents were heterozygous for the same 20 pairs of genes and these were genes carried on 20 different pairs of homologous chromosomes, it would be possible for their children to be of 3^{20} different genotypes, which would be almost 3.5 billion.

A qualitative trait affected by only one pair of genes may vary in a population because of a difference in the frequency of the two alleles at the same locus. Economic traits (quantitative) may also vary genetically for the same reason, but they may also vary because of a difference in the frequency of genes at several different loci. This is also demonstrated in Table 9.1.

Quantitative traits may show great variability because some of the many genes involved may express themselves phenotypically in either an additive or a nonadditive manner. Some traits may be affected by both types of gene action. Because selection and mating procedures are different for additive and nonadditive types of gene expression, it is important to know which type affects a certain trait. Information that gives some indication of which types of gene action affect a quantitative trait are presented in Table 9.2.

Table 9.2 Some factors that indicate the type, or types, of gene action affecting econmic traits

Characteristic	Type of phenotypic expression of genes		
	Additive	Nonadditive	Both additive and nonadditive
Heritability	High	Low	Medium
Amount of hybrid vigor	None	Considerable	Some
Inbreeding depression	None	Considerable	Some
Sex differences	Large	Small	Very small

Some economic traits, such as carcass quality and quantity, appear to be affected mostly by an additive type of gene expression. Others, such as fertility and livability, appear to be affected mostly by nonadditive types of gene action. Some traits, such as weaning weight in beef cattle, appear to be affected to a certain extent by both additive and nonadditive genes. It is possible that almost all quantitative traits determined by many pairs of genes may be determined to a certain extent by both additive and nonadditive gene action, but one may have a much more important effect than another.

How to use the knowledge of which type, or types, of gene action affect the various traits that we strive to improve in livestock through breeding methods will be discussed in more detail in subsequent chapters.

9.1.2   Environment

Phenotypic variations in economic traits due to environment are also of great importance. The environmental portion of the phenotypic variation may be denoted by the symbol σ_E^2. Environment includes all such factors as disease, nutrient supply, temperature effects, accidents, and others which the individual may encounter from the time of conception until its death.

Phenotypic variations due to environment are important because (1) they are not transmitted from parents to their offspring, (2) they overshadow variations due to heredity, (3) the proper environment is necessary for an individual to reach its genetic potential, and (4) rapid improvements can be made in the efficiency of livestock production by supplying uniform and superior environmental conditions to breeding animals and those used for commercial production.

9.1.3   Joint Action of Heredity and Environment

The interaction of heredity and environment means that animals of a certain genotype may perform more satisfactorily in one environment than they do in another. In other words, one environment permits the expression of the genetic characters in a breed or strain, while another does not. This is illustrated by the data presented in Fig. 9.2. These data show that the inbred Poland pigs were 23 pounds heavier at 154 days of age than were the inbred Landrace pigs when both were fed the same ration on pasture, but that the

Fig. 9.1   Extreme variation in the size of two littermate pigs. Both heredity and environment may be responsible for differences between individuals. (Courtesy of the University of Missouri.)

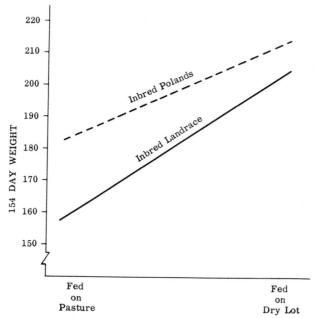

Fig. 9.2   Example of genetic × environmental interaction in inbred lines of swine concerning 154-day weights when fed on pasture and on dry lot.

difference was only 10 pounds when they were fed in dry lot. Thus, the Landrace pigs grew faster in comparison to the Poland pigs in dry lot than on pasture; which is another way of saying that the dry-lot conditions permitted the genes involved to achieve more complete expression. This seems reasonable, since the Landrace breed was originally developed under dry-lot conditions, whereas the Polands were developed to a greater extent on pasture.

Another example of the interaction of heredity and environment involves the performance of the Brahmans and their crosses in the southern and southwestern portion of the United States. Since they are more resistant than the British breeds to certain diseases, parasites, and high temperatures of that region, they perform more satisfactorily under those conditions. When compared with the British breeds under more desirable environmental conditions in the Midwest, however, they may not enjoy this advantage.

Still another example of the interaction of heredity and environment is the disease in humans and animals known as diabetes. This disease is characterized by the inability to metabolize glucose properly, so that the level of glucose in the blood stream becomes high, and glucose is excreted with the urine [1]. This disease was almost always fatal to humans before insulin was discovered as a means of treatment. Insulin is produced by the pancreatic gland and when released into the blood stream it causes a reduction in the blood-sugar level. The predisposition to develop diabetes seems to be inherited, but the disease may not develop unless the insulin-producing mechanism of the pancreas is overtaxed over a long period of time by the consumption of excess amounts of carbohydrates [4]. Under such conditions, insulin production becomes inadequate and diabetes develops. Thus, the right kind of internal environment is necessary to cause this inherited condition to become manifest.

Breeders should be interested in knowing if genetic-environmental interactions are important, and such knowledge should help answer the question of whether or not selection of animals for improvement in one set of conditions would also result in genetic improvement in another. Studies with dairy cattle indicate that the progeny of dairy sires rank similarly when their daughters are fed at different levels. In experiments with mice, however, evidence for genetic-environmental interactions was obtained [3]. In these experiments, two strains of mice derived from a single foundation stock were selected exactly in the same manner for weight at six weeks of age. One of these strains was fed ad lib, while the other was restricted to about 75 percent of the normal food intake between the ages of three and six weeks. The weight at six weeks increased under selection pressure in both strains. Exchanges of feeding levels were made for the two strains after five, seven, and eight generations of selection. The results showed that the improvement for rapid growth on a high plane of nutrition carried with it no improvement for rapid growth on a low plane. On the other hand, improvement of the genotype for rapid growth on a low plane of nutrition did carry with it a considerable improvement for growth on a high plane of nutrition.

In the past, the interaction of heredity and environment in causing variations was not considered to be very important, but now geneticists realize that this factor must be recognized. Genetic-environmental interactions are now being studied by some of the experiment stations, and more complete information should be available in the future. In the meantime, we must assume that interaction is important and must try to produce and select breeding stock under the same conditions in which the offspring will be produced.

9.2 IMPORTANCE OF HEREDITY AND ENVIRONMENT

Frequently discussed has been the question of whether heredity or environment is the more important in the expression of economic traits. Such a discussion here would be of little value, because it is now recognized that both are of very great importance. The best possible inheritance will not result

Fig. 9.3 These purebred calves are thin because they
have been on a maintenance ration of roughage during
the winter months. They have the inheritance necessary to
make rapid and efficient gains and good carcasses if
they are given the right environment to grow and fatten
for the market.

in a superior herd or flock unless the proper environment is also supplied
so that the animals can attain the limit set by their inheritance. Half-starved
and neglected purebreds are truly a disappointment to livestock men, in their
appearance as well as in their performance. Nevertheless, the best possible
environment will not develop a superior herd or flock unless the proper
inheritance is also present in the animals. The answer to a question illustrates
these statements. Why do not dairymen use beef cows for milk production
purposes, and why do not racing enthusiasts include draft horses in their
racing strings? The answer is, obviously, that the best possible feed, training,
or care cannot make a record-breaking milk-producer out of a beef cow or
a Kentucky Derby winner out of a draft horse. In these extreme examples,
it can readily be seen that the limit to performance is set by the animal's
own heredity, and the best possible environment will not cause that individual
to exceed its own genetic potential. At least this does not seem possible,
unless we can determine the physiological limitations set by inheritance
and can alter these by the use of hormones and other chemicals.

To make the most possible use of good inheritance, we must select
breeding animals that are superior because they possess more desirable
genes or combinations of genes. Superiority due to genes is the only thing
that is transmitted from the parents to their offspring. Superiority due to
environment will not be transmitted by the parents. This superior environ-

ment must be provided for the offspring if they are to be the equal of their parents.

Environment has received more attention in the past than heredity, but this is changing, and more attention is now being given to both. The proper environment is of great importance from the economic standpoint. In addition, it is becoming more and more obvious that animals must be kept in the kind of environment that allows them to show that they possess desirable inheritance for a particular trait. An extreme example here would be the selection of individuals for increased disease resistance. Individual animals must be exposed to the disease in question to determine which of them are resistant. Likewise, the controlled feeding of a boar to reduce his backfat will not also induce less backfat in his offspring if they are full fed a good ration to the usual market weight. A better estimate of the genotype of a boar for backfat thickness will be obtained by feeding him the same ration, and in the same way, that his offspring will be fed.

Breeders often mistake environmental effects for genetic effects when comparing livestock on different farms. The environment can be so different on two farms that the genetically superior animal may look inferior if he has been cared for improperly and so has not had the opportunity to show his full potential. For this reason, it is best to compare the performances of individuals within the same herd, where they have been fed and handled in the same manner, and not those of individuals in two different herds, where the environment might be quite different. Many swine-testing stations have been developed so that environment can be more nearly standardized for all animals at a central testing station. This is not the complete answer, however, because the pretest environments could have been quite different for various animals and might affect their performance while they are on test.

The importance of controlling environment as much as possible may be illustrated by the following example. All of the phenotypic variation in a trait is due to heredity (σ_H^2) and to environment (σ_E^2).* The portion of the variation due to heredity would be equal to the hereditary variance divided by the total variance, or

$$\text{Percent of hereditary variation} = \frac{\sigma_H^2}{\sigma_H^2 + \sigma_E^2} \times 100$$

Let us further assume that σ_H^2 is equal to 20 units and σ_E^2 is equal to 20 units. Thus, the percentage of the variance due to heredity would be

$$\frac{20}{20 + 20} \times 100$$

or 50 percent. Suppose, however, that we are able to reduce the environ-

---

*In this example it is assumed that interactions between heredity and environment are not important.

Fig. 9.4 A polled Hereford and a Holstein-Friesian bull. Both were grand champions of their breed. One is bred for meat and the other for milk. The best possible environment will not allow them to compete successively for the purpose each is selected for. (Courtesy of the American Polled Hereford Association and the Holstein-Friesian Association of America.)

mental variation to such an extent that it is only 10 units. In such a case, the portion of the variance due to heredity would be

$$\frac{20}{10 + 20} \times 100$$

or 67 percent.

When we correct weaning weights for every calf in a herd to the same age and same sex, as well as to the same age of dam, we are actually reducing the environmental variations between individuals in that herd, and a larger proportion of the remaining variance should be due to heredity. Thus, the superior individuals, after such corrections are made, would be more likely to be genetically superior, because we would increase our accuracy of picking those that possessed the more desirable genes or combinations of genes.

9.3 HERITABILITY ESTIMATES

Heritability estimates refer to that portion of the phenotypic variance in a population that is due to heredity. The percentage of heritability subtracted from 100 gives an estimate of the portion of the variance that is due to environment. Lush [6] has pointed out that heritability estimates are concerned with the differences between individuals or groups of individuals and not with their absolute values. More correctly, then, when we refer to the heritability estimate for a trait, we are referring to the portion of the differences for that trait in a population that is due to heredity. To further illustrate this point, let us assume that the heritability of backfat thickness in swine is 50 percent and that the average of the herd at a weight near 200 pounds is 1.40 inches. This does not mean that 0.70 inch of backfat is due to heredity and the remaining 0.70 inch to environment. It means that, of the differences between individuals in the herd in backfat thickness, approximately 50 percent are due to heredity and 50 percent are due to environment.

Lush has also pointed out that heritability may be used in either a narrow or a broad sense. It is important to understand the difference between the two. In the narrow sense, heritability estimates include mostly additive type of gene action or the average effects which the individual genes have in that population. This is approximately the same as the percentage of genetic progress made in the next generation when superior individuals are selected for parents. Heritability in the broad sense includes all of the effects of the entire heredity of each individual. Heritability in the broad sense includes, in addition to variations due to additive gene action, those which are due to dominance and epistasis. Most methods of estimating heritability include only a little more than the narrow (or additive) portion of the variation, but this varies with the method used to calculate the estimates.

9.3.1 Methods of Estimating Heritability

All heritability estimates are based on how closely relatives resemble each other [5]. From these calculations, an attempt is made to estimate the degree of correlation between the phenotype and the genotype of individuals in a population. It is not the purpose here to discuss in detail the methods of calculating heritability estimates, but we shall outline some of the different methods.

Identical twins have been used in genetic studies, especially in humans, to determine the relative influence of heredity and environment on various traits. Many statistical methods for using twins in genetic research have been developed, but all of them depend on the ability to distinguish between one-egg and two-egg twins. One-egg twins are derived from the same egg and thus have the same genetic makeup. Any difference between such twins should be of an environmental nature. Fraternal twins, or two-egg twins, develop from two different eggs and should be no more alike genetically than full brothers and full sisters who are not twins. Variations in two-egg twins would be due to both heredity and environment. Therefore, a comparison of two-egg and one-egg twins should give an estimate of the relative influence of heredity and environment on a particular trait. The formula used for obtaining estimates of heritability for certain human traits using identical and unidentical twins is [8]

$$H^2 = \frac{i^r - f^r}{1 - f^r}$$

where H^2 is the percentage share of hereditary determination of the observed intrapair difference in two-egg twins, i^r is the intrapair coefficient of correlation of identical (one-egg) twin pairs and f^r is that of fraternal (two-egg) twins. Studies of this kind have shown that intrapair differences are much greater in two-egg twins than in one-egg twins in almost every character studied.

More recently, identical twins alone have been used in genetic studies in dairy and beef cattle. In these studies, the heritability estimates have been calculated from the formula

$$H^2 = \frac{\text{(between-pair variance)} - \text{(within-pair variance)}}{\text{(between-pair variance)} + \text{(within-pair variance)}}$$

Heritability estimates for milk production in dairy cattle derived from records on identical and fraternal twins [2] range between 70 and 90 percent as compared to 36 to 50 percent in studies involving nontwin records from field data. The larger estimates from twin studies may be due to a number of factors. Possibly maternal and contemporary environmental effects may make twin members more alike than nontwins for a particular trait. In addition, the genetic variance as estimated from twin data, especially from one-egg twins, includes that which is due to dominance and epistasis. Little of this kind of genetic variance is included in heritability estimates obtained from nontwin data. In any event, it appears that heritability estimates from identical twin data are too high and are not a true indication of the progress one would expect to make in selection for a particular trait.

The resemblance between parents and their offspring is also used in calculating heritability estimates in farm animals. This method of calculation can take many forms, depending upon the nature of the records available. One method often used is to determine the intrasire regression of offspring on the dam. A heritability estimate by this method is largely heritability in the narrow sense, which is mostly additive gene action. Still another method often used is to determine the resemblance between sibs. This may be calculated from the intraclass correlations from the analysis of variance. An estimate calculated in this manner is likely to include some of the variations due to epistasis and dominance, as well as to additive gene effects. One convenience of this method is that it is not necessary to know the phenotype of the parents.

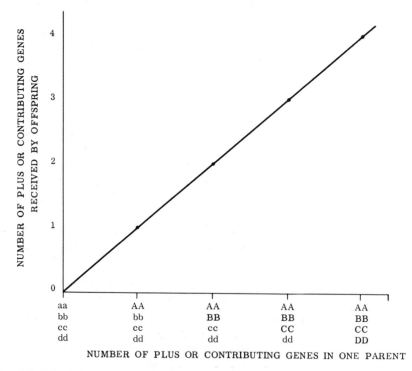

Fig. 9.5   A theoretical illustration of heritability involving additive gene action. The offspring receive their plus or contributing genes from the parents. In this example the heritability is 100 percent.

Heritability estimates may also be obtained by calculating the repeatability coefficient for a trait which is the correlation between different records by the same individuals. It gives an estimate of the upper limits of heritability and may be higher than the true heritability, if permanent environmental effects on the individual are important.

9.3.2 Value of Heritability Estimates

Heritability estimates tell something about the amount of progress that might be made in selection for a particular trait. For example, the heritability of rate of gain in beef cattle in the feed lot is about 60 percent, which means that this percentage of the total variation is due to genes and about 40 percent is due to environment. Thus, differences in bulls fed on a gain test under similar environmental conditions should be due largely to differences in inheritance, and those making the most rapid gains would be more likely to produce offspring which themselves would make rapid gains. When the heritability of a trait is high, the correlation between the phenotype and the genotype of the individuals, on the average, should also be high, and selection on the basis of the individual's own phenotype should be effective. High heritability estimates also indicate that additive gene action is important for that trait and the mating of the best to the best should produce more desirable offspring.

Quite often, heritability estimates for traits are as low as 10 to 15 percent or even lower. Litter size in swine is one such trait. A low heritability estimate tells us that there is a low correlation between genotype and phenotype and that if we used superior individuals for that trait for breeding purposes their offspring would not be as superior as when the heritability for a trait was high. To make progress in selection when the heritability of a trait is low, much more attention must be paid to the performance of the collateral relatives and the progeny.

Low heritability estimates also tell us that variations due to additive gene action are probably small. Evidence is also accumulating that when the heritability of a trait is low, nonadditive gene action such as overdominance, dominance, and epistasis may be important. This makes it necessary to use special methods of selection and mating for greater improvement in the herd or flock. These will be discussed more fully in a later chapter.

9.4 REPEATABILITY ESTIMATES

Repeatability estimates refer to the expression of the same trait, such as milk production, fleece weight, etc., at different times in the life of the same individual. Thus, there is no chance for segregation or independent assortment of the genes. On the average, the weaning weight of the calf from a first-calf heifer is about 47 percent accurate in predicting what her future records will be. Thus, if a cow weans a calf during her first lactation that is 100 pounds heavier than the average of all the heifers of her age in the herd, the cow should wean calves that average 47 pounds above the average of that group in later years.

Another definition of repeatability estimates is that they are the fraction of differences between single records of individuals that are likely to occur in future records of those same two individuals. For example, the repeatability of litter size at weaning in swine is about 16 percent. If one gilt weans ten pigs and another six in their first litters, one would expect an average difference in later litters of only 0.64 pigs (16 percent times four pigs difference).

A knowledge of repeatability estimates for the various traits may be used in selecting for future performance. When the repeatability estimate for a trait is high, culling on the basis of the first record should be effective in improving the over-all record of the herd the next year. In addition, offspring from the superior individuals in the herd should be given preference when selection is made for replacement stock.

Repeatability estimates also tell us something about how to allot animals on a feeding trial. If the repeatability of a trait is high, it becomes increasingly important to divide the offspring of each sire or each dam evenly among the different lots. Otherwise, if the offspring from one parent were in one lot and those of another parent in a second lot, lot differences ascribed to treatment might actually be largely due to hereditary differences. This procedure is of less importance if the repeatability of a trait is low.

Repeatability estimates also give an indication of how many records should be obtained on an individual before it may be culled from the herd or flock. The repeatability of weaning weight in beef cattle is about 47 percent, whereas it is only about 16 percent for litter size at weaning in swine. By using the following formula, we can estimate the repeatability of traits where a larger number of records is involved:

$$R = \frac{nr}{1 + (n - 1)r}$$

In this formula, R is the repeatability of more than one record, r is the repeatability of one record, and n is the number of records available for purposes of calculation. This formula was used to calculate the repeatability of weaning weight in cattle and litter weight at weaning in swine for one to five records, as shown in Table 9.3. These data indicate that one could cull beef cows for weaning weights of their calves on the basis of a single record with almost as much confidence as one could cull sows for litter size at weaning on the basis of five records.

Table 9.3 Repeatability estimates for weaning weight in beef cattle and litter size at weaning in swine with one to five records

	Repeatability in percent	
No. of records	*Weaning weights in beef cattle*	*Litter size at weaning in swine*
1	47	16
2	64	28
3	73	36
4	78	43
5	81	49

Lifetime averages which show the ability of certain individuals to repeat a high level of performance over a long period of time are very important in animal breeding. These records should be as accurate as possible and should be corrected for certain environmental factors before individuals in a herd are compared. Lush [7] has suggested the following formula for adjusting the records of cows with different numbers of records to the same basis, and he has called this the *probable producing ability* of an individual:

$$\text{Probable producing ability} = \text{herd average} + \frac{nr}{1 + (n-1)r} \times \text{her own average minus the herd average}$$

To illustrate how the probable producing ability of two cows for weaning weight of their calves may be calculated, let us assume that cow number 1 has weaned three calves with an average weaning weight of 475 pounds. Let us also assume that cow number 2 has weaned two calves with an average weaning weight of 490 pounds and that the herd average is 400 pounds. The R (repeatability) of two records is 0.64, or 64 percent, and for three records it is 0.73, or 73 percent. These figures are obtained from Table 9.3.

The probable producing ability of cow number 1 would be

$$\begin{aligned} \text{PPA} &= 400 + 0.73 \times 475 - 400 \\ &= 400 + 0.73 \times 75 \\ &= 454.75 \end{aligned}$$

The probable producing ability of cow number 2 would be

$$\begin{aligned} \text{PPA} &= 400 + 0.64 \times 490 - 400 \\ &= 400 + 0.64 \times 90 \\ &= 457.60 \end{aligned}$$

Thus, cow number 2 has a slightly higher PPA, although the two cows differ very little in this respect. The calculation of the probable producing ability of all cows in the herd with different numbers of records would make it possible to compare and cull them more accurately on a standard basis.

Lifetime averages are of value in selecting for traits in which the repeatability is low, but their value is decreased by the fact that the generation interval is increased and progress per year may be slow. A consistently high performance over a period of many years is a good indication that an animal possesses desirable genes for several traits. Whenever possible, both male and female replacement stock should be retained from such dams. Such a record would be useful in selecting for increased vigor, constitution, and longevity, which may be indicative of freedom from recessive or partially dominant genes with detrimental effects.

In conclusion, it has been pointed out in this chapter how variations in individuals are the raw material with which the animal breeder must work. It was further pointed out that the genetic variations within a herd or flock more truly are the raw material with which the breeder must work to make progress through the application of breeding methods. Methods used to measure the genetic portion of the variations in economic traits were also discussed. A more detailed discussion of how this information may be used for the improvement of each species of farm animals will follow in later chapters.

R E F E R E N C E S

1. Banting, F. G., and C. H. Best. "The Internal Secretion of the Pancreas," JLCM, 7: 251, 1922.

2. Brumby, P. J. "Monozygotic Twins and Dairy Cattle Improvement," ABA, 26: 1, 1958.

3. Falconer, D. W., and M. Latyszewski. "The Environment in Relation to Selection for Size in Mice," JG, 51: 67, 1952.

4. Lukens, F. D. W. "The Pathogenesis of Diabetes Mellitus," YJBM, 16: 301, 1944.

5. Lush, J. L. "Ways of Computing Heritability," PBRT, 1–2, May, 1954.

6. Lush, J. L. "Things Often Misunderstood or Overlooked about Heritability," PBRT, Chicago, 1–2, May, 1954.

7. Lush, J. L. *Animal Breeding Plans* (3rd ed.), (Ames: The Iowa State College Press, 1945).

8. Newman, H. H. "Aspects of Twin Research," SM, 52: 99, 1941.

QUESTIONS AND PROBLEMS

1. Explain the importance of phenotypic variation in quantitative traits to the animal breeder.

2. Which is the most important cause of phenotypic variations in farm animals, heredity or environment?

3. Does the genotype of an individual change during its lifetime? Does the phenotype change? Explain.

4. Under what conditions would two or more individuals be almost alike genetically?

5. How many different genotypes would be possible in the progeny of two parents heterozygous for the same 10 pairs of genes?

6. What kind, or kinds, of gene action may affect various quantitative traits in farm animals? How can this be determined? Why is such information important to the animal breeder?

7. What are some of the environmental factors that cause phenotypic variation in quantitative traits?

8. How much of the superiority of the parents due to environment is transmitted to their offspring? Why is this important?

9. Explain what is meant by the joint action (interaction) of heredity and environment? Why do you think such interactions are important in animal breeding?

10. Does the individual have a genetic limit to its potential performance? Give illustrations to prove your opinion.

11. What anatomical or physiological factors could set a genetic limit to the performance of an individual?

12. Why is it important to remove as much of the environmental phenotypic variation as possible in comparing animals for selection purposes? How can such environmental variations be reduced?

13. Define the term *heritability estimate* and explain the difference between heritability in the broad and narrow sense.

14. Which is more accurate in predicting progress in selection, a heritability estimate determined from a study of identical twins or one determined from a study of the resemblance between parents and offspring? Explain.

15. What does the degree of heritability of a quantitative trait indicate to the animal breeder?

16. Define a repeatability estimate in two different ways. Of what use are estimates of repeatability for quantitative traits to the animal breeder?

17. The repeatability estimate for weaning weight in beef cattle is almost twice as large as the heritability estimate for this trait. Why?

18. Assume that the repeatability of a quantitative trait in beef cattle is 0.50, or 50 percent. Calculate the repeatability (R) of 8 records for this trait.

19. Why are lifetime averages of great importance to the animal breeder? What are their disadvantages?

20. A farmer has a herd of swine which on the average weigh 195 pounds at six months of age. He selects for breeding purposes gilts that average 205 pounds and boars that average 225 pounds at six months of age. The offspring produced have an average six-month weight of 205 pounds. What is the apparent heritability of six-month weight in swine in this example?

21. A simple way to estimate heritability is to divide the dams mated to a sire into two groups; namely, a high-producing group and a low-producing group. The daughters of these two groups of dams are then tested for production. A dairy sire (X) is mated to two groups of cows. The high-producing group averages 650 pounds of butterfat and the low-producing group averages 500 pounds. The daughters of these two groups of dams average 585 and 570 pounds, respectively. What is the apparent heritability of butterfat production in this example?

22. Assume that the repeatability of weaning weight in beef cows is 0.47 (47 percent) when based on one record, 0.64 when based on two records, and 0.73 when based on three records. In a herd that averages 400 pounds, cow B, with three calves weaned weighing an average of 460 pounds and cow C, with two calves weaned weighing an average of 490 pounds, are compared. Which cow is probably superior for weaning heavy calves?

10

Principles of Selection

Selection is an important tool for rearranging genetic material to better fit individuals for a particular purpose. It may be defined as a process in which certain individuals in a population are preferred to others for the production of the next generation. Selection is of two general kinds, natural, or that due to natural forces, and artificial, or that due to the efforts of man.

10.1 NATURAL SELECTION

In nature, the main force responsible for selection is the survival of the fittest in a particular environment. Natural selection is of interest because of its apparent effectiveness and because of the principles involved.

Natural selection may be illustrated by considering the ecology of some of our wild animal species. Murie [3] studied the relationship between wolves in Mt. McKinley National Park in Alaska and other species of animals, especially the Dall or mountain sheep. Apparently, the wolves chase many sheep before they find one they can catch. Most of those killed by the wolves were the weaker animals, and included those that were either very young or very old. Thus, there was a tendency for nature to select against the weaker ones, and only the stronger survived to reproduce the species.

Some of the most interesting cases of natural selection are those involving man himself. All races of man that now exist belong to the same species, because they are interfertile, or have been in all instances where matings have been made between them. All races of man now in existence had a common origin, and at one time probably all men had the same kind of skin pigmentation—which kind we have no sure way of knowing. As the number of generations of man increased, mutations occurred in the genes affecting pigmentation of the skin, causing genetic variations in this trait over a range from light to dark or black. Man began to migrate into the various parts of the world and lived under a wide variety of climatic conditions of temperature and sunshine. In Africa, it is supposed, the dark-skinned individuals survived in larger numbers and reproduced their kind, because they were better able to cope with environmental conditions in that particular region than were individuals with a lighter skin. Likewise, in the northern regions of Europe men with white skins survived in a greater proportion, because they were better adapted to that environment of less intense sunlight and lower temperatures. But what of the Eskimos who lived in the polar regions of the North? They are also dark-skinned. Does this theory fit them? The most probable answer to this question is that the Eskimos are more recent migrants from Asia to the polar region, and compared to the Negro in Africa and the whites in Europe, they have not lived so long in that region. However, their dark skin could prevent sunburn from intense light rays reflected by the snow.

Recently, evidence has been obtained that there may be a differential selection for survival among humans for the A, B, and O blood groups [1]. It has been found that members of blood group A have more gastric carcinoma (cancer) than other types and that members of type O have more peptic ulcers. This would suggest that natural selection is going on at the present time among these different blood groups, and the frequency of the A and O genes might be gradually decreasing unless, of course, there are other factors that have opposite effects and have brought the gene frequencies into equilibrium.

Natural selection is a very complicated process, and many factors determine the proportion of individuals that will reproduce. Among these factors are differences in mortality of the individuals in the population, especially early in life; differences in the duration of the period of sexual activity; the degree of sexual activity itself; and differences in degrees of fertility of individuals in the population.

It is interesting to note that in the wild state, and even in domesticated animals to a certain extent, there is a tendency toward an elimination of the defective or detrimental genes that have arisen through mutations, through the survival of the fittest.

10.2 ARTIFICIAL SELECTION

Artificial selection is that which is practiced by man. Thereby, man determines to a great extent which animals will be used to produce the next generation of offspring. Even here, natural selection seems to have a part. Some research workers have divided selection in farm animals into two kinds, one known as automatic and the other as deliberate selection [2]. Litter size in swine may be used as an illustration of the meaning of these two terms. Here, automatic selection would result from differences in litter size even if parents were chosen entirely at random from all individuals available at sexual maturity. Under these conditions, there would be twice as much

Fig. 10.1 Above, market hogs in 1912, and below market hogs in more recent times. Changes in consumer demands for more lean and less fat together with selection pressure for the meat-type hog has wrought this change in type.

chance of saving offspring for breeding purposes from a litter of eight than from a litter of four. Automatic selection here differs from natural selection only to the extent that the size of the litter in which an individual is reared influences the natural selective advantage of the individual for other traits. Deliberate selection, in this example, is the term applied to selection in swine for litter size above and beyond that which was automatic. In one study by Dickerson and coworkers involving selection in swine [2] most of the selection for litter size at birth was automatic and very little was deliberate; the opportunity for deliberate selection among pigs was utilized more fully for growth rate, however.

Definite differences between breeds and types of farm animals within a species is proof that artificial selection has been effective in many instances. This is true, not only from the standpoint of color patterns that exist in the various breeds, but also from the standpoint of differences in performance that involve certain quantitative traits. For instance, in dairy cattle there are definite breed differences in the amount of milk produced and in butterfat percentage of the milk.

10.3 GENETIC EFFECT OF SELECTION

Selection does not create new genes. Selection is practiced to increase the frequency of desirable genes in a population and to decrease the frequency of undesirable genes. This may be illustrated by the following example, where A is the desirable gene and a the undesirable gene:

P_1 AA X aa

F_1 All Aa
(Freq. of A is 0.50)

F_2 Aa X Aa

     Progeny:

                1 AA
                2 Aa Freq. of gene A in F_2
                1 aa is still 0.50.

Let us assume that we cull all aa individuals in the F_2. If this were done, the remaining genes would be four A and two a. Thus, the frequency of the A gene would be increased to 0.67 and that of the a gene would be decreased to 0.33.

The increased frequency of the A gene when the aa individuals were culled would also increase the proportion of *AA* individuals in the population. If the frequency of the *A* gene were 0.50 (if it is assumed that requirements of the Hardy-Weinberg Law were met), the proportion of *AA* individuals would be 0.50 multiplied by 0.50 or 0.25. However, if the frequency of the *A* gene were increased to 0.67, the proportion of *AA* individuals would be 0.67 multiplied by 0.67 or 0.449.

If selection is effective, the genetic effects of selection are to increase the frequency of the gene selected for and to decrease the frequency of the gene selected against. If the frequency of the desirable gene is increased, the proportion of individuals homozygous for the desirable gene also is increased.

10.4 SYSTEMS OF SELECTION FOR DIFFERENT KINDS OF GENE ACTION

The different kinds of gene action that affect economic traits in farm animals were discussed in Chapter 4. It was pointed out that in certain instances such as coat color and horns only one pair of genes, or a relatively few genes, have a major effect on certain traits of great economic importance. Sometimes a single pair of genes may also have a major phenotypic effect on certain quantitative traits. An example is the gene for snorter dwarfism in beef cattle, where a pair of recessive genes (*dd*) may produce a dwarf, hiding or masking the phenotypic expression of many additive genes for fast growth and potential large mature size. It was also pointed out that in quantitative traits determined by many pairs of genes, some of these genes may express themselves in an additive manner, whereas others may express themselves in a nonadditive way. In fact, some quantitative traits may be affected by many pairs of genes, of which some have additive and some nonadditive phenotypic effects.

Because both qualitative and quantitative traits may be greatly affected by many different types of gene action, it seems important here to outline what methods may be used in selecting for or against them.

10.4.1 Selection for a Dominant Gene

In practice, we are very likely to be selecting for a dominant gene, because traits determined by such genes are usually desirable. Those individuals possessing a dominant gene will show it, but the problem here is one of distinguishing between the homozygous dominant and heterozygous dominant individuals. The heterozygous individuals must be identified by a breeding test before they can be eliminated. Selection for a dominant gene involves the same principle as selection against a recessive gene.

10.4.2 Selection Against a Dominant Gene

Selection against a dominant gene is relatively easy, providing the penetrance of the gene is 100 percent and it does not vary in its expression. Since each animal possessing a dominant trait should show this in its phenotype, eliminating the gene merely means that all animals showing the trait should be discarded. Whether or not this can be done at once, of course, depends upon the number of animals possessing the trait and whether one can afford to discard all of them at one time.

If the penetrance of the gene is low and the genes are variable in their expression, selection against a dominant gene would be much less effective. Selection for such a trait could not be based upon the individual's phenotype alone, but attention to the phenotype of the ancestors, progeny, and collateral relatives would also be necessary if selection were to be successful.

10.4.3 Selection for a Recessive Gene

Selection for a recessive gene is relatively simple, if penetrance is complete and genes do not vary too much in their expression. Selection under such conditions is merely a matter of keeping those individuals which show the recessive trait. A good example of such selection would be for the horned gene in cattle. To produce all horned cattle, one merely has to obtain horned breeding stock and mate them together. The only time polled individuals would be produced from such a mating is when a mutation from the horned to the polled gene occurs. This is so infrequent that it is seldom observed in an average-size herd.

10.4.4 Selection Against a Recessive Gene

Selection against a recessive gene is the same as selection for a dominant gene. In both instances the homozygous recessive individuals can be identified and discarded. Even when this is done, the recessive gene still remains in the herd, or population, being possessed by heterozygous dominant individuals. To eliminate the recessive gene entirely, the homozygous recessive and heterozygous dominant individuals both must be discarded, leaving only the homozygous dominant individuals.

Discarding or culling all homozygous recessive individuals reduces the frequency of the recessive gene but does not eliminate it. However, if the heterozygous dominant individuals in the population are preferred in selection, the frequency of the recessive gene may increase in spite of the fact that all homozygous individuals are eliminated. The frequency of the recessive gene could not exceed 0.50, however, because the frequency of this

Fig. 10.2 A Black Angus cow and her Red Angus calf. This is a recessive trait and both parents had to be carriers of the red gene. This is an example of a trait in which red individuals have been culled for many generations, but the recessive gene has not been eliminated from the breed.

gene if all individuals are heterozygous would be only 0.50. The amount the frequency of the recessive gene increases (toward the 0.50 level) when the heterozygous dominant individuals are favored in selection and all recessive individuals are discarded depends upon how much the heterozygous individuals are favored.

The following is a formula for determining the frequency of a gene in a population in which all of the homozygous recessive individuals are discarded, if it is assumed that the heterozygous individual is not favored in selection:

$$F_n = \frac{F_o}{1 + (N \times F_o)}$$

where

F_n is the frequency of the recessive gene after all homozygous recessive individuals have been discarded for n generations.

F_o is the original frequency of the recessive gene before the homozygous recessive individuals were discarded.

N is the number of generations of selection against the homozygous recessive individuals.

For example, let us assume that the frequency of a recessive gene in a population is 0.10. What would be the frequency of this recessive gene (F_n) after four generations of selection in which all homozygous individuals were discarded? The answer to this problem is

$$F_n = \frac{0.10}{1 + 4(0.10)} = \frac{0.10}{1.40} = 0.071$$

Data presented in Fig. 10-3 were calculated by using the above formula on a theoretical population in which the original frequency of the recessive was 0.50. Complete selection was then practiced against the homozygous recessive individuals for 20 generations. It will be noted that progress in selection against the recessive gene was very rapid at first, with a sharp decline in the frequency of the recessive gene during the first few generations

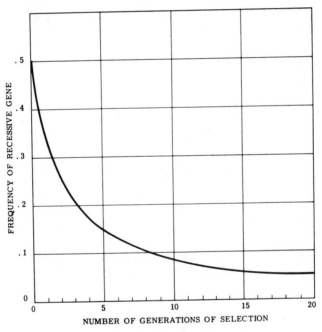

Fig. 10.3 Showing the decrease in the frequency of the recessive gene in a population when all homozygous recessive individuals are discarded and the heterozygote is not favored in selection.

of selection. But as selection continued, the rate at which the frequency of the recessive gene was lowered became less and less. This is what may be expected in a large population, where selection against a recessive gene is based on phenotype alone. Actually, the population will probably never be freed of the gene unless the heterozygous individuals are also identified and discarded along with those that are homozygous recessive.

Discarding all homozygous recessive individuals combined with the identification and elimination of all heterozygous dominant individuals should eliminate the recessive gene from the population, but this same recessive gene could reappear because of a new mutation. If homozygous recessive individuals appear in a herd or flock from parents of a dominant phenotype, it is immediately known that both parents are heterozygous for the recessive gene. Their elimination would also reduce the frequency of the recessive gene. The culling of other close relatives of the homozygous recessive individuals would further reduce the frequency of the recessive gene. All heterozygous dominant individuals must be identified and discarded. This can be done by conducting certain progeny tests especially designed to identify carriers of a recessive gene. These progeny tests are sometimes impractical in identifying females that are carriers of a recessive gene. These tests will be discussed in more detail in a later chapter.

10.4.5 Selection for Epistasis

As pointed out earlier, epistasis is the interaction between genes which are not alleles. These interactions may be of several different kinds. They may be either complementary or inhibitory, but we do not know for certain in what manner the genes may act, as far as their influence on the important economic traits in farm animals is concerned. We do have evidence, however, that epistasis may be of considerable importance in determining the performance of farm animals.

The example presented in Table 10.1 illustrates some of the principles involved in selection for epistasis. In this example we have assumed that the two groups of parents in the P_1 are homozygous for two pairs of alleles, but in opposite ways. In actual practice, it is very unlikely that each parent group would have a high frequency of a particular gene and a low frequency of the allele of that gene. However, we are assuming that they are all homozygous for purposes of illustration.

In this example parents in the P_1 would have less merit (1.40) than those in the F_1 because they did not possess the desired combination of genes $(A\_B\_)$. Thus, the average of the F_1 would be superior to the average of the P_1; this superiority is what is known as heterosis (or hybrid vigor), where

Table 10.1 Illustrating some principles involved in selection when epistasis is important

Assumptions: That a combination of A and B in the genotype adds 0.40 to the phenotype with the basic phenotype not including both genes having a value of 1.40.

Generation	Male parents		Female parents
P_1	AAbb 1.40	×	aaBB 1.40
		Average of P_1 is 1.40.	
F_1		All are AaBb. Average of F_1 is 1.80.	
F_2	AaBb 1.80	×	AaBb 1.80

Progeny:

1	AABB	1.80
2	AABb	1.80
1	AAbb	1.40
2	AaBB	1.80
4	AaBb	1.80
2	Aabb	1.40
1	aaBB	1.40
2	aaBb	1.40
1	aabb	1.40
	Average	1.62

Note: In the F_2 nine would have a phenotypic merit of 1.80 and seven would have a phenotypic merit of 1.40. Only one of the nine with a merit value of 1.80 would breed true. This example also includes the effect of complete dominance, since A is dominant to a and B to b.

the average of the F_1 exceeds the average of the P_1 generation. Other types of nonadditive gene action would also give heterosis.

Crossing the F_1 individuals to produce the F_2 would result in the production of nine different genotypes, but only two phenotypes. Nine of the individuals would have a high merit value of 1.80, whereas seven would have one only 1.40. If we selected the nine out of 16 individuals of the highest merit in the F_2, only one would breed true, and this would be the homozygous dominant genotype $AABB$. Such individuals mated together would produce all $AABB$ offspring, and the average of the progeny would be equal to that of the parents. Individuals of the eight other genotypes would be heterozygous for one or more pairs of genes and would not breed true, with the merit of their progeny, on the average, regressing toward a lower level.

Thus, when this form of epistasis is involved, selection of the superior individuals would cause us to make an error in selecting those that are genetically superior. When we consider the fact that many pairs of genes could interact in the individual to give an epistatic effect, we can see that it would be very improbable (but remotely possible) to develop a pure breeding strain for this epistatic effect.

The information presented in Table 10.1 also illustrates another important point. Note that the average of the F_2 individuals tended to regress from the average of the F_1, which means that some of the F_1 advantage over the P_1 (heterosis) was lost in the F_2 because of the segregation and recombination of genes.

Selection among families, lines, or breeds to find those that would give superior progeny would be the desired way of selecting for epistatic gene action. First, several different unrelated lines should be formed by inbreeding, which serves to make them homozygous for more of the pairs of genes they possess. Once these inbred lines are formed, they should be tested in crosses to find those that "nick," or combine, the best as indicated by the production of superior F_1 progeny. Once two or more lines are found that cross well, they can be retained as pure inbred lines and crossed again and again for the production of offspring for commercial purposes. This is the procedure followed in the production of hybrid seed corn.

The formation of many inbred lines, testing these in crosses to find those that combine the best, then combining these lines by crossing and inbreeding and selecting within the crosses should be helpful in developing superior inbred lines if epistasis is important. This procedure is probably too time-consuming and expensive to be of much practical value in farm animals. Furthermore, it is doubtful if one could ever fix all of the favorable epistatic effects in a single breed or line, regardless of the efforts applied to this end.

10.4.6 *Selection for Overdominance*

Overdominance is the interaction among genes that are alleles so that the heterozygous genotypes are superior, phenotypically, to the homozygotes. As was pointed out earlier, it is never possible to fix overdominance because it is due to heterozygosity.

Some of the principles involved in selection for overdominance are illustrated in Table 10.2. Two pairs of genes are used in this illustration, and it is assumed that each of the two pairs of heterozygous genes add to the merit of the individual over and above the merit of the homozygous parents. Actually, many pairs of heterozygous genes may be involved, and their phenotypic effects may not be equal.

Table 10.2 Illustrating some principles involved in selection when overdominance is important

Assumptions: That the basic phenotype is 1.40 and each combination of A^1A^2 and B^1B^2 adds 0.20 to the phenotype.

Generation	Male parents		Female parents
P_1	$A^1A^1B^1B^1$ 1.40	× Average of P_1 is 1.40	$A^2A^2B^2B^2$ 1.40
F_1		All are $A^1A^2B^1B^2$. Average of F_1 is 1.80.	
F_2	$A^1A^2B^1B^2$	× Progeny:	$A^1A^2B^1B^2$

	Progeny	
1	$A^1A^1B^1B^1$	1.40
2	$A^1A^1B^1B^2$	1.60
1	$A^1A^1B^2B^2$	1.40
2	$A^1A^2B^1B^1$	1.60
4	$A^1A^2B^1B^2$	1.80
2	$A^1A^2B^2B^2$	1.60
1	$A^2A^2B^1B^1$	1.40
2	$A^2A^2B^1B^2$	1.60
1	$A^2A^2B^2B^2$	1.40
	Average	1.60

Note: Only four out of 16 in the F_2 are superior, and this is because they are double heterozygotes. If they are selected for breeding, they would not breed true. Thus, selection should be between lines, families, or breeds to find those that cross best as measured by the merit of their offspring.

The average merit of the male and female parents in the P_1 is 1.40 as compared to 1.80 in the F_1 progeny. Thus, overdominance also results in the production of progeny superior to the average of the parents; this is hybrid vigor (heterosis). When the F_1 individuals were mated among themselves to give the F_2, there were nine genotypes and three phenotypes produced, with four of one phenotype being of superior merit (1.80). If we selected individuals from the F_2 with the greatest merit, all would be heterozygous for two pairs of genes and would not breed true when mated together. Note also that the average of the F_2 has fallen to 1.60 as compared to 1.80 in F_1 and 1.40 in the P_1. Thus, hybrid vigor in the F_2 has declined from that found in the F_1, as was the case with epistasis.

Although it is not possible to make the heterozygous individual breed true, it is possible always to produce them by mating individuals homozygous in opposite ways for the genes causing overdominance, as shown by the example presented in Table 10.2. A trait that illustrates this point is coat color inheritance in Shorthorn cattle. Roan cattle are heterozygous (*Rr*) and possess one gene for red and one for white. This is a type of inheritance in which neither gene is dominant to its allele. The mating of a large number of roan cattle should theoretically result in the production of a ratio of 1 red : 2 roan : 1 white. In spite of the fact that roan individuals do not breed true, it is possible to make matings in such a way that roan individuals may always be produced. This can be done by mating red (*RR*) with white (*rr*) individuals. Since the genotype may be distinguished by the phenotype, such matings can be made rather easily.

It is not possible to determine from their phenotype the genotype of individuals for certain economic traits. This is especially true where many pairs of genes with small effects may be involved. We do know, however, from a theoretical standpoint, that if the heterozygous individual does not breed true, we cannot hope to take advantage of overdominance by combining these genes into a single superior breed. We know also that, in order to always produce heterozygous individuals, we must cross lines or strains that are homozygous for the several pairs of genes that give the overdominant type of gene action. This may be illustrated in the following example:

$$a^1a^1b^2b^2c^1c^1d^1d^1 \qquad \times \qquad a^2a^2b^1b^1c^2c^2d^2d^2$$

$$\text{Inbred line one} \qquad\qquad \text{Inbred line two}$$

$$a^1a^2b^1b^2c^1c^2d^1d^2$$

$$\text{Linecross offspring}$$

Even though we cannot tell the genotypes of inbred line one and inbred line two from their phenotypes, we get some idea of how they complement each other by the performance of their crossbred, or F_1, progeny. This will allow us to practice a kind of selection known as *reciprocal recurrent selection*, which, theoretically at least, should be helpful in improving the nicking or combining ability of two inbred lines.

The procedure in practicing this kind of selection is to test many inbred lines in crosses to identify those that produce the superior crossbred offspring. Then two or more of these lines could be selected for superior combining ability by making reciprocal crosses between them, that is, by crossing boars of one line with sows of the other line, and vice versa, so that boars and sows of each line could demonstrate their crossing ability based on the performance of their crossbred progeny. The boars and sows from each line that demonstrated their ability to produce superior crossbred offspring could then be used to produce the pure line in the next generation. Several generations of such selection might improve the overall crossing or nicking ability of the two lines.

It can be seen, however, that the system of reciprocal recurrent selection is time-consuming and costly. Its effectiveness has not yet been thoroughly tested in farm animals, and its effectiveness may not be fully determined for a number of years. In the meantime, advantage may be taken of this kind of gene action and that of epistasis by crossing different breeds to produce crossbred offspring.

10.4.7 Selection for Additive Gene Action

Additive gene action affects many traits of economic importance. This type of gene action is indicated when the traits are highly heritable and show little or no heterosis or inbreeding depression.

The principles involved in selection for additive gene action are illustrated in Table 10.3. In this example, it is assumed that each contributing,

Table 10.3 Illustrating some principles involved in selection for additive gene action

Assumptions: That the residual genotype *aabb* gives a phenotypic value of 1.40, each plus gene (*A* or *B*) adds 0.10 to the phenotype, and that environment has no phenotypic effects.

Generation	Male parents		Female parents
P_1	AABB 1.80	×	aabb 1.40
		Average of P_1 is 1.60.	
F_1		All are *AaBb*. Average of F_1 is 1.60	
F_2	AaBb 1.60	×	AaBb 1.60
		Progeny: 1 *AABB* 1.80 2 *AABb* 1.70 1 *AAbb* 1.60 2 *AaBB* 1.70 4 *AaBb* 1.60 2 *Aabb* 1.50 1 *aaBB* 1.60 2 *aaBb* 1.50 1 *aabb* 1.40 Average 1.60	

Note: The best individuals in the F_2 are the best because they have more plus genes and not because of heterozygosity. Therefore, in selection for additive gene action find the best and mate the best to the best.

or plus gene (*A* or *B*), adds a value of 0.10 to the residual genotype *aabb*, which has a merit value of 1.40. Thus, the neutral genes (*a* and *b*) add nothing to the increased merit over the residual genotype.

The example used shows that the average value for the P_1, F_1, and F_2 generations are all the same (1.60). Therefore, this type of gene action does not give heterosis. In the F_2 generation, however, there is one individual with a merit value of 1.80 and four with a value of 1.70. To select for this kind of gene action the environment should be standardized as much as possible for all individuals to reduce phenotypic variation due to environment. The best for the trait should then be mated with the best. In other words, selection should be on the basis of individual merit and not on selection among different lines to obtain a "nicking," or heterosis effect. It is well, however, to select the best individuals with superior relatives, since this helps avoid mistaking environmental for genetic effects and *vice versa*.

10.5 SELECTION FOR A SINGLE QUANTITATIVE TRAIT

As mentioned earlier, quantitative traits are those affected by several pairs of genes, many of which have small individual phenotypic effects. The phenotype of such traits may be affected by additive or nonadditive gene action, or both. The phenotypic expression of such traits is also affected by environment. The amount of genetic progress (ΔG) made in one generation of selection for a quantitative trait depends upon the heritability (h^2) of the trait multiplied by the selection differential (S_d) for that trait. Thus the genetic progress expected in one generation of selection would be

$$\Delta G = h^2 \times S_d$$

Heritability estimates (h^2) for the various quantitative traits in each species of farm animals are given in later chapters which deal with improving these species through breeding methods. These estimates were obtained from many reports from experiment stations all over the world.

The selection differential (S_d) refers to the superiority, or inferiority, of those selected for parents (P_s) as compared to the average of the population ($\bar{P}$) from which the breeding animals were selected. The selection differential is also sometimes referred to as the *reach* and may be denoted by the following formula:

$$S_d = (P_s - \bar{P})$$

For example, if the average daily gain in a group of full fed calves is 2.00 ($\bar{P}$) pounds and in those kept for breeding is 2.50 (P_s) pounds, the selection differential would be 0.5 pound per day. If all animals were kept for breeding the selection differential would be zero and the expected genetic progress would be zero.

The selection differential may also be expressed in terms of standard deviation units, providing the frequency distribution curve for that trait is a normal bell-shaped curve. The formula would be

$$S_d = i\,\sigma_p$$

where i is the intensity of selection in standard deviation units and σ_p is the phenotypic standard deviation of the trait in the population from which the breeding individuals are selected. If the proportion of animals kept for breeding is known, the selection intensity i may be calculated from the formula $i = z/w$, where z represents the height of the curve where the group of animals selected for breeding are separated and where w represents the fraction of the population selected for breeding. This is sometimes referred to as selection differentials under *truncation*, which means that all individuals above a certain production level are kept for breeding. The value of z may be obtained from tables showing the ordinates and area of the normal frequency distribution curve.

Information presented in Table 10.4 shows the changes in the selection differential when different fractions of the population are saved for breeding for a certain trait. Data presented in this table show that as fewer animals are saved for breeding, the selection differential (in standard deviation units)

Table 10.4 Showing changes in the selection differential as units of the standard deviation when different proportions of the total population are saved for breeding

Fraction of all animals kept for breeding	Selection differential* as units of the standard deviation, or the selection intensity i
0.90	0.20
0.80	0.35
0.70	0.50
0.60	0.64
0.50	0.80
0.40	0.97
0.30	1.16
0.20	1.40
0.15	1.55
0.10	1.76
0.05	2.05
0.01	2.64
0.001	3.37

*The selection differential S_d equals $P_s - \bar{P}$ equals i times the phenotypic standard deviation σ_p. This assumes that the data fit a normal frequency distribution curve and that all animals above a certain value are kept for breeding (truncated).

increases. When all animals above a certain production level are kept for breeding, the selection differential S_d may be calculated by multiplying the selection intensity i for that level of selection times the phenotypic standard deviation σ_p for the trait in that population. As an example, let us assume that the selection intensity i is 0.20 and the phenotypic standard deviation σ_p for yearling weight in a group of cattle is 95 pounds. The selection differential S_d would be 95 multiplied by 0.20, or 19 pounds. On the other hand, if only 5 percent of the individuals were saved for breeding, the selection differential would be 95 multiplied by 2.05, or 194.75 pounds. Selection pressure would be much more intense in the latter case.

A number of factors may affect the size of the selection differential S_d. As shown in Table 10.4, the smaller the proportion of individuals in the total population kept for breeding, the larger the selection differential. Since fewer males than females are kept for breeding, the selection differential for males will almost always be larger than for females. In a herd expanding in numbers one would necessarily keep a larger number of females for breeding, whereas in a herd maintaining the same number of females year after year one would keep a smaller proportion of females for breeding, and the selection differential will tend to be larger. As a general rule, the selection differential will be larger in litter-bearing animals such as swine than in cattle, where a single birth usually occurs, because the number of offspring from which one may select replacements will tend to be larger. Estimated replacement rates for the different species of farm animals are given in Table 10.5. Even in a species such as cattle where single births usually occur, the selection differential will be larger in a herd where the percent calf crop weaned is large compared to a herd where it is low because a smaller proportion of the total calf crop will be needed for replacements. The sex ratio in a given

Table 10.5 The percentage of progeny required for breeding (replacements) when the herd number remains constant in the different species

Species	Percentage of total crop saved*	
	Males	Females
Beef cattle	4–5	40–50
Dairy cattle	4–5	50–60
Sheep	2–3	40–50
Swine	1–2	10–15
Horses	2–4	40–50
Chickens	1–2	10–15

*This assumes a 1 : 1 sex ratio and that the herd number remains constant.

herd or year may vary from the usual 1 : 1 ratio, and this can affect the size of the selection differential. In years when the proportion of individuals of one sex is large, the selection differential will tend to be higher than in a year when the proportion of individuals of that same sex is low.

Failure to accurately measure the economic traits can result in a smaller selection differential, because a mistake may be made in determining which individuals are actually superior for a particular trait. Thus some individuals may be saved for breeding that are inferior to some of those not saved for breeding. Failure to use records even though they are accurate can lower the size of the selection differential. For example, let us assume that we want to select for faster gain in the feed lot and individual A is outstanding in rate of gain but mediocre in conformation, whereas individual B is mediocre in rate of gain and outstanding in conformation. If we saved individual B because of his better conformation, we would have a smaller selection differential for rate of gain than if we selected individual A.

10.6 SELECTION PROGRESS OVER A PERIOD OF TIME

The amount of expected genetic gain made over a period of time (ΔG_t) through selection depends upon the size of the selection differential S_d, the degree of heritability of the trait h^2, and the length of the generation interval I_g. The formula used for computing the expected genetic gain over a period of time is as follows:

$$\Delta G_t = \frac{S_d \times h^2}{I_g}$$

The heritability estimate for a trait and the selection differential were discussed previously.

10.6.1 Generation Interval

The generation interval may be defined as the average age of the parents when their offspring are born. The generation interval varies with different species of animals and with the breeding and management systems followed to produce a new generation of breeding animals. The generation interval in swine can be reduced to one year if pigs are selected from the first litters of gilts bred to boars of the same age. Such parents produce only one litter. When this is practiced, gilts can be bred when they are seven to eight months of age and will produce litters by the time they are one year of age. If sows as well as boars are progeny-tested before they are used to produce breeding or replacement offspring, the generation interval may be two years or even longer. In cattle, the generation interval conceivably could be as short as 2.5 to 3.0 years, but on the average it is considerably longer than this if any progeny-testing is done or if the performance records of cows determine whether or not their offspring are kept for breeding purposes.

To illustrate how the length of the generation interval might affect the amount of progress made in selection over a period of time, let us compare two systems of breeding in swine, one where young boars and gilts are used for breeding purposes and the generation interval is one year, and one where replacements of breeding stock are made only after records are obtained on the parents from two previous litters. The generation interval in the latter case would be about two years. In a period of four years, we should have had the opportunity to produce four generations with the first selection system, but with the second only two generations would have been produced. It is obvious that if the heritability of the trait and the size of the selection differential are the same, we would expect to make more progress in selection in four than in two generations. The average lengths of the generation interval in some farm animals are shown in Table 10.6.

Table 10.6   Average length of the generation interval in years for different species of farm animals

	Average length of generation interval in years	
Species	Males	Females
Beef cattle	3.0–4.0	4.5–6.0
Dairy cattle	3.0–4.0	4.5–6.0
Sheep	2.0–3.0	4.0–4.5
Swine	1.5–2.0	1.5–2.0
Horses	8.0–12.0	8.0–12.0
Chickens	1.0–1.5	1.0–1.5

Note: The generation interval can be shorter than this in most species if an attempt is made to shorten it. If sows and boars produce only one pig crop, the generation interval would be about one year.

10.6.2   Genetic Correlations Among Traits

When we speak of genetic correlations among traits, we are referring to whether or not the same gene responsible for qualitative inheritance or some of the same genes responsible for quantitative inheritance affect two or more economic traits. Genetic correlations among traits are estimated by special statistical procedures or by selecting for one trait over a period of time and noting whether or not there is a change (correlated response) in traits not selected for as genetic improvement is made in the trait for which selection is practiced. Single-trait selection experiments must be carefully designed accurately to observe whether or not two or more traits are genetically correlated. Possible genetic correlations among economic traits and their relationship to genetic improvement over a period of time will be discussed more fully in later chapters, dealing with the selection and improvement of specific species of farm animals.

Pleiotrophy is probably the major cause of genetic correlations, although it is possible for linkage to have a similar transitory effect. Pleiotrophy is the process whereby one gene may affect two or more traits. Linkage means that the genes are carried on the same chromosome. Some genes may be so closely linked together on the same chromosome that they seldom, if ever, separate by crossing-over during synapsis in meiosis. Closely linked genes would tend to stay together over several generations, and the association of the traits determined by them would persist. Genes farther apart on the same chromosome would separate more readily by crossing over during synapsis in meiosis, and the relationship of the traits determined by such genes would break up, or become transitory. If pleiotrophy is the cause of genetic correlations this would suggest that the traits correlated would be affected by at least some of the same physiological pathways.

The genetic correlation between two traits may be very low, which means that probably very few of the same genes affect the two traits. Type and performance in beef cattle is a good example of this in that selection for type seems to have little influence on performance, or *vice versa*. Obviously, selection on the basis of one will not make an improvement in the other, and we might say that the two traits are inherited independently. If true, this means that it should be possible to get both in our animals, but to do so we must select for both.

Two or more traits may also be correlated from the genetic standpoint in a positive manner. By this is meant that selection for the improvement of one will also result in the improvement in the other, even though direct selection for its improvement has not been practiced. An example of this is rate and efficiency of gain in swine. Evidence has accumulated which indicates that if we select within a herd for fast-gaining individuals and make improvement in this trait, the efficiency of gain also improves. This indicates that physiologically, as well as genetically, the two traits are correlated or influenced by the same genes. If the genetic correlation between two traits is high enough, it may not be necessary to measure both, especially if the measurement of one requires added expense, time, and equipment. Thus, if there is a high enough genetic correlation between rate and efficiency of gain, we could measure only rate of gain and select for it and improve both at the same time. The rate of gain of individuals may be measured easily even if animals are fed in a group, but the efficiency of gain is more difficult to measure, since animals must be fed individually and careful attention to prevent feed wastage is required.

It is also possible for two traits to be genetically correlated in a negative manner. This means that selection for the improvement of one, if successful, results in a decline in the other to which it is genetically correlated. An example of such a correlation is butterfat percentage and milk yield in dairy cattle. Although there is evidence of other possible negative genetic relationships between economic traits in farm animals, more data are needed. Perhaps negative genetic correlations may explain why selection for dual-purpose animals has not been as successful as desired.

REFERENCES

1. Buckwater, J. A., *et al.* "Natural Selection Associated with the A, B, O Blood Group," S, 123: 840, 1956.
2. Murie, A. D. "The Wolves of Mt. McKinley," USDI, Fauna of the National Parks, Fauna Series No. 5, 1944.

QUESTIONS AND PROBLEMS

1. Define *selection.*

2. What are two general kinds of selection? Define the meaning of each.

3. Does man ever select breeding animals without nature giving him an assist? Explain.

4. Has selection been effective in the past? Explain.

5. What is the major genetic effect of selection? If selection is effective, what other genetic effect may result?

6. What would be the proportion of homozygous dominant individuals in a population if the frequency of the recessive allele were 0.70? If the frequency of the recessive allele were 0.20? What principle does this illustrate?

7. Why is it important to know how to select for different types of gene action?

8. Describe the procedures to follow in selection for a dominant gene; in selection against a dominant gene.

9. Describe the procedures to follow in selection for a recessive gene; to select against a recessive gene.

10. In actual practice, will the breeder be more likely to select for a dominant or a recessive gene? Explain.

11. Does culling all recessive individuals eliminate a recessive gene from a population? Explain.

12. Outline methods you would use to eliminate a recessive gene from a herd. How could you prevent the appearance of recessive individuals but not eliminate a recessive gene from a herd?

13. Assume that the frequency of a recessive gene in a population is 0.20. What would be the frequency of this recessive gene after all recessive individuals were discarded for eight generations? What conditions must be met to make these calculations valid?

14. Why is selection among families more desirable than selection among individuals when epistasis is important?

15. Is it possible to fix an epistatic trait in a population and make it breed true? Explain.

16. Explain why selection should not be based on individuality when over-dominance is important.

17. What is meant by reciprocal recurrent selection? What should be accomplished (theoretically) by using this system of selection?

18. Explain the procedures to use in selection for additive gene action.

19. What is heterosis? Which of the following types of gene action are responsible for heterosis: additive, epistasis, dominance, and/or overdominance?

20. What determines the amount of expected genetic progress through selection for one generation in a quantitative trait?

21. What is meant by the selection differential? What determines the size of the selection differential?

22. The average daily gain in a group of cattle was 2.25 lb. Bulls and heifers selected for breeding averaged 3.15 and 2.50 lb per day, respectively. If the heritability of daily gain in the feed lot is 0.60, what would be the expected genetic gain in the offspring? What would be the expected average phenotype for rate of gain in the progeny? What would be the expected genetic average in the offspring?

23. In a herd of cattle, the average yearling weight of all animals was 800 lb when corrected to a bull basis. Bulls and heifers kept for breeding weighed 1000 and 825 lb, respectively. The progeny of those individuals selected for breeding had an average yearling weight corrected to a bull basis of 845 lb. What is the apparent heritability of the trait?

24. In a herd where the average 210-day weight was 400 lb, 50 percent of the heifers and 5 percent of the bulls were kept for breeding. What is the selection differential for the bulls? For the heifers? For both bulls and heifers? (Assume that the standard deviation for weaning weight is 45 lb.)

25. Define generation interval. Define the generation interval in your own family.

26. The following are the dates of birth of the parents of calves born in the year 1970:

Calf no.	Year sire was born	Year dam was born
1	1965	1966
2	1963	1962
3	1965	1964
4	1963	1965
5	1966	1966
6	1966	1963
7	1966	1960
8	1967	1965
9	1967	1967
10	1965	1965

What is the generation interval for the sires? For the dams?

27. Using generation intervals listed in Table 10.6, what would be the expected genetic progress per year in a herd where the average yearling weight for bulls and heifers, corrected to a bull basis, were 1050 and 850 lb, respectively, if the average for the entire herd was 800 lb? Assume a heritability estimate for yearling weight of 60 percent.

28. Why is the generation interval usually shorter for bulls than for cows, and shorter for swine than cattle?

29. Does a shorter generation interval always increase genetic progress per year as compared to a longer generation interval? Explain.

30. What is meant by a genetic correlation among two traits? How may genetic correlations affect the amount of progress made in selection?

11

Selection of Superior
Breeding Stock

Any progress animal breeders hope to make through the application of breeding and selection methods will depend upon their ability to recognize and mate those animals possessing superior inheritance for a particular purpose. Superior inheritance is indicated by the phenotypic merit of the individual or upon its ability to combine well with others for the production of superior F_1 offspring. In any event, these superior animals produce the next generation if genetic advance is to be made.

Methods of estimating the breeding value of an individual from its own phenotype and that of its relatives are the subject of this chapter. The kinds of relatives an individual possesses are shown in Fig. 11.1.

11.1 SELECTION ON THE BASIS OF INDIVIDUALITY

Selection based on individuality means that an animal is kept or rejected for breeding purposes on the basis of its own phenotype for a particular trait, or traits. The progress made in selection depends upon how closely the geno-

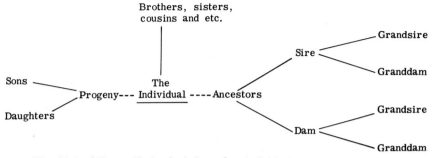

Fig. 11.1   Different kinds of relatives of an individual upon which selection may be based.

type is correlated with the phenotype. Sometimes this correlation is high, but there are times when it is low. The phenotype of the individual varies throughout its life because of environmental effects or the interaction between its genotype and environment. The genotype of an individual, however, is fixed at the time of fertilization and does not vary as does the phenotype.

11.1.1 Selection for Qualitative Traits

The phenotype of the individual (individuality) is often used to estimate its breeding value for qualitative traits such as coat color and horns or lack of horns. Selection for such traits on the basis of the individual's phenotype is more effective in some instances than in others. The genotype of the homozygous recessive individual, where only one pair of genes is involved, may be determined from its phenotype. For example, we know that a red Angus carries two red genes (genotype *bb*) because this red gene is recessive to the black gene (*B*). The genotype of the individual carrying a dominant gene cannot be determined from its phenotype because we cannot distinguish, phenotypically, between the homozygous dominant and the heterozygous dominant individual. For example, we cannot tell if a black Angus is *BB* or *Bb* from its phenotype, or color. Thus, selection on the basis of individuality for qualitative traits may be useful but it is not always completely accurate. Information on the phenotypes of the close relatives as well as that of the individual makes these estimates of the genotype more accurate. The same is true for quantitative traits.

11.1.2 Selection for Quantitative Traits

Quantitative traits are those affected simultaneously by many pairs of genes and various elements of the environment so that there is no sharp distinction among phenotypes of the individuals within a group. Such traits may be affected mostly by additive gene action or mostly by nonadditive gene action, or both.

In selection for quantitative traits the breeder attempts to estimate the genotype of the individual from its phenotype. If such a trait were 100 percent heritable, the phenotype and genotype of the individual for that trait would be identical. However, no quantitative trait is 100 percent heritable, because environment always affects the phenotype to a certain extent.

The phenotypic merit of the individual for important economic traits (quantitative) is determined by comparing the individual's phenotype with that of the average of all individuals within a group from which it is selected, as described in Chapter 10. To be effective, the comparison must be made under carefully controlled environmental conditions with other animals of nearly the same age and at the same time. Accurate records are also required.

The individual's record is of little value unless it shows where that individual ranked relative to others under similar conditions. For example, let us assume that an individual gained 2.50 lb per day on a standard gain test. This would estimate its phenotype for gain, but unless it can be compared with the gain of others on the same test it tells us little or nothing about the individual's relative phenotype. If the average of all individuals on this same gain test was 1.80, however, we can get an estimate of where that individual ranked as compared to the average of all individuals on the test. The relative phenotypic merit of this individual would be 2.50 minus 1.80, or 0.70 lb per day above the average of all individuals on this particular test. If another individual on the same test gained only 1.50 lb per day, his phenotypic value would be 1.50 minus 1.80, or −0.30 lb per day gain. This means that his gain was 0.30 lb per day below the average of all individuals on the test.

Some animal breeders prefer to express an individual's phenotypic value for a trait as a percentage of the average of all individuals on that particular test. This is sometimes called the trait ratio and may be calculated for any trait as follows:

$$\frac{\text{Individual's record for a trait}}{\text{Group average for the same trait}} \times 100$$

Thus, the gain ratio of the individual that gained 2.50 lb per day in the previous example would be 2.50 divided by 1.80 times 100, or 139. This would mean that this individual gained 39 percent faster than the average of the group from which it was selected. The individual that gained only 1.50 lb per day would have a gain ratio of 1.50 divided by 1.80 times 100, or 83. It would rank about 17 percent below the average of all individuals on this specific gain test. A ratio of this kind may be calculated for any trait or index including several traits, in the same manner, and may be used to compare individuals in different years, on different farms, or in different herds. It avoids some of the differences between years, farms, or herds whether these are due to differences in environment or genetic merit, providing genetic and environmental interactions are not important.

As stated previously, the phenotype of an individual for a quantitative trait is not a true estimate of its genotype, because no trait is 100 percent heritable, but it is also affected by environment. For example, if daily gain used in the previous example is 50 percent heritable, this suggests that one-half of the 0.70 daily gain superiority of the individual is likely due to environment. This environmental portion of its phenotypic superiority will not be transmitted to its offspring, but only that which is due to the average effects of genes. For the individual that had a daily gain 0.30 lb below the average of the group from which it was selected, this environmental inferiority 'will not be transmitted to its offspring. Therefore, in general, there is a tendency for the average phenotype of the offspring of a phenotypically superior individual to regress toward the average of the population, whereas the average phenotype of the progeny of phenotypically inferior individuals will tend to rise toward the average of the population.

One should remember that we are speaking of average heritabilities and average phenotypes in these examples. This means that there will be a general tendency for these examples to be true, but there will be individual exceptions. Even from superior or inferior parents, there will be a tendency for one or more progeny to be superior or inferior, but the phenotypic average of the progeny of the superior parent should exceed that of the inferior parent.

The probable breeding value (PBV) of an individual expressed as a deviation from the average for a particular trait may be determined from

$$\text{PBV} = \bar{P} + b_1(P_i - \bar{P})$$

where b_1 is the regression coefficient for the genotype of the individual on its own phenotype. Regression coefficients for different degrees of heritability for a certain trait are given in Table 11.1. P_i is the phenotypic value of the individual selected and $\bar{P}$ is the phenotypic average for all individuals tested at the same time from which the individual was selected. The PBV calculated in this manner is the estimated genetic superiority of the individual (ΔG) over the average of the group from which it was selected and is always nearer the group average than its phenotypic value because of environmental effects.

The PBV of the individual that gained 2.50 lb per day as compared to 1.80 lb per day for all individuals tested at the same time, if it is assumed that daily gain is 50 percent heritable, would be $1.80 + 0.50\ (2.50 - 1.80)$, or 2.15 lb per day. Data presented in Table 11.1 show that as the heritability of a trait increases, the accuracy of predicting the PBV of an individual from its phenotype also increases.

Table 11.1 Regression coefficients (b_1) for genotype on phenotype and accuracy of selection at different degrees of heritability when selection is based on the individual's own phenotype (one record)

Heritability estimate of a trait	Regression coefficient (b_1)	Accuracy of selection*
0.10	0.10	0.32
0.20	0.20	0.45
0.30	0.30	0.55
0.40	0.40	0.63
0.50	0.50	0.71
0.60	0.60	0.77
0.70	0.70	0.84
0.80	0.80	0.89
0.90	0.90	0.95
1.00	1.00	1.00

*Accuracy of selection was determined from $\sqrt{b_1}$

11.1.3 Selection for Quantitative Traits Affected Largely by Nonadditive Gene Action

Nonadditive gene action tends to lower the heritability and to increase the expression of hybrid vigor. Selection on the basis of individuality for such traits is usually less effective than for highly heritable traits. The superior individual for such a trait is often highly heterozygous (overdominance) and does not breed true. The more homozygous individuals tend to be less desirable, phenotypically.

The offspring of heterozygous parents tend to regress toward the mean of the population because on the average the offspring are less heterozygous than the parents. Superior parents for such traits may also be superior because of environment, which is another factor responsible for their offspring regressing toward the mean of the population.

Selection for nonadditive gene action involves selection among lines, families, or breeds to find those that "nick" or combine the best on the basis of the performance of their F_1 progeny. This kind of selection will be discussed in more detail later in connection with progeny tests.

11.2 SELECTION ON THE BASIS OF PEDIGREES

A pedigree is a record of an individual's ancestors related to it through its parents. In the past, most of the information included in pedigrees has consisted only of the names and registration numbers of the ancestors, and little has been indicated as to their phenotypic or genotypic merit. More recently, data to indicate the phenotypic merit of ancestors are being included in pedigrees, especially in swine. Swine pedigrees now often include information on the size of the litter at birth and weaning, and the weight of the litter at weaning, as shown in Fig. 11.2. In addition, some information is now being included in the pedigree on carcass quality of ancestors, as a result of the Certified Meat Hog Program sponsored by the different breed associations. This is a step in the right direction and may result in improvement in highly heritable traits such as carcass quality and quantity.

Even with performance and other records included along with the names and registration numbers of ancestors in the pedigree of an individual, more information is needed. Records of the ancestors could be more useful in predicting the breeding value of an individual if they gave some indication of how that ancestor ranked for certain traits in comparison with its contemporaries. For example, if the selection differential for the ancestor could be presented in the pedigree (calculated from the average performance record of the contemporaries subtracted from the performance record of that ancestor), or if the performance record of the ancestor could be expressed as a percentage of the average performance of its contemporaries (trait ratios), the ancestor's records would have greater predictive value.

11.2.1 The Pedigree in Selection for Qualitative Traits

A study of pedigrees, if full information is available on the phenotypes and genotypes of the ancestors, may be of importance in detecting carriers of a recessive gene. Such information has been used with success in combating dwarfism (a recessive trait) in beef cattle. Cattle were sold on the basis of whether or not they were "pedigree clean" or "pedigree dirty." The latter term refers to an animal having some ancestors that produced dwarfs and who, therefore, were carriers of the dwarf gene. This would indicate that the animal whose pedigree was being studied could also be a carrier of this gene.

A disadvantage of the use of the pedigree in selection against a recessive gene is that there are often unintentional and unknown mistakes in pedigrees that may result in the condemnation of an entire family of breeding when actually it may be free of such a defect. On the other hand, the frequency of a recessive gene in a family may be low and records may be incomplete, so that an animal appears to have a "clean" pedigree. Then, later, it will be found that the gene is present, and this family, once thought to be free of the recessive gene, will be called a "dirty" family.

A definite disadvantage of pedigree selection as used in dwarfism in beef cattle is that all animals with the same or similar pedigree are condemned. This occurs in spite of the fact that individuals in such a line are

Pedigree of Brilliant Miss Royalty 895658 11-STAR PR 439
Raised 11 consecutive litters all PR qualified.
Av. 11.72 Farrowed
    10.18 Raised
    387.4 lbs. at 56 days

Transformer's Masterpiece
326965 3-STAR PR 21
Has 28 dau. with 52 litters.
Av. 11.15 pigs Farrowed;
    9.67 pigs Raised;
    358.35 lbs. at 56 days

Transformer 288723
1-STAR PR 22
9 dau.: 19 litters
Av. 11.15 Farrowed
    9.67 Raised
    358.35 lbs. at 56 days

Kasters Queen
707876

Steam Roller 256561
2-STAR PR 12
17 dau.: 23 litters
Av. 10.8 Farrowed
    10 Raised
    370.48 lbs. 56 days

Harper's Sue 3rd
639384

Master Roller
277319

Dolls Gal
666442

Century Hi Roller
Sire of 4 PR boars

Lady Flash
Harpers Hi Hope
Highway Sue
Gardners Roller
Betsy Clan
Wicks Clan
Nebraska Doll 1st

Mt. Ararat Brilliant
Beauty 710908
5-STAR PR 164
5 PR litters
Av. 11.8 pigs Farrowed
    9.8 pigs Raised
    356.5 lbs. at 56 days.

Litter mate to Mt. Ararat
Bright Beauty
3-STAR PR 116

Mt. Ararat Quicksilver
283503

American Beauty 654280
1-STAR PR tested sow
Av. 9 pigs Farrowed
    8 pigs Raised
    376 lbs. at 56 days

The Mercury
267327

Cesor's Brilliant
Princess 651608

Century of Earlham 259101
8 tested dau.: 12 PR litters
Av. 11.58 Farrowed
    9.75 Raised
    372 lbs. at 56 days

Rambler Beauty 630074
3-STAR PR 28
4-PR litters
Av. 10.75 Farrowed
    8 Raised
    348.75 at 56 days

The Clipper Deluxe
Powder Puff
Main Answer
Lady Brilliant 4th
Century Hi Roller
Sire of 4 PR boars

Lady Brilliant 4th
The Clipper Deluxe

Fair Lady

Fig. 11.2 Pedigree of Brilliant Miss Royalty (895658) which includes information on the performance of relatives. A pedigree of this kind is of much more use in selection than one containing only names and numbers. (Courtesy of the American Hampshire Swine Record Association.)

free of the recessive gene, as proved by progeny tests (Fig. 11.3). Nevertheless, the individual still has a questionable pedigree and will be discriminated against by many breeders, either because they are not familiar with the mode of inheritance affecting such a trait or because they are afraid to trust progeny-test information.

Fig. 11.3 This young bull was proved free of the dwarf gene by the progeny test even though his sire was a known carrier of the dwarf gene.

Still another disadvantage of pedigree selection is that a pedigree may often become popular because of fashion or fad and not because of the true merit of the individuals it contains. The popularity of the pedigree may also change from time to time, and the value of such a pedigree may decrease and individuals from this line of breeding may even be discriminated against (Figure 11.4). If the popularity of a pedigree is actually based on merit, there is less danger of its value decreasing in a short period of time.

11.2.2 The Pedigree in Selection for Quantitative Traits

The use of records of the performance of ancestors to increase the accuracy of determining the probable breeding value of an individual can help increase the accuracy of these predictions under certain conditions. To be of value for this purpose, the records of the ancestors must give some idea of their merit as compared to that of their contemporaries, and the heritability of the trait must be something less than 100 percent. Since the heritability of a trait is never so high, good records on the performance of ancestors make predictions of the individual's breeding value more accurate.

How much attention should be paid to the performance records of an

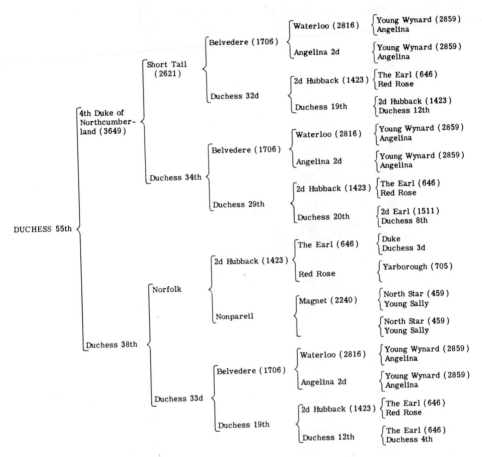

Fig. 11.4 Pedigree of Duchess 55th. A very popular pedigree at one time, but the family later lost popularity because of barrenness in the cows.

ancestor depends upon the following factors: (1) The degree of relationship between the ancestor and the individual. An individual is related by 0.50 to each parent, 0.25 to each grandparent, and 0.125 to each great-grandparent, providing no inbreeding is involved. (2) The degree of heritability of the trait. (3) Environmental correlations among animals used in the prediction. (4) How completely the merit of the ancestors used in the prediction is known.

Statistically the *accuracy of selection* is represented by the correlation of the genotype of the individual with the phenotypic average of the relatives used in selection. *The probable breeding value* (PBV) of an individual is calculated from the regression of the genotype of the individual on the phenotypic average of the relatives times the selection differential for the individual or its relatives. These terms will be used often in the remainder of this chapter.

The accuracy of selection using the phenotype of the individual and/or that of various ancestors is given in Table 11.2. These calculations are based

Table 11.2 Accuracy of selection at different degrees of heritability when selection is based on the individual's records or on its records plus those of its ancestors

Records used	Accuracy of selection at different degrees of heritability					
Heritability estimate	0.10	0.30	0.50	0.70	0.90	1.00
Individual's own records	0.316	0.548	0.707	0.837	0.949	1.000
Individual's own records plus those of one parent	0.347	0.581	0.730	0.847	0.950	1.000
Individual's own records plus those of one parent, those of one grandparent, and one greatgrandparent (One line of descent)	0.375	0.614	0.756	0.861	0.953	1.000
Records of the sire and dam plus those of all four grandparents.	0.265	0.434	0.534	0.609	0.674	0.707

on the assumption that the phenotypes of all individuals are equally well-known and that no environmental correlations exist between various relatives for a certain trait. Also, it is assumed that no inbreeding is involved.

Data presented in Table 11.2 show that the accuracy of selection increases as the heritability of the trait increases, that attention to records of certain ancestors increases the accuracy of selection, and that attention to the individual's own performance record gives a greater accuracy in selection than performance records on the sire, the dam, and all four of its grandparents. For highly heritable traits the records of ancestors add little to an individual's own record.

Regression coefficients for predicting the PBV of the individual based on its own phenotype or the phenotype of any one of its ancestors are given in Table 11.3. Predictions of the PBV of the individual based on its own phenotype or the phenotype of one of its ancestors may be calculated from $\bar{P} + b_i (P_j - \bar{P})$, where b_i is the regression coefficient for trait i, P_j represents the phenotype of the jth individual or ancestor and $\bar{P}$ represents the average phenotype of all individuals in the contemporary group from which the individual, or the ancestor, was selected. These regression coefficients show that information on the performance of the individual is more valuable in predicting its PBV than information on any one ancestor.

Table 11.3 Regression coefficients that may be used to predict the PBV of individuals from information on the phenotype of the individual or that of certain of its ancestors

Regression coefficients		Heritability of the trait					
		0.10	*0.30*	*0.50*	*0.70*	*0.90*	*1.00*
Individual's record	b_1	0.100	0.300	0.500	0.700	0.900	1.000
Parent's record	b_2	0.050	0.150	0.250	0.350	0.450	0.500
Grandparent's record	b_3	0.025	0.075	0.125	0.175	0.225	0.250
Great-grandparent's record	b_4	0.012	0.038	0.062	0.088	0.112	0.125

Note: The PBV of an individual is equal to $\bar{P}_i + b_i(P_j - \bar{P}_i)$, where b_i is the regression coefficient for trait i, P_j the phenotypic record of the individual j, and $\bar{P}_i$ the phenotypic average of that individual's contemporaries.

Information summarized in Table 11.4 may be used to predict the PBV of the individual from various combinations of its own records and those of certain ancestors. Since two or more relatives are used to predict the PBV of the individual, partial regression coefficients are used (Becker, 1970). In making these predictions, the partial regression coefficient for the individual is multiplied by the selection differential for the individual and this is added to the product of the partial regression coefficient and selection differential for each ancestor used to predict the individual's PBV. This emphasizes the importance of having performance records on the individual or the ancestors that may be compared to the average of their contemporaries to obtain the proper selection differential.

Information presented in Table 11.4 may also be used to calculate the PBV of the individual based on various combinations of records where the phenotypes are expressed as a percentage of the average performance (trait ratios) of all individuals in the group from which that individual was selected. This may be done by merely substituting the appropriate ratio for the corresponding record of the individual, or its ancestors, and using 100 wherever a contemporary average is indicated.

Probably one of the greatest deficiencies of pedigree selection is that good performance records on each ancestor as compared to its own contemporaries are lacking. If such information is not available, it is important that the merit of the sire, dam, or other ancestor be carefully assessed by comparing it to some base such as the breed average. Furthermore, pedigree selection provides no basis for selection among individuals that are all descendants of the same ancestor.

Pedigrees do have the advantage that they are cheap to use, they may be used to select for traits not expressed early in life such as cancer eye and longevity, and they may be used to select for traits expressed only in one sex such as milk and egg production.

Table 11.4 Predicting the breeding values of individuals from information on the phenotype of the individual combined with that of certain ancestors

A. Prediction based on the individual's own record plus that of one parent.

		Heritability of the trait			
Partial regression coefficients		*0.10*	*0.30*	*0.50*	*0.70*
Individual's record	b_1	0.098	0.284	0.467	0.658
Parent's record	b_2	0.045	0.107	0.133	0.120

$$PBV = \bar{P}_{ic} + b_1(P_i - \bar{P}_{ic}) + b_2(P_p - \bar{P}_{pc})$$

B. Predictions based on the individual's own record plus that of one parent, one grandparent, and one great-grandparent all in one line of descent.

Individual's record	b_1	0.097	0.282	0.464	0.657
Parent's record	b_2	0.044	0.101	0.124	0.113
Grandparent's record	b_3	0.020	0.037	0.034	0.019
Great-grandparent's record	b_4	0.009	0.014	0.010	0.004

$$PBV = \bar{P}_{ic} + b_1(P_i - \bar{P}_{ic}) + b_2(P_p - \bar{P}_{pc}) + b_3(P_{gp} - \bar{P}_{gpc}) + b_4(P_{ggp} - \bar{P}_{ggpc})$$

C. Prediction based on the individual's own record plus that of both parents.

Individual's record	b_1	0.096	0.267	0.429	0.603
Sire's record	b_2	0.045	0.110	0.143	0.139
Dam's record	b_2	0.045	0.110	0.143	0.139

$$PBV = \bar{P}_{ic} + b_1(P_i - \bar{P}_{ic}) + b_2(P_s - \bar{P}_{sc}) + b_2(P_d - \bar{P}_{dc})$$

D. Prediction based on the performance records of the parents and grandparents.

Sire's record	b_2	0.048	0.134	0.214	0.301
Dam's record	b_2	0.048	0.134	0.214	0.301
Maternal					
Grandsire's	b_3	0.023	0.055	0.071	0.070
Granddam's	b_3	0.023	0.055	0.071	0.070
Paternal					
Grandsire's	b_3	0.023	0.055	0.071	0.070
Granddam's	b_3	0.023	0.055	0.071	0.070

$$PBV = \bar{P}_{ic} + b_2(P_p - \bar{P}_{pc}) + b_2(P_d - \bar{P}_{dc}) + b_3(P_{mgs} - \bar{P}_{mgsc}) + b_3(P_{mgd} - \bar{P}_{mgdc})$$
$$+ b_3(P_{pgs} - \bar{P}_{pgsc}) + b_3(P_{pgd} - \bar{P}_{pgdc})$$

Note : The PBV is predicted by multiplying the partial regression coefficient b_x by the selection differential for that particular individual or ancestor and then adding this to the average of their contemporaries. Where the letter *c* is included in the subscript, it indicates the individual's contemporaries. For example, $\bar{P}_{sc}$ means the average of the contemporaries of the sire.

11.3 SELECTION ON THE BASIS OF PROGENY TESTS

Selection on this basis means that the breeder makes a decision to keep or cull a sire or dam based on the average merit of their offspring as compared to the average merit of the progeny of contemporary sires and dams. Progeny tests may be used in selection for both qualitative and quantitative traits.

11.3.1 Progeny Tests in Selection for Qualitative Traits

Probably the most effective use of progeny tests in selection for qualitative traits is to determine if an individual of the dominant phenotype is homozygous or heterozygous. To produce a pure breeding line or herd for a dominant trait or to eliminate all individuals carrying a recessive gene in the line or herd, one must discard all homozygous recessive individuals as well as the heterozygous individuals that, although possessing the dominant phenotype, are carriers of the recessive gene. The recessive individuals are identified from their phenotypes, but the heterozygous and homozygous individuals have similar phenotypes. The genotypes of these two dominant phenotypes must be determined through progeny tests unless it is known that one parents is recessive. A dominant individual is heterozygous if one of its parents or one of its offspring is a recessive homozygote. One can never be absolutely certain that an individual is homozygous dominant after it is progeny-tested, but the more dominant offspring the individual produces without producing any that are recessive when certain test matings are made, the higher the probability that the individual is homozygous dominant.

The kinds and numbers of test matings to make in animals producing only one young per birth to determine if they are homozygous dominant are shown in Table 11.6, and for animals with multiple births in Table 11.7.

Table 11.6 Kinds and numbers of matings with all offspring of the dominant phenotype necessary to prove an individual of the dominant phenotype homozygous dominant at the 95 and 99 percent levels of probability in animals giving birth to one young

Kinds of matings	Number of matings that produce dominant phenotype offspring and no recessive offspring*	
	95%	*99%*
Mated to homozygous recessive individuals	5	7
Mated to known heterozygous individuals	11	16
Male mated to unselected daughters of another known heterozygous dominant male	23	35
Sire mated to his own unselected daughters†	23	35

*Only one recessive offspring is required to prove the individual heterozygous dominant.
†The sire would be tested for any recessive gene he might be carrying and not just one specific recessive gene.

Table 11.7 Kinds and numbers of matings with all offspring of the dominant phenotype necessary to prove an individual of the dominant phenotype homozygous dominant at the 95 and 99 percent levels of probability in animals giving birth to litters

Kinds of matings	Numbers of matings	No. of dominant phenotype offspring* and no recessives necessary to prove the tested parent homozygous dominant at the probability level of:	
		95%	99%
Mated to homozygous recessive individuals	To 1 or more homozygous recessive individuals	5 total	7 total
Mated to known heterozygous individuals	To 1 or more known carriers of the recessive gene	11 total	16 total
Male mated to unselected daughters of another known heterozygous male	At least 5 unselected daughters	11 each daughter	
	At least 7 unselected daughters		16 each daughter
Sire mated to his own unselected daughters	At least 5 unselected daughters	11 each daughter†	
	At least 7 unselected daughters		16 each daughter†

*Only one recessive offspring is required to prove the tested individual heterozygous.
†Each daughter would also be tested for all recessive genes she might have received from her sire. The sire would also be tested for any recessive gene he might be carrying. The sire has to be mated to different daughters, whereas each daughter is tested when she has a sufficient number of offspring when mated to her sire.

It is much simpler to progeny-test for recessive genes in litter-bearing animals than in those that give birth to a single young. In species where only one young is born it is practical to progeny-test males. In litter-bearing animals both sires and dams may be progeny-tested. In all species where a sire can be mated to his own daughters, this has the advantage of progeny-testing him for all detrimental genes he might possess and not for just a specific gene. Other matings (those not between related animals) test only for a specific gene. Thoroughly progeny-tested males that prove to be free of detrimental recessive genes could be used on a wide-scale basis for artificial insemination or for establishing new inbred lines.

Mating a sire to his own daughters has the disadvantages of a high degree of inbreeding (25 percent) and a large number of daughters being required, especially in species giving birth to a single young. Fewer daughters are required, however, for progeny-testing a sire from a litter-bearing species,

Fig. 11.5 Sam 951, probably the outstanding sire of the Charolais breed of cattle in America. He was mated with over 200 of his own daughters, and no recessive defects were uncovered by these matings indicating he is free of detrimental recessive genes. (Courtesy of Litton Charolais Ranch.)

but the daughters to which he is bred may also be progeny-tested for all recessive genes they may possess which came from their sire.

In species of animals that produce only one young at birth, it would be impossible under practical conditions to progeny-test a female by mating her to 23 to 35 of her own sons, because she would never produce this number of sons in her lifetime. It would be possible, although sometimes impractical, to test a sire by mating him to 23 to 35 of his own unselected daughters. He might be so old, however, by the time he had been mated to this many different daughters that he might be dead or out or service. Also, if such a progeny test were conducted, the progeny would be inbred by at least 25 percent and might lack vigor. The use of artificial insemination when the sire was young might produce enough daughters in one breeding season to make such a progeny test possible at a relatively young age. In animals such as swine which normally produce several young at birth, such progeny tests can be conducted and the disadvantages encountered with cattle are not so important.

Mating a male to his own unselected half-sisters is a method of progeny-testing to prove him free from any recessive gene he might have received from the parent that makes him related to the females to which he is mated. If he is mated to 23 different half-sisters and no homozygous recessive offspring are produced, one would say he was free of any recessive gene he might have received from their common parent at the 95 percent level of probability. He is tested free of these recessive genes at the 99 percent level of probability if he is mated to 35 different unselected half-sisters and produces no homozygous recessive offspring. *Such a test, however, does not test the male for recessive genes that he might have inherited from the other parent.* The progeny produced from such matings would be only 12.5 percent inbred, and many times enough half-sisters would be available to progeny-test a sire by the time he is three or four years of age, especially if he were used on a relatively large scale by natural mating or by artificial insemination. A boar can also be mated to unselected half-sisters with the number of matings and progeny required per mating for the test being the same as those for sire × daughter matings presented in Table 11.7. Again, however, this tests a boar for possible recessive genes he received from the parent common to him and his half-sisters but not for genes that might come from the other parents.

If the frequency of a recessive gene in a population is high, the random mating of a sire to females in that population gives some indication of whether or not he is carrying a recessive gene. If the frequency of a gene is low, the probability of proving a sire free of a recessive gene is very low. Under practical conditions, the frequency of most detrimental recessive genes is low. For example, if the frequency of a recessive gene in a population is 0.25, which is high, the probability of any one individual selected at random's being heterozygous for that recessive gene is 2(0.25 × 0.75), or 0.375, and the probability of producing a homozygous recessive is (0.25 × 0.25), or 0.0625. If the frequency of a recessive gene in a population is 0.05, which is relatively low, the probability of an individual selected at random's being heterozygous is 2(0.05 × 0.95) or 0.095, or less than ten percent. However, all dominant phenotype parents that produce a homozygous recessive offspring are heterozygous regardless of the frequency of the recessive gene in the population. If the heterozygous individuals are favored in selection, the frequency of the recessive gene would remain at a high frequency, and there would be more heterozygous dominant and more homozygous recessive individuals produced in that population than in populations where there is no selective advantage for the heterozygote.

Mating a sire to his full-sibs will also test him for any recessive gene he might have received from both parents, providing that enough different full sisters are available for such a test. Enough full sibs (23 to 35 in cattle) will never be available for such a test. In some instances in swine, enough full sisters for such a test (five to seven) might be available, but even then the progeny would be inbred by at least 25 percent.

In progeny-testing for a specific detrimental recessive gene, it sometimes is not practical to use homozygous recessive individuals as testers, because they may not live to maturity or they may be of very low fertility. As an alternative, individuals known to be carriers of a recessive gene may be used for testing purposes. Such a test requires more matings and more offspring to prove the individual tested homozygous dominant at a certain level of probability as compared to using homozygous recessive individuals as testers.

11.3.2 Progeny Tests in Selection for Quantitative Traits

Progeny tests may be used to predict more accurately the PBV of a parent for a quantitative trait. The principle involved in the progeny test is that each progeny receives a sample one-half of its inheritance from each of its parents, and this is a sample one-half of the parent's breeding value. By increasing the number of progeny tested for a certain parent and calculating the average of these progeny, it is possible to obtain an estimate of the repeated parent's breeding value (usually a sire) based on this relationship.

Two or more offspring from the same parent will vary from each other genetically, because they will not receive the same genes from a parent unless that parent is homozygous for all genes it possesses. Complete homozygosity for all genes may occur in theory, but probably never occurs in practice. The segregation of genes in the gametes of a particular parent that is not homozygous for all the genes it possesses will result in different genes being transmitted to its various offspring. For a given pair of alleles, some offspring will receive the same gene from the parent, but since many pairs of alleles are involved, the probability of two offspring receiving exactly the same set of genes from a parent is very low. This segregation of genes in the gametes is sometimes called Mendelian error and complicates selection. However, the more progeny obtained from a particular parent, the better the estimate of the kinds of genes that parent possesses.

The various offspring of a parent will also vary from each other because of differences in environment. It is very unlikely that two individuals ever have the same environment throughout their lives or even in the same period of their lives. If a number of individuals are compared in as standard an environment as possible, the average differences among individuals due to environment are reduced, although they are not entirely eliminated.

Progeny tests are conducted to compare the performance of the progeny of two or more parents. Usually sires rather than dams are progeny-tested, because sires generally produce more progeny in a given season or year. The more carefully the progeny tests are conducted, the more accurate will be the determination of the PBV of the parents being compared.

Several precautions should be taken to make progeny tests more accurate. Some of these are as follows: (1) Dams mated to all sires on a given progeny test should be selected randomly. Breeding the superior dams to a certain sire will tend to cause that sire's PBV to be overestimated. (2) Standardize rations and feeding practices. Feed animals the same ration and in the same manner. Feeding some parent groups different rations, different amounts of ration, or feeding one group on pasture and another in dry lot will bias the progeny test. (3) Do not feed all progeny of a single sire in the same pen. Some pens may be more favorable, or less favorable, for performance, and this tends to increase environmental variations among the different sire groups. If progeny of a single parent are fed in the same pen and several progeny groups are fed in the same way, the pens could be rotated among the different sire groups at regular intervals to reduce pen effects. (4) Compare different parent groups raised in as nearly the same environment or location as possible. (5) Compare parent groups born during the same year or same season of the year when possible. (6) Include all healthy progeny of a particular parent in the test, if possible, whether they are inferior or superior. This tends to average Mendelian and environmental errors for each sire group. (7) The larger the number of progeny tested per parent, within limits, the more accurate the estimate of that parent's PBV. This is because environmental effects and Mendelian segregation errors are averaged, and this gives a better estimate of the kinds of genes a parent possesses. However, the breeder may want to test fewer progeny per parent in order to progeny-test and compare more parents.

The accuracy of selection (which is the correlation of the genotype of the parent with the average phenotype of its progeny) may be calculated as follows:

$$\frac{h}{2}\sqrt{\frac{n}{1+(n-1)t}}$$

where h is the square root of the heritability for a trait, n is the number of progeny per parent used in the average, and t is approximately $\frac{1}{4}h^2$ or $h^2/4$ if the progeny group is composed of half-sibs and there is no environmental correlation between sibs. To illustrate how the figures presented in Table 11.8 were derived, let us calculate the accuracy of selection for a parent

Table 11.8 Accuracy of selection* on the basis of the progeny average using different numbers of progeny and different heritability estimates

Number of progeny	Accuracy at different levels of heritability									
	0.10	0.20	0.30	0.40	0.50	0.60	0.70	0.80	0.90	1.00
1	0.158	0.224	0.274	0.316	0.354	0.387	0.418	0.447	0.474	0.500
2	0.220	0.309	0.374	0.426	0.471	0.511	0.546	0.577	0.606	0.632
3	0.267	0.369	0.442	0.500	0.548	0.588	0.624	0.655	0.682	0.707
4	0.305	0.417	0.495	0.555	0.603	0.643	0.678	0.707	0.733	0.756
5	0.337	0.456	0.537	0.598	0.646	0.685	0.717	0.745	0.769	0.791
6	0.365	0.490	0.572	0.633	0.679	0.717	0.748	0.775	0.797	0.817
7	0.390	0.519	0.602	0.661	0.707	0.743	0.773	0.798	0.819	0.837
8	0.413	0.544	0.627	0.686	0.730	0.765	0.793	0.816	0.836	0.853
9	0.433	0.567	0.650	0.707	0.750	0.783	0.810	0.832	0.850	0.866
10	0.452	0.587	0.669	0.726	0.767	0.799	0.824	0.845	0.862	0.877
15	0.527	0.664	0.741	0.791	0.826	0.852	0.872	0.888	0.902	0.913
20	0.582	0.716	0.787	0.831	0.861	0.883	0.900	0.913	0.924	0.933
25	0.625	0.754	0.818	0.858	0.884	0.903	0.917	0.928	0.937	0.945
30	0.659	0.782	0.842	0.877	0.900	0.917	0.930	0.939	0.947	0.953
35	0.688	0.805	0.860	0.892	0.913	0.928	0.939	0.947	0.954	0.960
40	0.712	0.823	0.874	0.904	0.923	0.936	0.946	0.953	0.959	0.965
45	0.732	0.839	0.886	0.913	0.930	0.942	0.952	0.958	0.964	0.968
50	0.749	0.851	0.896	0.921	0.937	0.948	0.956	0.962	0.967	0.971
75	0.811	0.893	0.927	0.945	0.956	0.964	0.970	0.974	0.978	0.981
100	0.848	0.917	0.944	0.958	0.967	0.973	0.977	0.981	0.983	0.985

*Perfect accuracy of 1.00 is the ultimate, and it is assumed that environmental correlations and inbreeding are zero.

when the trait is 0.30 heritable and 20 progeny are used in the test. The accuracy of selection would be

$$\frac{0.5477}{2}\sqrt{\frac{20}{1+(20-1)\frac{0.30}{4}}} = 0.2739\sqrt{\frac{20}{1+(19)0.075}}$$

$$= 0.2739 \times 2.8718, \text{ or } 0.787$$

Data summarized in Table 11.8 were calculated by using this formula. each figure presented shows the accuracy of selection using different numbers of progeny per parent for a trait at different degrees of heritability. For example, with a heritability estimate of 0.50, 15 progeny per parent are required for an accuracy of 0.826. With a heritability estimate of 0.20, 40 to 45 progeny per parent are required to progeny-test with the same degree of accuracy.

The relative accuracy of selection on the basis of progeny tests as compared to selection on the basis of individual performance is shown in Table 11.9. These relative values were calculated from the correlation of the PBV

Table 11.9 The relative accuracy of progeny tests as compared to selection on the basis of individual performance*

Number of progeny	Heritability of a trait									
	0.10	0.20	0.30	0.40	0.50	0.60	0.70	0.80	0.90	1.00
1	0.50	0.50	0.50	0.50	0.50	0.50	0.50	0.50	0.50	0.50
2	0.70	0.69	0.68	0.67	0.66	0.66	0.65	0.65	0.64	0.63
3	0.85	0.83	0.81	0.79	0.78	0.76	0.75	0.73	0.72	0.71
4	0.97	0.93	0.90	0.88	0.85	0.83	0.81	0.79	0.77	0.76
5	1.07	1.02	0.98	0.95	0.91	0.88	0.86	0.83	0.81	0.79
6	1.16	1.10	1.04	1.00	0.96	0.93	0.89	0.87	0.84	0.82
7	1.23	1.16	1.10	1.05	1.00	0.96	0.92	0.89	0.86	0.84
8	1.31	1.22	1.15	1.09	1.03	0.99	0.95	0.91	0.88	0.85
9	1.37	1.27	1.19	1.12	1.06	1.01	0.97	0.93	0.90	0.87
10	1.43	1.31	1.22	1.15	1.09	1.03	0.99	0.95	0.91	0.88
15	1.67	1.49	1.35	1.25	1.17	1.10	1.04	0.99	0.95	0.91
20	1.84	1.60	1.44	1.31	1.22	1.14	1.08	1.02	0.97	0.93
25	1.98	1.69	1.49	1.36	1.25	1.17	1.10	1.04	0.99	0.95
30	2.09	1.75	1.54	1.39	1.27	1.18	1.11	1.05	1.00	0.95
35	2.18	1.80	1.57	1.41	1.29	1.20	1.12	1.06	1.01	0.96
40	2.25	1.84	1.60	1.43	1.31	1.21	1.13	1.07	1.01	0.97
45	2.32	1.88	1.62	1.44	1.32	1.22	1.14	1.07	1.02	0.97
50	2.37	1.90	1.64	1.46	1.33	1.22	1.14	1.08	1.02	0.97
75	2.57	2.00	1.69	1.49	1.35	1.25	1.16	1.09	1.03	0.98
100	2.69	2.05	1.72	1.52	1.37	1.26	1.17	1.10	1.04	0.99

*The accuracy ratio was determined from $r_{G\bar{P}_0}/h$.

of the parent with the average phenotype of the progeny ($r_{G\bar{P}_o}$) divided by the square root of the heritability of the trait. Expressed in a formula, this would be $r_{G\bar{P}_o}/h$, which is equal to the relative accuracy of the progeny test as compared to that based on individual performance. The relative accuracy of the progeny test as compared to individual performance alone when the heritability estimate of a trait is 0.30 and the number of progeny tested per parent is 10 would be 0.669 (from Table 11.8) divided by 0.5477 (the square root of h^2), or 1.22. This means that such a progeny test would be 1.22 times more accurate than selection on the basis of individual performance alone.

Data summarized in Table 11.9 show that progeny tests as compared with individual selection are relatively more accurate at lower levels of heritability. Progeny tests including at least five or more progeny are required to give as high a degree of accuracy in selection as a record of the individual's own performance. If a higher degree of accuracy is required, more progeny must be included in the test. Properly designed progeny tests will eliminate much of the environmental error among the different progeny groups.

Regression coefficients that may be used to predict the PBV of a parent are presented in Table 11.10. The PBV of the parent is determined by multiplying the regression coefficient (b) for a particular number of progeny at a certain level of heritability by the selection differential of the progeny ($\bar{P}_o - \bar{P}_{co}$). This is added to the average of the contemporary progeny ($\bar{P}_{co}$). For example, let us assume that the average gain of 10 progeny ($\bar{P}_o$) was 2.15 lb per day and the average gain of all contemporary groups ($\bar{P}_{co}$) was 2.00. If the heritability of the trait was 0.50, the PBV of the parent would be $2.00 + 1.176(2.15 - 2.00)$, or 2.18. The value of 1.176 is the regression coefficient (b) obtained from the appropriate column in Table 11.10. Selection differentials expressed as a ratio to the group average can be used by substituting 100 in the formula for the group average.

Progeny-testing parents among litter-bearing animals includes some factors not involved in testing those parents normally giving birth to one young. This is particularly true in swine, where eight or more pigs may be weaned per litter. It is often impractical to feed an entire litter on a performance test, so a smaller sample of the litter is used, often consisting of two males and two females. Even smaller samples of the litter may be used in some cases. When several litters from the same sire are compared, those within a litter are related by 0.50, whereas progeny of a sire in different litters (half-sibs) are related by 0.25. This assumes that no inbreeding is involved. Although pigs within the same litter are more alike genetically for many quantitative traits than are pigs in different litters, littermates also tend to be more alike phenotypically because they have a more similar environment from conception to weaning.

Diallel crossing has been proposed to equalize or reduce the importance of maternal influences in progeny tests. The diallel cross consists of determining the breeding value of two or more males by breeding them at different times to the same two or more groups of females and comparing the averages of the progeny of each sire. For example, let us assume that we are going to breed ten sows (numbered one through ten) to two different boars (A and B) in a progeny test. In the spring we could breed sows one through five to

Table 11.10 Regression coefficients for predicting the PBV of a parent from information on the phenotypes of its half-sib progeny

Number of progeny	Regression coefficient* when the heritability is					
	0.10	0.20	0.30	0.40	0.50	0.60
1	0.050	0.100	0.150	0.200	0.250	0.300
2	0.098	0.190	0.280	0.364	0.444	0.522
3	0.142	0.272	0.392	0.500	0.600	0.692
4	0.186	0.348	0.490	0.616	0.728	0.828
5	0.228	0.416	0.576	0.714	0.834	0.938
6	0.266	0.480	0.654	0.800	0.924	1.028
7	0.304	0.538	0.724	0.876	1.000	1.106
8	0.340	0.592	0.786	0.942	1.066	1.170
9	0.376	0.642	0.844	1.000	1.126	1.228
10	0.408	0.690	0.896	1.052	1.176	1.276
15	0.556	0.882	1.098	1.250	1.364	1.452
20	0.678	1.026	1.238	1.380	1.482	1.558
25	0.782	1.136	1.340	1.470	1.562	1.630
30	0.870	1.224	1.418	1.538	1.622	1.682
35	0.946	1.296	1.478	1.590	1.666	1.722
40	1.012	1.356	1.528	1.632	1.702	1.752
45	1.072	1.406	1.570	1.666	1.730	1.776
50	1.124	1.450	1.604	1.694	1.754	1.796
75	1.316	1.596	1.718	1.786	1.830	1.860
100	1.438	1.680	1.780	1.834	1.870	1.892

*The regression coefficient $= \dfrac{h^2}{2}\left(\dfrac{n}{1 + (n-1)t}\right)$

where t is $0.25h^2$, n is the number of progeny per sire group, and h^2 is the heritability of the trait. The PBV for the parent is $\bar{P}_{co} + b(\bar{P}_o - \bar{P}_{co})$, where b is the regression coefficient for a certain number of progeny at a certain heritability, $\bar{P}_o$ is the phenotypic average of the progeny, and $\bar{P}_{co}$ is the phenotypic average of the contemporaries of the progeny.

boar A and sows six through ten to boar B. For fall litters the five sows numbered six through ten would be bred to boar A and those numbered one through five would be bred to boar B. Thus, each boar would be bred to each of the ten sows during the year, and differences between dams bred to the same boar would be eliminated. The environmental differences due to dam effects would also be reduced.

The diallel cross may be used under practical conditions to progeny-test swine and poultry, where each female may produce several young that are full sibs or half-sibs two or more times during the year. This test is much less practical in cattle, however, where only one young is produced per year (monotocous species) and because of a long gestation period.

Although carefully conducted progeny tests are sometimes more accurate in predicting the PBV of a parent than the parent's own performance, progeny tests greatly lengthen the generation interval and lower the amount of progress made in selection for additive gene effects over a period of years. This occurs because the progeny have to be old enough that their phenotype can be measured before selections can be made. In cattle, progeny-testing for yearling weight would require an extra two years compared with selection on individual performance and three and one-half years to complete a progeny test for maternal traits. Adequate numbers of progeny per parent must also be tested in order to make the progeny test meaningful, and it is desirable to progeny-test and compare many sires to find one or more that are superior. Simply progeny-testing a parent does not change that parent's genotype; it is how the progeny of a parent rank in comparison to the progeny of other parents that is important. Progeny tests are meaningless unless selection is practiced among parents so tested. The genetic gains from each cycle of testing and selection depend upon the fraction of progeny-tested parents selected for breeding use. Differences among averages of progeny are likely to be disappointingly small.

Progeny-testing several parents, of course, increases the expense of testing, and under some conditions it is not practical. However, when a parent completes a progeny test and is found to have a superior breeding value, it should be used for breeding until a known superior replacement is identified.

11.4 SELECTION ON THE BASIS OF COLLATERAL RELATIVES

Collateral relatives are those not directly related to an individual as ancestors or progeny. Thus, they are the individual's brothers, sisters, cousins, uncles, aunts, etc. The more closely they are related to the individual in question, the more valuable is the information they can supply for selection purposes.

Information on collateral relatives, if complete, gives an idea of the kinds of genes and combinations of genes the individual is likely to possess. Information of this kind is now being used in Meat Hog Certification Programs, where a barrow and a gilt from each litter may be slaughtered to obtain carcass data. This is done because otherwise the animal itself has to be slaughtered if information on its own carcass quality is to be obtained. Information on collateral relatives is also used in selecting dairy bulls, since milk production can be measured only in cows even though the bull possesses and transmits genes for milk production to his offspring. Records on collateral relatives can also be used in the selection of poultry for egg and meat production and for all-or-none traits such as mortality, disease resistance, number of nipples, number of vertebrae, or fertility.

The half-sib or full sib test is actually a progeny test of one or both parents but differs in usage by doing away with the waiting period involved when information is used as a progeny test. Half-sibs all have one parent in common, whereas full sibs have both parents in common.

The more homozygous the parent, the more likelihood the various offspring of that parent will receive the same genes from that parent. For example, if parent A is homozygous for four pairs of genes ($AABBCCDD$), each offspring will receive the same genes ($ABCD$) from that parent. If parent B were heterozygous for these same four pairs of alleles ($AaBbCcDd$), all of the offspring of that parent would not receive the same combination of genes from that parent. In fact, such a parent could transmit 16 different combinations of genes to its progeny (2^n, where n is the number of pairs of heterozygous genes). The performance of several sibs helps to determine, within limits, the kinds of genes the parent possesses for a quantitative trait.

Selection on the basis of sib tests means that an individual is kept for breeding or is rejected (culled) on the basis of the average phenotype of its brothers and sisters. These may be maternal half-sibs, paternal half-sibs, or full sibs. The principles involved in the use of sib tests to estimate the PBV of an individual are similar to those used in pedigree and progeny selection.

The accuracy of selection on the basis of the phenotypes of sibs depends upon the degree of heritability (h^2) of the trait, the closeness of the relationship (R) of the sibs and the individual being selected, the number of sibs (n) used to determine the sib average, and the degree of correlation (t) between the phenotypes of the sibs. The accuracy of selection may be calculated from

$$Rh\sqrt{\frac{n}{1 + (n-1)t}}$$

In using this formula we are assuming that no inbreeding is involved, and if the tests are designed in such a way that the environmental correlations among the phenotypes of the sibs are zero, t equals Rh^2.

The accuracy of selection of the individual on the basis of varying numbers of half-sibs and at different levels of heritability is shown in Table 11.11. Each figure was calculated by using the formula above. These data show that the accuracy of selection increases as the records on a larger number of half-sibs is considered and as the heritability of the trait increases. The accuracy of selection never exceeds 0.50, regardless of the number of half-sibs tested and the degree of heritability of the trait.

The relative accuracy of selection on the basis of the performance of half-sibs as compared to selection on the basis of the individual's own performance is shown in Table 11.12 for different numbers of half-sibs and different degrees of heritability. These data show that nearly 30 half-sibs are required to give the same accuracy as information on the individual's own record when the heritability is as low as 0.10, and 100 or more when heritability is higher than 0.10. Also, adding the record of another half-sib is

Table 11.11   Accuracy of selection of the individual on the basis of the records of its half-sibs using different numbers of half-sibs and different heritability estimates

Number of half-sibs	Accuracy of selection when the heritability is						
	0.10	*0.20*	*0.30*	*0.40*	*0.50*	*0.60*	*0.70*
1	0.079	0.112	0.137	0.158	0.177	0.194	0.209
2	0.110	0.154	0.187	0.213	0.236	0.255	0.273
3	0.134	0.185	0.221	0.250	0.274	0.294	0.312
4	0.152	0.209	0.247	0.277	0.302	0.322	0.339
5	0.169	0.228	0.268	0.299	0.323	0.342	0.359
6	0.183	0.245	0.286	0.316	0.340	0.359	0.374
7	0.195	0.259	0.301	0.331	0.354	0.372	0.387
8	0.206	0.272	0.314	0.343	0.365	0.383	0.397
9	0.216	0.283	0.325	0.354	0.375	0.392	0.405
10	0.226	0.294	0.335	0.363	0.384	0.400	0.412
15	0.263	0.332	0.370	0.395	0.413	0.426	0.436
20	0.291	0.358	0.393	0.415	0.430	0.441	0.450
25	0.312	0.377	0.409	0.429	0.442	0.451	0.459
30	0.330	0.391	0.421	0.439	0.450	0.459	0.465
35	0.344	0.403	0.430	0.446	0.456	0.464	0.469
40	0.356	0.412	0.437	0.452	0.461	0.468	0.473
45	0.366	0.419	0.443	0.456	0.465	0.471	0.476
50	0.375	0.426	0.448	0.460	0.468	0.474	0.478
75	0.406	0.447	0.463	0.473	0.478	0.482	0.485
100	0.424	0.458	0.472	0.479	0.483	0.486	0.489

Note: Accuracy of selection is the correlation between the genotype of the individual and the phenotypic average of its half-sibs or $(r_{G\bar{P}_{hs}})$.

affected by the law of diminishing returns. However, in instances where information cannot be obtained from the individual, such as in bulls for milk production and roosters for egg production, records on half-sibs can be used effectively in selection.

Full sibs may be used in selection, but the fact that they have a similar maternal environment from conception to weaning lowers the accuracy of their use for such a test. For example, for full sibs the t in the formula is not likely to be Rh^2, but it is more likely to be $Rh^2 + c^2$, where c^2 is the added contribution of maternal effects. For litter size and weight in swine, c^2 has been estimated to be about 0.10. If c^2 is 0.10, the accuracy of selection is 0.326 for 10 full sibs when h^2 is 0.10 instead of 0.415. When h^2 is 0.50, the accuracy of selection is 0.549 instead of 0.620.

The accuracy of selection on the basis of the performance of full sibs as compared to selection on the basis of individuality at different levels of heritability is shown in Table 11.13. The relative accuracy of selection on

Table 11.12 Relative accuracy of selection on the basis of the performance of half-sibs as compared to selection on the basis of individual performance

Number of half-sibs	Relative accuracy of selection when the heritability is						
	0.10	0.20	0.30	0.40	0.50	0.60	0.70
1	0.25	0.25	0.25	0.25	0.25	0.25	0.25
2	0.35	0.35	0.34	0.34	0.33	0.33	0.33
3	0.42	0.41	0.40	0.40	0.39	0.38	0.37
4	0.48	0.47	0.45	0.44	0.43	0.42	0.41
5	0.53	0.51	0.49	0.47	0.46	0.44	0.43
6	0.58	0.55	0.52	0.50	0.48	0.46	0.45
7	0.62	0.58	0.55	0.52	0.50	0.48	0.46
8	0.65	0.61	0.57	0.54	0.52	0.49	0.47
9	0.69	0.63	0.59	0.56	0.53	0.51	0.48
10	0.71	0.66	0.61	0.57	0.54	0.52	0.49
15	0.83	0.74	0.68	0.63	0.58	0.55	0.52
20	0.92	0.80	0.72	0.66	0.61	0.57	0.54
25	0.99	0.84	0.75	0.68	0.63	0.58	0.55
30	1.04	0.88	0.77	0.69	0.64	0.59	0.56
35	1.09	0.90	0.79	0.71	0.65	0.60	0.56
40	1.13	0.92	0.80	0.72	0.65	0.60	0.57
45	1.16	0.94	0.81	0.72	0.66	0.61	0.57
50	1.19	0.95	0.82	0.73	0.66	0.61	0.57
75	1.28	1.00	0.85	0.75	0.68	0.62	0.58
100	1.34	1.03	0.86	0.76	0.68	0.63	0.58

Note: The relative accuracy of selection was calculated from the ratio of $r_{G\bar{P}_{hs}}/h$.

the basis of the performance of full sibs as compared to selection on individuality is shown in Table 11.14.

Information presented in Table 11.14 shows that selection on the basis of the individual's own performance is relatively more accurate than selection on the basis of full sib records when the trait is highly heritable. However, when traits are low in heritability and records on six or more full sibs are available, selection on the basis of full sibs is more accurate.

The PBV of the individual may be predicted by multiplying the regression coefficient of the genotype of the individual (**d**) on the average phenotypes of the sibs by the selection differential of the sibs ($\bar{P}_{sibs} - \bar{P}_{co}$). This may be expressed as

$$\text{PBV of the individual} = \bar{P}_{co} + Rh^2 \frac{n}{1 + (n-1)t} (\bar{P}_{sibs} - \bar{P}_{co})$$

The relationship coefficient (R) is 0.25 for half-sibs and 0.50 for full sibs, if we assume that there is no inbreeding. The value of t is approximately

Table 11.13 Accuracy of selection of the individual on the basis of the records of its full sibs using different numbers of full sibs and different heritability estimates

Number of full sibs	Accuracy of selection when the heritability is						
	0.10	0.20	0.30	0.40	0.50	0.60	0.70
1	0.158	0.224	0.274	0.316	0.354	0.387	0.418
2	0.218	0.302	0.361	0.408	0.447	0.480	0.509
3	0.261	0.354	0.416	0.463	0.500	0.530	0.556
4	0.295	0.392	0.455	0.500	0.535	0.562	0.584
5	0.323	0.423	0.484	0.527	0.559	0.584	0.604
6	0.346	0.447	0.507	0.548	0.577	0.600	0.618
7	0.367	0.468	0.526	0.564	0.592	0.612	0.629
8	0.385	0.485	0.541	0.577	0.603	0.622	0.637
9	0.401	0.500	0.554	0.588	0.612	0.630	0.644
10	0.415	0.513	0.565	0.598	0.620	0.637	0.649

Note: Accuracy of selection is the correlation between the genotype of the individual and the phenotypic average of its full sibs ($r_{G\bar{P}_{fs}}$). It is assumed that no inbreeding is involved and possible environmental correlations are ignored.

Table 11.14 Relative accuracy of selection on the basis of the performance of full sibs as compared to selection on the basis of individual performance

Number of full sibs	Relative accuracy of selection when heritability is						
	0.10	0.20	0.30	0.40	0.50	0.60	0.70
1	0.50	0.50	0.50	0.50	0.50	0.50	0.50
2	0.69	0.67	0.66	0.65	0.63	0.62	0.61
3	0.83	0.79	0.76	0.73	0.71	0.69	0.66
4	0.93	0.88	0.83	0.79	0.76	0.73	0.70
5	1.02	0.95	0.88	0.83	0.79	0.75	0.72
6	1.10	1.00	0.93	0.87	0.82	0.78	0.74
7	1.16	1.05	0.96	0.89	0.84	0.79	0.75
8	1.22	1.09	0.99	0.91	0.85	0.80	0.76
9	1.27	1.12	1.01	0.93	0.87	0.81	0.77
10	1.31	1.15	1.03	0.95	0.88	0.82	0.78

Note: The relative accuracy of selection was calculated from the ratio of $r_{G\bar{P}_{fs}}/h$.

$0.25h^2$ for half-sibs and $0.50h^2$ for full sibs except when c^2 is considered. The letter n represents the number or sibs making up the phenotypic average and h^2 the heritability of the trait.

Where the trait may be measured in the individual as well as in the sibs, performance records for both may be used to predict the accuracy of selection and the PBV of the individual. The accuracy of selection when such records are used is

$$\text{Accuracy of selection} = \frac{h[1 + (n-1)R]}{\sqrt{n[1 + (n-1)t]}}$$

The prediction of the PBV of the individual then becomes

$$\text{PBV} = \bar{P}_{\text{sibs}} + h^2\left[\frac{1-R}{1-t}(P_i - \bar{P}_{\text{sibs}}) + \frac{1+(n-1)R}{1+(n-1)t}(\bar{P}_{\text{sibs}} - \bar{P}_{c\ \text{sibs}})\right]$$

$\bar{P}_{c\ \text{sibs}}$ is the average of all such families on the test. The combination of records on the individual and its sibs for selection is more advantageous than records on the individual's own performance when R and t are greatly different.

As one resorts to various selection aids to increase selection accuracy over that resulting when the PBV is estimated from the individual's own performance, he usually pays a price in the form of reduced selection differentials, fewer unique selection opportunities, and frequently in delays needed to secure this information. Many times additional time and expense are involved. Thus, increased accuracy of selection by using information on relatives may not always be compatible with a greater rate of progress from selection and a practical application of these principles. These points must be kept in mind when the breeder strives to improve his animals through selection.

11.5 SELECTION FOR SPECIFIC COMBINING ABILITY

Selection for specific combining ability means that selection is practiced to take advantage of hybrid vigor when nonadditive gene action is important.

Selection on the basis of individuality usually is not the most efficient method of selection for traits affected largely by nonadditive gene action. Increased merit in such traits usually depends upon heterozygosity through crossbreeding, resulting in the expression of hybrid vigor. If dominance is important, however, selection on the basis of individuality will be effective in improving traits within a breed. It is less effective if epistasis and over-dominance are important. The relationship between hybrid vigor and the different types of nonadditive gene action will be discussed in the chapter on outcrossing and crossbreeding.

As pointed out previously, the heterozygous individual does not breed true. Heterozygous parents, if large numbers are involved, produce approximately one-half heterozygous offspring, the other one-half being either homozygous dominant or homozygous recessive. Thus, on the average, the off-spring of heterozygous parents are inferior to the parents.

Although heterozygous parents mated together do not produce all heterozygous offspring, matings can be made to produce all heterozygous offspring, if it is assumed that no new mutations occur. For example, a homozygous dominant parent (*AA*) when mated to another that is homo-zygous recessive (*aa*) would produce all heterozygous offspring. Thus, heterozygosity depends upon the parents being homozygous in opposite ways, or for different alleles.

In quantitative inheritance where many genes may affect the same trait, it is not possible to determine from the phenotype which animals are homo-zygous in opposite ways for many genes. Certain methods may be used, however, to identify those individuals that are homozygous in opposite ways.

The method probably most often used in finding lines or breeds that "nick" best when crossed is the method used in the production of hybrid seed corn. The first step is to form several different inbred lines by close inbreeding. Inbreeding increases the homozygosity of all pairs of genes that the individuals in an inbred line possess. If inbreeding were 100 percent within an inbred line, all individuals within that line should be homozygous for all the gene pairs they possess, regardless of the phenotypic expression of those genes. The breeder, of course, has no sure way of knowing what pairs of genes are homozygous within an inbred line. In practice this is not necessary.

The next step after inbred lines are formed is to test them in crosses to determine which lines combine to produce the best linecross progeny. In general, the two inbred lines producing the most superior progeny when crossed are the ones that are homozygous in opposite ways for many pairs of genes, giving greater heterozygosity in the progeny. These inbred lines are kept pure to cross again and again in later years to produce progeny that may be the source of commercial animals most of which are sold and not kept for breeding purposes.

Reciprocal recurrent selection is a system of selection for increasing the combining ability of two or more lines or breeds that have already demon-strated from past crosses that they "nick" or combine well. The principles involved assume that individuals in the two lines are not completely homo-zygous in opposite ways for all pairs of genes but that one allele may be present at a high frequency in one line and at a low frequency in the other line. Crossing the lines and selecting the individuals to reproduce each pure line on the basis of the performance of their crossbred progeny theoretically should make the two lines more homozygous in opposite directions. Recip-rocal recurrent selection is described in Chapter 10 as a method of selection between lines, families, or breeds to take advantage of overdominance.

In farm animals, selection is usually practiced for more than one trait.

Since one trait may be affected mostly by nonadditive gene action and another by additive gene action, or one trait may be affected by both, the question arises as to how one can develop a selection and mating system to make the best use of both additive and nonadditive gene action. The practical answer seems to be to select and improve the highly heritable traits (additive gene action) within the pure lines or breeds by finding the best and mating the best to the best, followed by crossing these improved lines or breeds to take advantage of hybrid vigor due to nonadditive gene action.

11.6 METHODS OF SELECTION

In the preceding sections of this chapter, the importance of the merit of the individual and its relatives in identifying genetically superior breeding animals was discussed. Our discussion now will be directed toward the methods of selection that may be practiced, using the information obtained from records on the individual and/or its relatives. It should be kept in mind when selection methods are being discussed that the amount of progress made, regardless of the method used, also depends upon the size of the selection differential (selection intensity), the heritability of the trait, the length of the generation interval, as well as some other factors.

From the practical standpoint, the net value of an animal is dependent upon several traits that may not be of equal economic value or that may be independent of each other. For this reason, it is usually necessary to select for more than one trait at a time. The desired traits will depend upon their economic value, to a great extent, but only those of real importance should be considered. If too many traits are selected for at one time, less progress will be made in the improvement of any particular one. Assuming that the traits are independent and their economic value and heritability are about the same, the progress in selection for any one trait is only about $1/\sqrt{n}$ times as effective as it would be if selection were applied for that trait alone. For example, if four traits were selected for at one time in an index, the progress for one of these traits would be on the order of $\frac{1}{2}$ (not $\frac{1}{4}$) as effective as if it were selected for alone.

Several methods may be used for determining which animal should be saved and which should be rejected for breeding purposes. Three of these methods that are generally used have been discussed by Hazel and Lush.

11.6.1 Tandem Method

In this method, selection is practiced for only one trait at a time until satisfactory improvement has been made in this trait. Selection efforts for this trait are then relaxed, and efforts are directed toward the improvement of a second, then a third, and so on. This is the least efficient of the three methods to be discussed, from the standpoint of the amount of genetic progress made for the time and effort expended by the breeder.

The efficiency of this method depends a great deal upon the genetic association between the traits selected for. If there is a desirable genetic association between the traits, so that improvement in one by selection results in improvement in the other trait not selected for, the method could be quite efficient. If there is little or no genetic association between the traits, which means that they are inherited more or less independently, the efficiency would be less than if the traits were genetically associated in a desirable manner. Since a very long period of time would be involved in the selection practiced, the breeder might change his goals too often or become discouraged and not practice selection that was intensive and prolonged enough to improve any desirable trait effectively. A negative genetic association between two traits, in which selection for an increase in desirability in one trait results in a decrease in the desirability of another, would actually nullify or neutralize the progress made in selection for any one trait. Therefore, the efficiency of such a method would be low.

11.6.2 Independent Culling Method

In the use of this method, selection may be practiced for two or more traits at a time, but for each trait a minimum standard is set that an animal must meet in order for it to be saved for breeding purposes. The failure to meet the minimum standard for any one trait causes that animal to be rejected for breeding purposes. To show the disadvantage of this method, let us use an example in swine, where minimum standards are set so that any pig saved for breeding must be from a litter of 8 weaned, must weigh 180 pounds at 5 months of age, and must have no more than 1.3 inches of backfat at 200 pounds live weight. Let us assume that pig A was from a litter of 9 pigs weaned, weighed 185 pounds at 5 months, and had 1.3 inches of backfat. For pig B, let us assume that it was from a litter of five weaned, weighed 225 pounds at five months, and had 0.95 inch of backfat at 200 pounds. If the independent culling method of selection were used, pig B would be rejected, because it was from a litter of only five pigs. However, it was much superior to pig A in its weight at five months and in backfat thickness, and much of this superiority could have been of a genetic nature. In actual practice, it is possible to cull some genetically very superior individuals when this method is used.

The independent culling method of selection has been widely used in the past, especially in the selection of cattle for show purposes, where each animal must meet a standard of excellence for type and conformation regardless of its status for other economic traits. It is also used when a particular color or color pattern is required. It is still being used to a certain extent in the production of show cattle and in testing stations. It does have an important advantage over the tandem method in that selection is practiced for more than one trait at a time. It sometimes is also disadvantageous, because an animal may be culled at a young age for its failure to meet minimum standards for one particular trait, when sufficient time to complete the test might reveal superiority in other traits.

11.6.3 The Selection Index

This method involves the separate determination of the value for each of the traits selected for, and the addition of these values to give a total score for all of the traits. The animals with the highest total scores are then kept for breeding purposes. The influence of each trait on the final index is determined by how much weight that trait is given in relation to the other traits. The amount of weight given to each trait depends upon its relative economic value, since all traits are not equally important in this respect, and upon the heritability of each trait and the genetic associations among the traits [3].

The selection index is more efficient than the independent culling method, for it allows the individuals that are superior in some traits to be saved for breeding purposes even though they may be slightly deficient in one or more of the other traits. If an index is properly constructed, taking all factors into consideration, it is a more efficient method of selection than either of the other two which have been discussed, because it should result in more genetic improvement for the time and effort expended in its use.

Selection indexes seem to be gaining in popularity in livestock breeding. The kind of index used and the weight given to each of the traits is determined to a certain extent by the circumstances under which the animals are produced. Some indexes are used for selection between individuals, others for selection between the progeny of parents from different kinds of matings, such as line-crossing and crossbreeding, and still others for the selection between individuals based on the merit of their relatives, as in the case of dairy bulls, where the trait cannot be measured in that particular individual. Examples of these indexes and how they may be used will be given in later chapters dealing with each species of farm animals.

11.6.4 Calculation of a Selection Index

An index may be developed by rather simple means or by more complex methods where several genetic parameters may be used in their calculation. The example used here to illustrate the calculation of a selection index will include (1) phenotypic and genotypic correlations among traits, (2) genetic and phenotypic variances for the traits, and (3) the relative economic values for the traits included in the index. Information on each of these factors used to construct an index may be obtained from reports in the literature, or, preferably, they may be determined from data on the herd or flock in which they are to be used at a particular time or in a particular set of circumstances.

Although a selection index may include several traits, it is more desirable from the practical standpoint to limit the number of traits to a few of the greatest economic importance and of the highest heritability. The example used to illustrate the calculation of an index will include only two traits, weaning weight and weaning type score in beef cattle. The various genetic parameters were taken from the literature and the economic values were estimated. The index calculated is meant for the illustration of calculations only and is not suggested for use in a herd of beef cattle.

Two normal simultaneous equations including symbols for the various factors needed are

(1) $V_P(X_1)b_1 + \text{Cov}_P(X_1X_2)b_2 = V_A(X_1)a_1 + \text{Cov}_A(X_1X_2)a_2$

(2) $\text{Cov}_P(X_1X_2)b_1 + V_P(X_2)b_2 = \text{Cov}_A(X_1X_2)a_1 + V_A(X_2)a_2$

The symbols used in these two equations represent the following:

$X_1 = $ average weaning weight.

$X_2 = $ average weaning type score.

$V_P(X_1) = $ phenotypic variance of weaning weight (X_1).

$V_A(X_1) = $ additive genetic variance of weaning weight (X_1).

$V_P(X_2) = $ phenotypic variance of weaning type score (X_2).

$V_A(X_2) = $ additive genetic variance of weaning type score (X_2).

$\text{Cov}_P(X_1X_2) = $ phenotypic covariance of weaning weight and weaning type score.

$\text{Cov}_A(X_1X_2) = $ additive genetic covariance of weaning weight and weaning type score.

$a_1 = $ economic value assigned to weaning weight.

$a_2 = $ economic value assigned to weaning type score.

$b_1 = $ partial regression coefficient for weaning weight.

$b_2 = $ partial regression coefficient for weaning type score.

Estimates of the various parameters used in the construction of the index are

Weaning weight (X_1)

$$V_A(X_1) = 394$$
$$V_P(X_1) = 2233$$

Weaning type score (X_2)

$$V_A(X_2) = 14$$
$$V_P(X_2) = 44$$

Covariances

$$\text{Cov}_A(X_1 X_2) = 51$$
$$\text{Cov}_P(X_1 X_2) = 282$$

Economic values

Weaning weight or $a_1 = 15$

Weaning type score or $a_2 = 10$

The two normal equations are set up with the values used in this illustration.

(1) $2233b_1 + 282b_2 = 394 \ (15) + 51 \ (10)$
(2) $282b_1 + 44b_2 = 51 \ (15) + 14 \ (10)$

Next, the calculations are made solving the equations for b_1 and b_2.

(1) $2233b_1 + 282b_2 = 6420$
(2) $282b_1 + 44b_2 = 905$

1. Divide 282 in equation (1) by 44, which gives 6.4091.
2. Multiply equation (2) by 6.4091, which gives:

$$1807.37b_1 + 282b_2 = 5800.24$$

3. Subtract equation (2) from equation (1) and solve for b_1:

$$
\begin{aligned}
2233.00b_1 + 282b_2 &= 6420.00 \\
1807.37b_1 + 282b_2 &= 5800.24 \\
\hline
425.62b_1 + \quad 0 \ \ &= \ \ 619.76
\end{aligned}
$$

$$b_1 = \frac{619.76}{425.63} = 1.4561$$

4. Substitute b_1 (1.4561) in equation (1) and solve for b_2:

$$2233(1.4561) + 282b_2 = 6420$$
$$282b_2 = 6420 - 3251.47$$
$$282b_2 = 3168.53$$
$$b_2 = 11.236$$

5. The selection index would be

$$\text{Index} = 1.4561 X_1 + 11.2360 X_2$$

6. The index may be simplified as follows:

$$\text{Index} = X_1 + \frac{11.2360}{1.4561} X_2$$

$$\text{Index} = X_1 + 7.72 X_2$$

11.7 COMPARISON OF DAMS WITH DIFFERENT NUMBERS OF RECORDS

Two methods are available to the animal breeder for improving the average genetic level of performance of females in his herd or flock. One is to select replacement females from those with superior individual records and from relatives with superior records, as discussed previously. Another method is to cull the poorly producing females on the basis of their performance records.

Females within a herd or flock, and this is especially true of cattle, often vary widely in age and in the number of parturitions in which they have produced young. The most probable producing ability (MPPA) is useful in ranking the dams in a herd when there is a difference in the number of records of different females. The formula for this calculation is

$$\text{MPPA} = \bar{H} + \frac{NR}{1 + (N - 1)R}(\bar{C} - \bar{H})$$

where

$\bar{H}$ is 100, the herd average trait ratio.
N is the number of parturition records per dam.
R is the repeatability of the trait ratio.
$\bar{C}$ is the average trait ratio for all young the dam has produced.

As an example to illustrate the calculation of the MPPA, let us use weaning weight in beef cattle and assume that a cow has produced six calves during her lifetime. Also let us assume that the average weaning weight of her calves (corrected for sex) is 525 lb, whereas the average of the herd is 450 lb. Thus, the weaning weight ratio of this cow's calves would be 525/450, or 117. The records needed to calculate the MPPA for this cow would be $\bar{H}$ is 100, N is 6, R is 0.45, and $\bar{C}$ is 117. The MPPA for this cow would be

$$\text{MPPA} = 100 + \frac{6(0.45)}{1 + (6 - 1)0.45}(117 - 100), \text{ or } 114.1$$

By calculating the MPPA for each cow in the herd, the cows lowest in performance could be identified and culled, which would tend to raise the average weaning weight of calves in this herd thereafter. In addition, replacement females could be kept from dams with the highest MPPA.

REFERENCES

1. Becker, W. A. *Manual of Procedures in Quantitative Genetics* (Pullman: Washington State College Bookstore, 1964).

2. Hazel, L. N. "The Genetic Basis for Constructing Selection Indexes," G. 28: 476, 1943.

3. Hazel, L. N., and J. L. Lush. "The Efficiency of Three Methods of Selection," JH, 33: 393, 1942.

4. Johannson, I., and J. Rendel. *Genetics and Animal Breeding* (San Francisco: W. H. Freeman and Company, 1966).

5. Lush, J. L. *Animal Breeding Plans* (3rd ed.) (Ames: Iowa State College Press, 1945).

QUESTIONS AND PROBLEMS

1. What relatives of an individual are his direct relatives? His collateral relatives?

2. Do the phenotype and genotype of the individual vary at different periods of life? Explain.

3. Why is selection on the basis of phenotype usually accurate for recessive traits but not for dominant traits?

4. What different kinds of gene action may affect two or more quantitative traits?

5. Would one be able to determine the genotype of an individual from its phenotype if a quantitative trait were 100 percent heritable? Explain.

6. Why is it important to have some figure that shows how an ancestor, or other relative, ranked in performance as compared to its contemporaries?

7. A bull calf gained 4.00 lb per day on a gain test as compared to an average of 2.70 lb per day for other bull calves on the same test. What would be his gain ratio?

8. Another bull calf in another herd gained 4.00 lb per day on a gain test, whereas the average of all calves on the same test was 2.20 lb per day. What is the gain ratio for this bull calf?

9. Which bull calf of the two mentioned in questions 7 and 8 would you think should be superior genetically? Why?

10. What would be the PBV of the bull calf in question 7? In question 8?

11. How can one tell if a quantitative trait is affected mostly by additive or nonadditive gene action?

12. Why isn't selection on the basis of individuality effective for a trait determined almost entirely by nonadditive gene action?

13. Even though the pedigrees of some purebred animals possess some performance records, what do these records usually lack, as a general rule, to make them useful in selecting the individual?

14. What are some advantages and disadvantages of selection for qualitative traits on the basis of pedigrees?

15. What factors determine the amount of attention to pay to performance records of ancestors in selection?

16. Statistically, what is meant by accuracy of selection? By the probable breeding value of an individual?

17. Which gives the greatest accuracy in selection, records on the performance of the individual or on the performance of its parents, grandparents and great-grandparents combined?

18. Why are records on the grandparents less useful in selection than records on the parents?

19. Assume that you want to progeny-test a herd sire to determine if he is pure polled or heterozygous polled. What kind and how many matings would you recommend to prove his genotype?

20. What are the advantages and disadvantages of progeny-testing a sire by mating him to his own daughters?

21. What are some advantages of progeny-testing a sire by mating him to his half-sisters?

22. Assume that the frequency of a recessive gene in a herd is 0.15. What proportion of offspring in this herd would one expect to be heterozygous for this recessive gene?

23. Outline the methods you would use if you had 100 females available to progeny-test five males.

24. Which gives the greatest accuracy in selection, selection on the basis of an individual's own records or those of its progeny? Explain.

25. Under what conditions do progeny tests give the greatest accuracy in selection?

26. Assume that in a progeny test for rate of gain the average gain for 100 progeny of five different sires (20 progeny per sire) was 2.20 lb per day. If daily gain were 50 percent heritable, what would be the PBV of sire number one if his progeny gained 2.70 lb per day?

27. What is meant by a diallel cross? Why is it sometimes recommended as a method of progeny-testing a sire? What species of farm animals should it be used for in progeny testing?

28. What are some disadvantages for progeny tests when a quantitative trait is involved?

29. Which is the most effective in determining the individual's PBV, selection on the basis of performance of that individual or selection on the basis of the performance of its full sibs?

30. What is meant by reciprocal recurrent selection? What kinds of traits should this kind of selection be used for as a means of genetic improvement?

31. Under what conditions is selection on the basis of half-sibs most accurate?

32. Which gives the greatest accuracy, selection based on the performance of half-sibs or full sibs? Why?

33. Distinguish between tandem selection, independent culling method of selection, and the selection index. Which is the most effective and why?

34. What traits and how many should be included in a selection index?

35. What are some disadvantages of using the independent culling method of selection?

12

Principles of Inbreeding

Inbreeding is another tool, in addition to selection, that the animal breeder may use for the improvement of farm animals. Inbreeding is a mating system in which progeny are produced by parents more closely related than the average of the population from which they come.

Most livestock producers are familiar with the effects of inbreeding, and they avoid it as much as possible. It is avoided because past experience has shown that inbreeding is usually associated with the appearance of genetic defects and a general over-all decline in vigor and performance. In humans, it is believed that children from the marriage of first cousins are doomed to be deformed physically or mentally, and this belief is so strong that we have certain moral and legal laws which prohibit the marriage of close relatives.

The desire on the part of the animal breeder to avoid inbreeding has resulted in many good sires being shipped to market when their daughters become old enough for breeding. Actually, the production records of the daughters of a sire are a means of indicating his breeding ability, and sires that prove to be superior should be used for breeding purposes as long as they retain the ability to reproduce. Many herds today could profit from the use of superior sires by taking advantage of inbreeding if the owners understood the effects and possible consequences of this system of mating.

Studies with plants and animals show that inbreeding is not always detrimental. Many plants such as oats, peas, and beans are self-fertilized and thus are highly inbred as compared to cross-pollinated plants such as corn. Laboratory rats have been inbred by full brother × full sister matings for many generations without a decline in vigor and have produced as large litters when mated to littermate males as when mated to unrelated males [4]. Many of our present-day breeds of farm animals were established by making use of inbreeding, and many outstanding herds of swine and cattle in the past were developed by this system of mating.

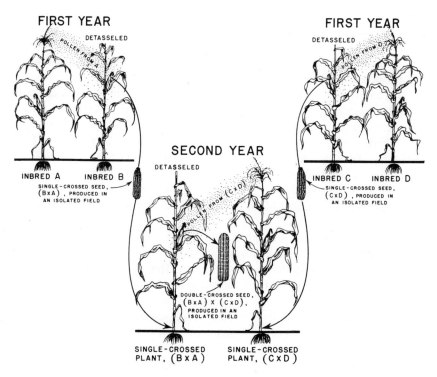

Fig. 12.1 Diagram illustrating how hybrid seed corn is produced. The principle involved is to inbreed corn, which normally is cross-polinated, to form several inbred lines. These are then tested to find which ones "nick" the best in crosses, and then they are used as diagrammed above for seed corn production. Something similar could be done with livestock. (Courtesy of the Field Crops Department, University of Missouri.)

12.1 GENETIC EFFECTS OF INBREEDING

The genetic effect of inbreeding is that it makes more pairs of genes in the population homozygous regardless of the kind of gene action involved. All phenotypic effects of inbreeding result from this one genetic effect, so it is very important to understand how homozygosity is brought about.

To illustrate the genetic effect of inbreeding, we shall use a single pair of genes and let D be the dominant and d the recessive allele. We shall also assume for this example that we are dealing with plants that are self-fertilized and that there is no selection for or against the dominant or the recessive genes. We shall also assume that the parent generation we are working with contains 1600 individuals, all of which are heterozygous for the two genes (Dd). This is shown in Table 12.1. The crossing of heterozygous individuals will give a genotypic ratio of 1DD: 2Dd: 1dd in the first generation. Since the ratio is the important thing to consider, we shall keep the number of individuals in each generation at 1600 but shall change the ratio as homozygosity increases.

A study of this example will show that in succeeding generations an important change has taken place in the percentage of heterozygotes in the population. In the parent generation, all of the individuals were heterozygous, whereas in the first generation this was true of only 50 percent of the individuals. Thus, we can say that after one generation of self-fertilization, the number of individuals in the population that were homozygous was increased by 50 percent. In this generation, one-fourth of all individuals are now homozygous dominant (DD) and one-fourth are homozygous recessive (dd).

Now let us continue the inbreeding for more generations by self-fertilization. Individuals of the genotype DD when self-fertilized will produce nothing but offspring of this same genotype. The same is true of parents of genotype dd. All individuals of genotype Dd, however, will produce offspring again at the genotypic ratio of $1DD: 2Dd: 1dd$. In the second generation of inbreeding we now have the following genotypes: $600DD: 400Dd: 600dd$. Seventy-five percent of the individuals are now homozygous dominant and homozygous recessive. Continuing this self-fertilization, we find that the homozygosity of the population produced increases with each generation but at a decreasing rate.

Other important points are demonstrated in this example. First, note that in the parents the frequencies of both the D and the d genes were 0.50. In the fourth generation, the frequency of each gene was still 0.50, showing that inbreeding did not change the frequency of the genes in the population. If we had discarded the individuals of genotype dd or if they had died, we would have caused a decrease in the frequency of the d gene. Such an effect

Table 12.1 Example showing how inbreeding increases the number of pairs of homozygous genes*

Generation number	Genotype	Percent homozygous gene pairs	Frequency of d gene	Frequency of dd genotype
0	160Dd	0	0.50	0
1	400DD · 800Dd · 400dd	50	0.50	0.25
2	400DD + 200DD · 400Dd · 200dd + 400dd	75	0.50	0.375
3	600DD + 100DD · 200Dd · 100dd + 600dd	87.5	0.50	0.437
4	700DD + 50DD · 100Dd · 50dd + 700dd	93.8	0.50	0.468
5	750DD + 25DD · 50Dd · 25dd + 750dd	96.9	0.50	0.484

*Adapted from *General Genetics* by A. M. Srb and R. D. Owen (San Francisco : W. H. Freeman and Company, 1952).
Note: This example assumes self-fertilization with no selection practiced against the recessive gene. Inbreeding in animals where self-fertilization cannot be practiced would progress in the same way, except that homozygosity would be increased more slowly.

would have been due to selection, however, and not to inbreeding as such.

A second point to note in this example is that even though we used only one pair of genes, all other pairs of genes that segregate independently are made homozygous at the same rate, regardless of their phenotypic effects. Still a third point is that even though the rate of inbreeding is much slower in animals than in plants, because self-fertilization cannot take place in animals, the genetic effects are still the same, except that homozygosity increases at a slower rate, depending upon the degree of relationship between parents that produce inbred offspring.

12.2 CONSEQUENCES OF HOMOZYGOSITY

The fact that inbreeding increases the number of pairs of genes that become homozygous regardless of the phenotypic expression of these genes and how many genes are involved allows us to make certain conclusions regarding this important genetic effect.

Inbreeding does not increase the number of recessive alleles in a population but merely brings them to light throught increasing homozygosity. This is well illustrated in Table 12.2, where it is shown that even though the fre-

Fig. 12.2 Dwarf parrot-mouth lamb. A hidden recessive gene in a herd brought out by the mating of two carriers of the recessive gene. (Courtesy of the University of Missouri.)

quency of homozygous recessive individuals (*dd*) increased markedly as inbreeding progressed, the frequency of the *d* gene did not change from the parental to the fourth generation. In other words, the proportion of recessive genes in the original population was the same as in the fourth generation, but they were hidden by being paired with a dominant gene. Undoubtedly, there are many hidden recessive genes in outbred populations. Dr. H. J. Muller [6] estimates that the human population now carries an average of eight hidden recessive and semidominant genes per person, and the load may be more than this. The situation probably exists in farm animals. Progeny-test records on six bulls selected at random [5] showed that each was carrying from one to four detrimental recessive genes.

Inbreeding or increased homozygosity does not uncover dominant genes. This is true because individuals that are heterozygous or homozygous dominant will both show the dominant effect of the gene, and in most cases there is little or no difference in the phenotype of individuals of the two genotypes. Inbreeding, however, does increase the possibility that animals carrying genes with a dominant effect will be homozygous (*DD*) rather than heterozygous (*Dd*). Since from the physiological standpoint dominant genes usually have a favorable effect whereas recessive genes have an unfavorable one, culling the less desirable animals should result in an increase in the frequency of the dominant or favorable genes in the population. This is the reason that the animal breeder who uses inbreeding must be prepared to cull a large proportion of the less desirable individuals from his herd. This requires that a large number of animals be produced in developing a line, and for this reason it is very costly and not too practical in some instances.

Inbreeding fixes characters in an inbred population through increased homozygosity whether or not the effects are favorable or unfavorable. For example, in Table 12.1 it was pointed out that 47 percent of the population was homozygous (*dd*) in the fourth generation of self-fertilization. Inbreeding among such individuals can do nothing but produce *dd* offspring, and as inbreeding progresses, unless selection against this genotype is practiced, the trait becomes fixed in a larger proportion of the individuals in the inbred line. Many times inbreeding may increase at a rapid rate, and recessive genes may be fixed or made homozygous in many members of the line before much attention is paid to their presence. One inbred line of swine at the Missouri Station [1] had to be discarded because of the presence of a recessive gene for hemophilia that caused many in the line to bleed to death at castration or at farrowing time. The early identification of such defects in an inbred line together with rigid culling and a moderate rate of inbreeding are necessary to prevent the fixation of recessive traits in many of the members of an inbred line. Once they are made homozygous or are fixed in that line, the only cure is to cull known homozygous recessive and heterozygous individuals and their close relatives or to outbreed and introduce the more favorable allele for that trait. The breeder must then backcross to the original inbred line and try to retain the desirable traits of the original line as well as the desirable ones introduced by the outcross.

Since inbreeding makes more pairs of genes homozygous, the offspring of inbred parents are more likely to receive the same genes from their parents than are offspring of noninbred parents. This means that the offspring are more likely to resemble the parent from which they receive a dominant gene, and because of this they are more likely to resemble each other. This is another way of saying that inbred parents are more likely to be prepotent than noninbred parents, at least for traits conditioned by dominant genes. Superior inbred lines should possess more pairs of homozygous dominant genes than inferior inbred lines, since dominant genes are thought to produce more favorable effects.

Fig. 12.3 This animal is afflicted with hemophilia due to an autosomal recessive gene in the homozygous state. Inbreeding uncovered this defect in a line of swine. (Courtesy of the University of Missouri.)

12.3 *PHENOTYPIC EFFECTS OF INBREEDING*

Inbreeding if accompanied by selection may increase the phenotypic uniformity among animals within an inbred line for traits such as coat color and horns that are conditioned by genes with large monofactorial effects. This is not true, however, of some of the quantitative traits that are affected by many pairs of genes with different modes of expression. Many of these genes seem to act in a physiological manner upon the efficiency of metabolism of many chemical compounds. Studies to date, mostly in swine, point out rather conclusively that increased inbreeding or homozygosity is accompanied by a decline in those traits that are closely related to physical fitness. Among such traits are fertility, viability, and growth rate. Type and conformation may be affected to a certain extent, especially by the occurrence of crooked legs and less bloom shown in inbred animals because they are less vigorous than those which are not inbred.

Inbred animals within an inbred line are more likely to be alike genotypically than phenotypically for traits of economic importance. This is because many inbred animals are not as able to cope with their environment as are those which are not inbred. Many pigs in highly inbred litters are quite variable in size and weight at 154 days of age, whereas pigs from crossbred litters may be more uniform. The variation within inbred litters should be due more to environmental than to genetic causes. When mated to nonrelated animals, inbred animals should breed better than they look, whereas non-inbred animals are often disappointing from the breeding standpoint because they look better than they breed.

12.4 PHYSIOLOGICAL BASIS OF INBREEDING EFFECTS

Little is known about the physiological basis of inbreeding effects, because few experiments have been designed to study this in farm animals. Many of the adverse effects of inbreeding in other animals, however, are known to be due to the action of recessive genes, and much is known about the physiology of their action in some instances.

As a general rule, the action of recessive genes is unfavorable to the well-being of the individual. This varies from those genes that are lethal in the recessive state to those that have such a slight effect that it can hardly be noticed or may not be noticed at all. Undoubtedly, many of the adverse effects of inbreeding are due to several pairs of recessive genes, each of which has only a slight detrimental effect on the same trait. Probably the action of most, if not all, such genes is through the failure to produce required enzymes or through the production of abnormal proteins and other compounds.

As mentioned previously, inbred animals are usually less able to cope with their environment than are noninbred animals. This is shown by greater death losses early in life, the lack of the ability to reproduce efficiently in many instances, and a slower growth rate and a smaller mature body size. Similar effects are often produced by poor nutrition or exposure to many diseases. Yet, the above adverse effects are often noted even when efforts are made to supply inbred animals with rations of proper quantity and quality and to control infectious diseases. This suggests that the adverse effects of inbreeding may be due to some physiological insufficiency and perhaps to a deficiency or lack of balance of hormones from the endocrine system.

Thomas Bates used inbreeding in the development of the Duchess family of Shorthorn cattle. Of 58 Duchess cows, 24 were barren and the remaining 34 produced only 110 calves during their lifetime. The actual reason for their sterility and low fertility was not known, but it must have had a physiological basis and undoubtedly involved genes and the endocrine system.

Studies with inbred mice have shown definite differences between inbred lines in certain metabolic and physiological traits. For instance, a study of four inbred strains showed a variation in the daily secretion rate of L-thyroxine from 3.5 micrograms per 100 grams of body weight in one line to 6.45 in another [2].

A study was made with inbred Peppin Merino sheep in which inbred animals were injected with crude pituitary extracts and the effects on growth rate and wool production observed [3]. This treatment produced a highly significant increase in growth rate in inbred lambs over the first 10 weeks of age when compared with untreated inbred lambs. Continued treatment from 10 to 23 weeks of age caused no further change in growth rate. Differences between treated and untreated lambs in body weight and size were maintained long after the cessation of treatment. The injections caused a significant increase in wool production while they were being given but not afterwards. Similar treatments produced no detectable response in noninbred lambs. It was concluded that the effect of inbreeding may be attributed at least in part to a reduction in pituitary activity.

Many more studies are needed in which an attempt is made to determine the physiological reason for adverse inbreeding effects. Such studies, although they may be costly, would be helpful in giving us a clearer understanding of the normal physiology of farm animals. Studies are also needed to determine if nutrient requirements of inbred animals are greater than those which are not inbred. Studies such as these might be helpful in increasing the efficiency of production of inbred animals for seed stock purposes.

12.5   EFFECT OF INBREEDING ON DIFFERENT KINDS OF GENE ACTION

As far as the genotypes are concerned, increased homozygosity following inbreeding affects all independent pairs of segregating genes in the same way. On the other hand, increased homozygosity following inbreeding may have a quite different effect on the phenotype, depending upon the kind of gene action involved. These effects are illustrated by hypothetical examples in Tables 12.2 and 12.3.

12.5.1 Dominance and Recessiveness

Most geneticists agree that much of the decline in vigor that accompanies inbreeding is due to the uncovering of detrimental recessive genes through increased homozygosity. These recessive genes are hidden by dominant genes in the noninbred population. If we use a single pair of genes as an example and assign values to the dominant and recessive phenotypes, as shown in Table 12.2 we can see that the average values for the population decrease toward those of the recessive genotypes as the degree of homozygosity increases. We also are assuming here, of course, that there is no selection against the recessive genotypes. Eventually, however, if complete homozygosity were attained, and this is not very likely, there would be no further decrease in the values, because there would be no further uncovering of recessive genes.

In all probability, many pairs of recessive genes may affect the size and vigor of plants and animals, and some may have a greater effect than others. In spite of this, increased homozygosity of many pairs of recessive genes would have a similar effect to that shown in Table 12.2.

12.5.2 Overdominance

This type of gene action can also be responsible for adverse effects of inbreeding in farm animals as homozygosity increases. This is illustrated in Table 12.2. Overdominance, it will be remembered, means that the heterozygous genotypes are superior to either of the homozygotes. Thus, as inbreeding increases, there is a decrease in the number of heterozygous and an increase in the number of homozygous individuals in the population. This will result in a deterioration in a trait affected by this kind of gene action, unless rigid selection is practiced for the more desirable individuals. If we select the superior animals for breeding purposes, however, there would be a tendency to favor those that are the most heterozygous and cull those that are more homozygous. This would result in a slower increase in the degree of homozygosity in the population than one might expect.

12.5.3 Epistasis

This type of gene action may also be responsible for the deterioration of a trait when inbreeding is practiced as shown in Table 12.3. Because epistasis involves the interaction of genes that are not alleles, we must use at least two different pairs of genes to illustrate inbreeding effects on this type of gene action. The mating of individuals of the genotype *AaBb* could result in offspring of nine possible different genotypes.

Table 12.2 Hypothetical example showing the influence of inbreeding or increased homozygosity on different types of gene action (self-fertilization)*

Number of generation	Genotypes	Population average in units		
		Dominance†	Overdominance‡	Additive§
0	1600Dd	180	180	180
1	400DD, 800Dd, 400dd	170	160	180
2	400DD + 200DD, 400Dd, 200dd + 400dd	165	150	180
3	600DD + 100DD, 200Dd, 100dd + 600dd	163	145	180
4	700DD + 50DD, 100Dd, 50dd + 700dd	161	143	180
5	750DD + 25DD, 50Dd, 25dd + 750dd	161	141	180

*Adapted from *General Genetics* by A. M. Srb and R. D. Owen (San Francisco: W. H. Freeman and Company, 1952).
†Assume that *dd* gives 140 units and *DD* and *Dd* 180 units.
‡Assume that *dd* and *DD* give 140 units and *Dd* 180 units.
§Assume that *dd* gives 160 units, *Dd* 180 units, and *DD* 200 units or that each, plus gene *D*, adds 20 units to the residual genotype (*dd*) of 160 units.

Table 12.3 Hypothetical example showing the influence of inbreeding or increased homozygosity on epistatic gene action

Number of generation	Genotypes	Average of population in units
0	1600 *AaBb*	200
1	100 *AABB* (200) 200 *AABb* (200) 100 *AAbb* (150) 200 *AaBB* (200) 400 *AaBb* (200) 200 *Aabb* (150) 100 *aaBB* (150) 200 *aaBb* (150) 100 *aabb* (150)	178
2	300 *AABB* (200) 100 *AABb* (200) 150 *AAbb* (150) 100 *AaBB* (200) 200 *AaBb* (200) 50 *Aabb* (150) 200 *aaBB* (150) 200 *aaBb* (150) 300 *aabb* (150)	172

The following assumptions were made in this example:
That each new generation was propagated by self-fertilization.
That combinations of *A*- *B*- resulted in 200 units and all other combinations in 150 units.

The example shown in Table 12.3 assumes that combinations of the genes *A* and *B* are favorable, whereas all other combinations are unfavorable. This example shows that there is a decline in the population average following inbreeding if we start with parents that are completely heterozygous for the two genes *A* and *B*.

Under actual conditions, we have no way of knowing what kinds of epistatic gene action affect the various traits. In this example, if we selected the most desirable individuals for breeding purposes and self-fertilized them, we would move toward fixing the desired combination of genes in the population, although the progress made would be very slow. From the theoretical standpoint, developing inbred lines, crossing them, and then developing a new inbred line from the cross should be helpful in fixing a larger number of favorable combinations of genes with epistatic effects. Although this seems possible, it is not too probable, and it is doubtful that combinations of many epistatic genes with favorable effects could be fixed in a single inbred line.

12.5.4   Additive Gene Action

In additive gene action, there are no dominant or recessive genes, nor are there interactions between the various alleles or pairs of genes. Inbreeding would cause both plus and neutral genes to become more homozygous, but if selection were not practiced as in the example shown in Table 12.2, there would be no decline in the trait as inbreeding increased. Actually, it seems more likely that if superior individuals were selected when this was the only kind of gene action involved, the merit of the population should improve until genetic variation was exhausted. When this point was reached, no further improvement would result.

If additive gene action were the only kind affecting the important economic traits in farm animals, it would be possible to build superior purebred lines that would breed true for the important traits. Unfortunately, however, different traits of economic importance are affected by both additive and nonadditive gene action, and in many instances both types affect the same trait. Thus, the development of superior inbred lines is by no means simple, and it is necessary to try to fix superior genes in inbred lines and then cross them to get combinations of genes that will give hybrid vigor or heterosis.

12.6   POSSIBLE USES OF INBREEDING

The most important factor limiting the usefulness of inbreeding in livestock breeding is the decline in vigor that almost always follows or accompanies its use. This is doubly important, because the traits affected the most adversely by inbreeding are those of the greatest importance from the economic standpoint. In spite of these disadvantages of inbreeding, there are certain instances where it may be used to advantage in livestock production.

Inbreeding may be used to determine the actual genetic worth of an individual. This is particularly true when a sire can be mated to 23 to 35 of his own daughters. As mentioned before, such a practice is advantageous because it will test for all recessive genes the sire may be carrying, but at the same time it should give some indication of the desirable genes he may possess. Such a mating system results in 25 percent inbreeding in one generation and may be too extreme or severe under most conditions. Another disadvantage of this practice is that the male may be dead or no longer in service by the time the test is completed, and he may also fail the test at its completion. This method of testing could be used to advantage in a large herd where artificial insemination could be practiced and where the main objective was genetic improvement in the stock.

Inbreeding could be used in a practical way to select against a recessive gene that is of economic importance. Since inbreeding brings out the hidden recessive genes, the homozygous recessive individuals as well as many of the heterozygotes could be identified and culled. This would require severe culling, however, and might be too costly under most conditions.

Inbreeding may be used to form distinct families within a breed, especially if selection is practiced along with it. Selection between such inbred families for traits of low heritability would be more effective than selection based on individuality alone, especially if there were distinct or definite family differences. Family selection is more effective than individual selection, because it tends to reduce some of the environmental variations that breeders often mistake for those of a genetic nature.

Inbreeding should be used only for the production of seed stock. But even when the breeder uses it for this purpose, he has to determine how much he can sacrifice in the way of lower production and performance to increase the purity of his breeding animals. This becomes more important when we consider the fact that purchasers of breeding stock will pay no more, as a general rule, for inbred stock, because they do not recognize the value of its increased prepotency. On the contrary, the purchaser is more likely to discriminate against inbred sires and dams, because they may be smaller in size for their age, and their conformation may not be as desirable as that of noninbred animals. In addition, many livestock producers have the mistaken opinion that inbred parents transmit less desirable genes to their offspring than those that are not inbred.

The most practical use of inbreeding at the present time seems to be to develop lines that can be used for crossing purposes, as is done for hybrid seed corn. When two or more inbred lines are found that nick well in crosses, they are more likely than are noninbred animals to do so in future crosses because of their purity. The most practical use of inbreeding is to use inbred sires in three-line rotation crosses on crossbred females for commercial production. Inbred females should not be used for commercial production because of their reduced performance as mothers.

From the research standpoint, inbreeding is of value to determine the type, or types, of gene action that affects the various economic traits in farm animals. If inbreeding effects are very great, the trait is affected by nonadditive gene action. If inbreeding effects are very small or nonexistent, the trait is affected mostly by additive gene action. As mentioned in a previous chapter, different methods of selection are required for the two different kinds of gene action.

12.7 HOW TO DEVELOP AND USE
INBRED LINES

The production and widespread use of hybrid seed corn leaves no doubt about the important role inbred lines can play in the increased efficiency of production of plants. The principles of inbreeding that apply to plants should also apply to animals, although the problem of developing inbred lines in animals involves much more time and expense. If short cuts or special methods for developing inbred lines in animals could be developed, this would be a great boon to the efficiency of animal production.

The discovery of parthenogenesis in a strain of Beltsville Small White turkeys has led to some interesting developments in the production of inbred lines [7]. Parthenogenetic turkeys are those born without fathers. Selection for a higher rate of parthenogenesis in this line has been effective, so parthenogenetic embryos can be expected in six to twelve percent of all unfertilized eggs and about ten percent of these embryos survive to hatching. Parthenogenetic turkeys are males and carry the diploid number of chromosomes, all of which come from the mother. About one in five of these males that survive to maturity produces some viable sperm, so they can be used for breeding purposes. Fatherless turkeys are being used for research in several areas of biology, including research on the development of inbred lines of turkeys.

Theoretically, parthenogenetic male turkeys should be homozygous for all loci, and this seems to be confirmed by experimental results. If this is true, any detrimental recessive gene carried by the dam could become homozygous in the embryo and could cause the appearance of recessive genetic defects or death of the homozygous individuals. This suggests that those parthenogens that survive to breeding age may be relatively free of lethal and/or semilethal genes and if used for breeding could decrease the detrimental genetic load in a flock over a period of years.

By backcrossing a male parthenogen turkey with his own dam, the offspring produced are highly inbred (homozygous), equaling that resulting from self-fertilization in plants (F_x of 0.50). Some inbred lines are being developed in this way, and lines should be produced that are relatively free of detrimental or lethal recessive genes.

Parthenogenesis probably does not occur in farm mammals, or at least it has not been proven to occur. Therefore, the systems of developing inbred lines of turkeys cannot be used in other species. The problems involved and possible ways to reduce these problems in developing inbred lines in farm animals will now be discussed.

12.7.1   Problems in the Development of Inbred Lines

The development of inbred lines in animals is much more expensive and time-consuming than in plants. Plants produce many more progeny (seed) per individual than animals, and the expense of growing each plant is much less. Many plants can also be grown at any one time during the season, with the generation interval seldom, if ever, exceeding one year, whereas in animals it may be several years. Inbred lines among plants can be developed much more rapidly than among animals, because most plants can be self-fertilized. Some plants are self-fertilized by artificial means. Still others are self-sterile and possess a built-in mechanism for maintaining heterozygosity.

Experiments with both plants and animals show that upon inbreeding some inbred lines survive only a short while because of the appearance of lethal or detrimental traits, causing the line to be incapable of reproducing and maintaining itself. Other lines may not show lethal or highly detrimental recessive traits, but on the average the performance of the line may decline because of a decline in vigor. It is probably best to discard lines that decline greatly in vigor, because this indicates that they possess undesirable, although not necessarily lethal, genes. Even within inbred lines that perform well, death losses may be higher before weaning than in noninbred individuals. Those individuals that die before weaning are probably those, in many cases, which carry defective recessive genes, or at least the most undesirable combination of genes. It appears that Mother Nature is trying to eliminate those carrying defective genes. In general, those lines that perform well and possess normal vigor should be further inbred to develop the inbred lines.

Inbreeding, because it involves the mating of relatives, naturally limits the genetic diversity within a particular line to that possessed by the original parents, which by necessity are few in number. In a large herd or flock where inbreeding is practiced, several distinct inbred lines, or sublines, will be developed, especially if parent × offspring or sib matings are made. The genes each inbred line will possess will largely depend upon the sample of genes present in the original individuals within that line. Genetic drift and selection, both artificial and natural, will also help determine the genetic composition of a particular inbred line. In any event, as inbreeding proceeds the different lines will become more and more distinct in their genotypes and phenotypes.

The more intense the inbreeding used to develop an inbred line, the more limited the number of individuals used to establish the line and the more limited the number and kinds of genes the line possesses. For example, if one develops an inbred line by parent × offspring or full brother × full sister matings, fewer individuals will be mated to produce that line than if half-sibs are used for its development. However, the rate of increase of inbreeding is less in half-sib matings as compared to full sib or parent × offspring matings.

The main question involved in developing inbred lines in farm mammals is whether or not enough extra performance is gained in crossing different inbred lines as compared to crossing different breeds to justify the time and expense of developing inbred lines. Research has not yet given a satisfactory answer to this question.

12.7.2 How to Develop Inbred Lines

The only certain way to know what will happen when inbreeding is practiced is to inbreed and observe the results. Some precautions may increase the chances that an inbred line will perform satisfactorily. It seems reasonable that one should start an inbred line using parents that have good performance and no obvious genetic defects. Also, their relatives should be superior in these respects. Even when this is done, however, there is no certainty that a superior inbred line will be produced.

The following are some suggestions for developing inbred lines. Inbreed as rapidly as possible at first for the early exposure of any detrimental or lethal recessive genes that may be present in the original breeding stock. Two methods may be used that will give the maximum amount of inbreeding (25 percent) in the first generation. One of these is to make parent × offspring matings by mating a sire to his own daughters or a dam to her own sons. The offspring of such matings will carry a little less than 75 percent of the inheritance of the parent that is mated to its own offspring. Thus, one parent is responsible for most of the recessive genes that are paired in the offspring. If that parent is genetically superior and has produced superior

Fig. 12.4 A typical boar of the Uark Poland China inbred line. The current inbreeding for this line is about 75 percent. Although there has been some decline in vigor, the performance of individuals within this inbred line is still satisfactory. (Courtesy of Dr. P. R. Noland, University of Arkansas.)

progeny in the past, this indicates that such a parent possesses superior genes, although it does not prove it. It also gives some indication of the number and kind of recessive genes that a parent possesses when it is mated with several offspring (see Chapter 11).

The mating of full sibs also results in 25 percent inbreeding in the first generation progeny. In such a mating, two different individuals (both grandparents) instead of one contribute to the inbreeding, but each is responsible for only one-half of the increase in homozygosity (12.5 percent). In the parent × offspring mating, one individual (who is both a parent and a grandparent to the inbred progeny) is responsible for all of the increase in homozygosity (25 percent). In the parent × offspring mating a specific recessive trait is more likely to be expressed if it is carried by the parent than in the full-sib matings, because the amount of homozygosity due to a single parent is twice as high (25 percent as compared to 12.5 percent). Progeny from full sibs may eventually express a wider variety of recessive traits, because inbreeding traces back to two different unrelated ancestors, and they are more likely to possess a greater variety of recessive detrimental genes than a single ancestor.

Regardless of the kind of mating made, the more intense the inbreeding, the more likely recessive traits are to be expressed if the genes for these traits are present in the original population. This may be desirable, however, because more intense inbreeding will identify more quickly those lines carrying detrimental recessive genes so they can be discarded earlier.

The breeder should watch for recessive traits in inbred progeny and for a sharp decline in vigor in the inbred offspring. Recessive traits may be internal or external in their phenotypic expression. Internal traits may never be observed unless the breeder performs post mortems and is really looking for them. If recessive defects are found to be present in an inbred line, it may be advisable to discard the line. If several lines are started, one or more may not show defects, and efforts can be concentrated on these to develop a superior line.

Once two or more superior inbred lines of animals are formed by inbreeding and inbreeding has reached 40 to 50 percent, it is often desirable to propagate the line by half-brother × half-sister matings. This will slow the rate of increase in inbreeding, and more individuals can be used to further develop the line. This will also give an opportunity to use more intense culling and selection in the further development of that line.

12.7.3 How to Use Inbred Lines

The main reason for developing inbred lines is to use two or more of them for crossing. The ones crossed year after year should be those that give the best performing linecross progeny. Such parents are said to have superior combining ability.

To find lines with superior crossing ability they have to be tested in crosses to determine the comparative performance of the progeny of different line crosses. The more lines that are tested in crosses the more time and expense involved. In farm mammals, all possible test crosses may not be practical.

Theoretically, lines should not be tested in crosses until they have an inbreeding coefficient of nearly 100 percent and homozygosity of all genes has been attained. This is not practical in most species of animals, so the testing of the various lines in crosses should begin when inbreeding is 40 to 50 percent. The more homozygous (more inbred) the lines, the greater the repeatability of performance one can expect in recurrent crosses of these same lines. In other words, if two lines cross or combine well the first time they are crossed and both lines are highly homozygous for different alleles, they are likely to repeat their superior crossing performance in later years.

12.7.4 *How to Improve Inbred Lines*

Once inbred lines are formed, matings must continue to be made within the inbred line to keep it pure. A particular inbred line may have certain faults that need to be corrected, and this can be done only by introducing new genes from outside breeding stock if the line is highly homozygous. Several methods are used in the improvement of inbred lines of corn, and some of the same principles may be used to improve inbred lines of animals.

Convergent improvement may be used to improve two or more inbred lines, each of which is superior for one trait that is mediocre in the other. For example, let us assume that line A is superior for growth rate but inferior in carcass quality, whereas line B is inferior in growth rate but superior in carcass quality. The following steps may be used to improve each line:

1. Cross lines A and B to produce A × B progeny.
2. Backcross AB with A and with B.
3. Select within the AB × A and the AB × B backcrosses for both superior growth rate and carcass quality and continue to inbreed each line.
4. Continue to backcross and select within each line for each trait.

The objective of such a procedure is to strengthen a weak trait in an inbred line, but at the same time an attempt is made to retain the good characteristics of that line as well as its identity. It is a proposed method of incorporating new desirable additive genes into an inbred line.

Another method that could be used to improve inbred lines of farm mammals is similar to *gamete selection*, suggested for the improvement of inbred lines of corn. In animals, an inbred line could be improved by mating several outbred females possessing desirable traits with males from the inbred line. If some of these females appear to add desirable genes to the inbred line as indicated by the superior phenotype of their progeny, their inheritance could be gradually incorporated into the inbred line by continued backcrossing of her descendants with the members of the original inbred line. New inheritance could also be introduced into an inbred line by using outbred males in a similar manner.

12.7.5   The Future Use of Inbreeding

Considerable inbreeding was used in the early development of present-day breeds of livestock. Only a few producers of purebreds now make use of this system of breeding, although some use a mild form of linebreeding which will be discussed in detail in a later chapter. Most producers of purebreds follow a system of outbreeding and try to maintain a certain amount of heterozygosity in their breeding animals, because they perform better, on the average, than inbred animals. However, the development of superior inbred lines and crossing those with the best combining ability could considerably increase the efficiency of livestock production.

Inbreeding probably will never be popular in small herds of purebred animals. It may become popular in some larger purebred herds, but this may never be commonplace. Large commercial companies who have the capital, facilities, and the scientific knowledge are more likely to form and merchandise inbred lines of livestock in the future. Several commercial companies are already developing inbred lines for this purpose.

REFERENCES

1. Bogart, R., and M. E. Muhrer. "The Inheritance of a Hemophilia-like Condition in Swine," JH, 33: 53, 1942.

2. Chai, C. K. "Endocrine Variation (Thyroid Function in Inbred and F_1 Hybrid Mice)," JH, 49: 143, 1958.

3. Doney, J. M. "The Effects of Inbreeding on Four Families of Peppin Merinos; 3; The Influence of Crude Pituitary Extract on Inbred Lambs," AJAR, 10: 97, 1959.

4. King, H. D. "Studies on Inbreeding," JEZ, 26: 3, 55; 29: 134, 1918–19.

5. Mead, S. W., P. W. Gregory, and W. M. Regan. "Deleterious Recessive Genes in Dairy Bulls Selected at Random," G, 31: 574, 1946.

6. Muller, H. J. "Our Load of Mutations," AJHG, 2: 111, 1950.

7. Olsen, M. W. "Potential Use of Parthenogenetic Development in Turkeys," JH, 60: 346, 1969.

QUESTIONS AND PROBLEMS

1. Define *inbreeding*.

2. Why do breeders, as a general rule, try to avoid inbreeding?

3. Is inbreeding always bad?

4. Explain the principles involved in the production of hybrid seed corn.

5. Will the principles used for the production of hybrid seed corn work with animals? Explain.

6. What is the main genetic effect of inbreeding? Does inbreeding, as such, increase the frequency of recessive genes? Explain.

7. The frequency of recessive genes in a population often decreases as the amount of inbreeding increases. Why?

8. Why are dominant genes often increased in frequency when inbred lines are formed?

9. Why aren't dominant genes uncovered by inbreeding?

10. How could inbreeding be used to eliminate recessive genes from a population?

11. Why does inbreeding tend to increase breeding purity of inbred individuals?

12. Why is there a danger of fixing detrimental genes in an inbred line?

13. Why do inbred individuals usually breed better than they look, whereas outbred individuals tend to look better than they breed?

14. What are some of the phenotypic effects of inbreeding?

15. What is the possible physiological basis of inbreeding effects?

16. How does inbreeding affect traits which themselves are affected by overdominance? Is it possible to develop an inbred line that will not decline in performance if overdominance is important? Explain.

17. If a trait were affected 100 percent by additive gene action, what would be expected to happen, phenotypically, as inbreeding increased? Why?

18. Discuss some of the possible uses of inbreeding on a practical basis.

19. Explain how fatherless turkeys may be used in the development of inbred lines. Can the same thing be done with farm mammals? Explain.

20. Inbreeding usually results in greater death losses of the young between birth and weaning. What is likely to be the genetic difference between those that die and those that survive to weaning in an inbred line?

21. Suppose you mated a sire to his own daughters and the offspring were excellent. What does this indicate to you genetically?

22. Why does intensive inbreeding limit the genetic diversity within an inbred line?

23. What is the main practical problem in developing inbred lines of farm animals for commercial production?

24. Outline in detail the procedure you would recommend for developing inbred lines of farm mammals.

25. Discuss the differences, genetically, in developing an inbred line by parent $\times$ offspring matings as compared to full-sib matings. Which mating system would you prefer for developing an inbred line and why?

26. What is the main reason for developing inbred lines?

27. At what stage in their development (amount of inbreeding) would you start testing inbred lines in crosses to find those that combine best?

28. What methods could be used to improve an inbred line once it has been developed?

29. Do you think that inbreeding will be used on a larger scale in the future? Why?

30. Why is it recommended that inbred females not be used for livestock production on a commercial basis?

13

How to Measure
Inbreeding
and Relationships

Because inbreeding has such a very definite genotypic and phenotypic effect on many important economic traits in farm animals, it is important to know how to measure the amount of inbreeding in a pedigree. This chapter will outline methods of calculating inbreeding and relationship coefficients. The inbreeding coefficient is a measure of the decrease in the proportion of heterozygous genes over what was present before the inbreeding was practiced.

Wright [7] developed a formula for figuring the inbreeding coefficient by considering the probability that gametes from the sire and dam both carried the same gene. His formula has been modified by Lush [4], and this will be discussed as a means of measuring the amount of inbreeding of an individual or a herd. These coefficients may be calculated without preparing arrow diagrams of the pedigree, but the use of these diagrams make calculations easier, so their preparation will be discussed first.

13.1 PREPARING AN ARROW DIAGRAM OF A PEDIGREE

The pedigree of Real Prince Domino 33rd is shown in Fig. 13.1 and will be used to explain the procedure for preparing an arrow diagram. To simplify the procedure, the name of each individual in the pedigree should be replaced by either a number or a letter. In extended pedigrees, numbers may be more desirable than letters because there may be more ancestors than letters in

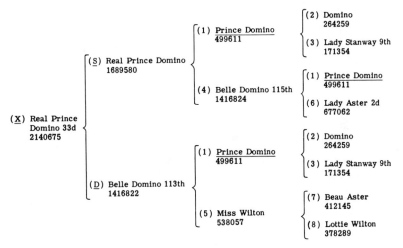

Fig. 13.1 Pedigree of Real Prince Domino 33rd (2140675), which includes letters and numbers to replace the names of the ancestors to be used in the pedigree in Fig. 13.2.

the alphabet. A good method is to list the individual whose pedigree is being studied as X, his sire as S and his dam as D. Numbers then may be used for the remainder of the ancestors, but the same number should be given the same individual each time he or she is found in the pedigree. The pedigree of Real Prince Domino 33rd, with letters and numbers replacing the names of individuals in the pedigree, is shown in Fig. 13.2

The next step is to begin the arrow diagram by placing the letter X on a piece of paper along with the sire S and the dam D in the usual position in the pedigree. Then arrows should be drawn from S to X and from D to X, with the arrows pointing to X, as shown in Fig. 13.3. After this is done, the first common ancestor in the pedigree should be located. A common ancestor is one that appears in both the dam's and the sire's pedigree. Now draw arrows from the common ancestor (1) to the sire, with the arrows pointing toward the sire. Do the same for the dam. If other individuals are between the common ancestor and the sire or dam, they must be included in the arrow pathway in the proper position. The completed arrow diagram of Real Prince Domino 33rd as shown in Fig. 13.3 contains only one common ancestor, which is individual number 1, or Prince Domino. He traces to individual X twice through the sire and once through the dam. Although 2 and 3 are common ancestors of S and D, they are not used in calculating F_x since they are common ancestors only through individual 1.

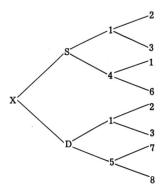

Fig. 13.2 Pedigree of Real Prince Domino 33rd (2140675), with letters and numbers replacing the names of his ancestors.

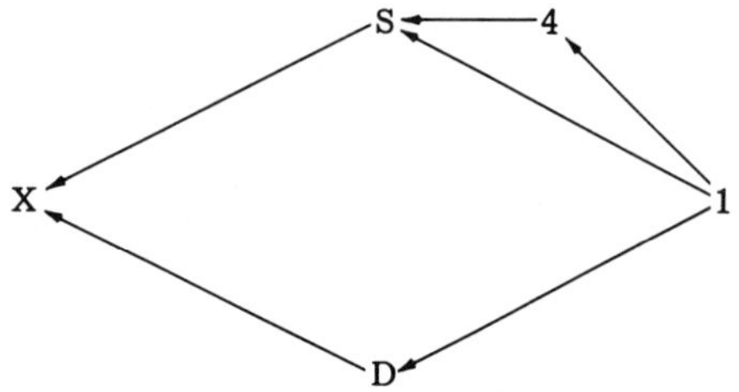

Fig. 13.3 Arrow diagram of the pedigree of Real Prince Domino 33rd.

13.2 CALCULATING INBREEDING COEFFICIENTS

The formula for calculating inbreeding coefficients is as follows:

$$F_x = \tfrac{1}{2} \sum [(\tfrac{1}{2})^n (1 + F_a)]$$

where

F_x refers to the inbreeding coefficient of individual X.

$\sum$ is the Greek symbol meaning to sum or add all paths.

n is the power to which one-half must be raised, depending upon the number of arrows connecting the *sire* and *dam* through the common ancestor.

F_a is the inbreeding coefficient of the common ancestor.

If the common ancestor is not inbred, the formula to use in calculating the inbreeding coefficient becomes

$$F_x = \tfrac{1}{2} \sum [(\tfrac{1}{2})^n]$$

Our first examples will involve calculating inbreeding coefficients for different kinds of matings in which the common ancestor is not inbred; thus the shorter formula will be used.

13.2.1 Half-Sib Matings

The following pedigree and arrow diagram show a half-sib mating, the sire and the dam of individual X having had the same sire (C). The only *common ancestor* in this pedigree is individual C, because he appears in the pedigree of both the sire and the dam of individual X. The arrow diagram shows that there is only one pathway from C to X through the sire and only one through the dam. This pathway may now be straightened out for illustrative purposes, and it becomes $X \longleftarrow S \overset{1}{\longleftarrow} C \overset{2}{\longrightarrow} D \longrightarrow X$.

We number the arrows running from sire (S) through the common ancestor (C) to the dam (D). We do not count the arrows running from individual X to the sire and dam. The number of arrows connecting the sire

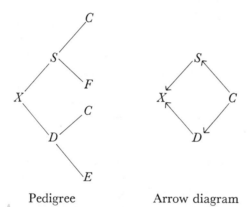

Pedigree   Arrow diagram

and dam with the common ancestor is two, and this is the n in the formula. Our calculation of the inbreeding coefficient now proceeds by letting

$$F_x = \tfrac{1}{2}(\tfrac{1}{2})^2, \quad \text{or} \quad \tfrac{1}{2}(0.25), \quad \text{or} \quad 0.125$$

The inbreeding coefficient of individual X, then, is 0.125, and this can be expressed as 12.5 percent by multiplying the inbreeding coefficient by 100.

13.2.2 Full-Sib Matings

The method for calculating the inbreeding coefficient for a full-sib mating is very similar to that described for half-sib matings, except that an additional path and common ancestor are involved. The following pedigree-and-arrow diagram illustrates how calculations are made for such a mating. The two pathways are

$$X \longleftarrow S \overset{1}{\longleftarrow} C \overset{2}{\longrightarrow} D \longrightarrow X = (\tfrac{1}{2})^2 = 0.25$$
$$X \longleftarrow S \overset{1}{\longleftarrow} F \overset{2}{\longrightarrow} D \longrightarrow X = (\tfrac{1}{2})^2 = 0.25$$
$$\overline{\text{Total} \hspace{5cm} 0.50}$$

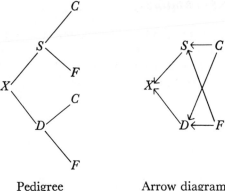

Pedigree   Arrow diagram

The inbreeding coefficient of individual X is one-half of the sum of these two paths, or $F_x = \frac{1}{2}(0.50) = 0.250$, or 25 percent inbreeding. Note that in this last pedigree there are two common ancestors (C and F). We merely calculate the figure for all of the pathways, which totals two in this pedigree, from the common ancestors, as denoted by the arrow diagram, and then add or sum all paths, as $\sum$ indicates. Then, when all are added, we take one-half of the total to get the inbreeding coefficient.

13.2.3 Sire × Daughter Matings

The inbreeding coefficient is calculated for parent × offspring matings in the same manner as for half- and full sibs with only slight variations. The following is a pedigree of an individual from a mating of a sire to his own daughter. The inbreeding coefficient from such a mating is 0.25, providing the sire himself is not inbred.
The pathway is

$$X \longleftarrow S \xrightarrow{1} D \longrightarrow X = (\tfrac{1}{2})^1 = 0.50$$

Thus, $F_x = \frac{1}{2}(0.50)$, or 0.25, or 25 percent inbreeding.
Inbreeding coefficients for dam × son matings are calculated in a

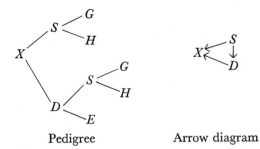

Pedigree   Arrow diagram

similar manner, except that the arrow diagrams run from the dam as the common ancestor.

13.2.4 *Sire* × *Daughter Mating with the Sire Inbred*

The following pedigree is one in which a sire × daughter mating is made, but the sire himself is inbred. The first step in calculating the inbreeding coefficient for such an individual is to complete the arrow diagram as shown.

The first common ancestor in this pedigree, of course, is individual S, which is the sire of individuals X and D. At this point, one might ask what to do about the other common ancestors such as A, B, C, and E? The answer to this is that we take care of these individuals by first calculating the inbreeding coefficient of individual S or the sire, as was done in the previous example for full-brother × full-sister matings. After this is done, for each path going from individual S to individual D, which is just one in this case, we multiply the path by $(1 + F_a)$, or one plus the inbreeding coefficient of individual S.

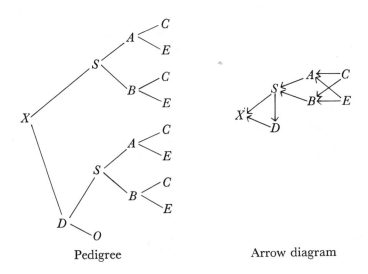

Pedigree   Arrow diagram

The calculation of the inbreeding coefficient for individual S is as follows:

$$S \longleftarrow A \overset{1}{\longleftarrow} C \overset{2}{\longrightarrow} B \longrightarrow S = (\tfrac{1}{2})^2 = 0.250$$
$$S \longleftarrow A \overset{1}{\longleftarrow} E \overset{2}{\longrightarrow} B \longrightarrow S = (\tfrac{1}{2})^2 = 0.250$$
$$\overline{\text{Total} \qquad\qquad\qquad\qquad\qquad\qquad 0.500}$$

The inbreeding coefficient of individual S, or F_s, would be $\tfrac{1}{2}$ (0.500), or 0.250, or 25 percent. We now proceed to calculate the inbreeding coefficient (F_x) of individual X. Only one pathway is involved, and this is

$$X \longleftarrow S \overset{1}{\longrightarrow} D \longrightarrow X = (\tfrac{1}{2})^1, \quad \text{or} \quad 0.50$$

Since individual S, which is the common ancestor, is inbred, we must use the complete formula as given earlier, The computations then are $F_x = \tfrac{1}{2}$ $[0.50(1.25)] = \tfrac{1}{2}(0.625)$, or 0.3125. Thus, individual X is 31.25 percent inbred.

13.3   COEFFICIENTS OF RELATIONSHIP

The coefficient of relationship between two individuals is an expression of the probability that they possess duplicate genes, because of their common line of descent, over and above those found in the base population. It is evident that an increase in inbreeding causes the relationship of individuals within an inbred line to increase. On the other hand, it is possible for two different inbred lines to be inbred the same amount but still not be related. For example, individuals in a Landrace line of hogs might be inbred 25 percent and in a Poland line a similar amount. They do not have duplicate genes in common because of their common descent, although they do possess some of the same genes because they belong to the same species.

It is often of practical importance to know something about the degree of relationship between two individuals. For instance, an animal may be offered for sale that has a pedigree similar to that of another animal that sold earlier at another sale for a high price, and the breeder wants to know the maximum that he should bid for this animal. Or he may have a bull for sale at private treaty and would like to use the situation described as a sales point. Another instance when relationship coefficients might be of value is when a livestock breeder may have the chance to buy two animals, one of which shows excellent type and comes at a high asking price and the other of which is lacking in some one point but not to the extent to warrant disregarding the animal entirely. If they have a high coefficient of relationship, one would probably perform just as well in the breeder's herd as the other. He could purchase the cheaper animal and produce as good stock with it as with the one that was more expensive.

From the theoretical standpoint, relationships could have another use for the livestock breeder. For traits such as carcass quality that cannot be measured very well until after the death of the individual, the slaughter of a relative should give some indication of the carcass quality of the individual in question. The value of the relative in this respect would be proportional to the degree the two individuals were related. A full brother or sister would be worth more than a half sister or brother in this respect. Full brothers and sisters within an inbred line would also be more closely related than would be full brothers and sisters that are not inbred. Relationship coefficients would give a good indication of the value of records of relatives from this standpoint.

Relationships are of value in calculating PBV of individuals as shown in Chapter 11. Relationships are of two different kinds. Collateral, that is, those not related as ancestors or descendants; the other is direct, that is, those related as ancestors or descendants. The two kinds will be discussed separately.

13.3.1 Relationship Coefficients between Collateral Relatives

Methods of calculating relationships are very similar to those used for calculating inbreeding coefficients, and arrow diagrams are of value in this respect. The formula is as follows:

$$R_{xy} = \frac{\sum [(\frac{1}{2})^n (1 + F_a)]}{\sqrt{(1 + F_x)(1 + F_y)}}$$

where

R_{xy} is the relationship coefficient between animals X and Y.

$\sum$ is the Greek symbol meaning to sum or add.

n is the number of arrows connecting individual X with Y through the common ancestor for each path.

F_x is the inbreeding coefficient of animal X.

F_y is the inbreeding coefficient of animal Y.

F_a is the inbreeding coefficient of the common ancestor.

If individuals X and Y and their common ancestor are not inbred, the formula becomes

$$R_{xy} = \sum [(\frac{1}{2})^n]$$

The following is an example in which the relationship coefficient between half brothers and half sisters is calculated. In this example, we shall let individual X be the male and individual Y the female, although relationship coefficients may be calculated for animals of the same sex. Since none of the individuals involved are inbred, we can use the simple form of the formula for calculating the relationship coefficient.

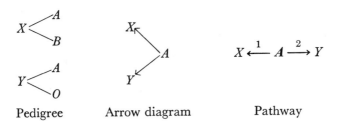

Pedigree   Arrow diagram   Pathway

The relationship coefficient between individual X and Y, or R_{xy}, is $(\frac{1}{2})^2$, or 0.250. This means that these two individuals are related by about 25 percent, or they probably have an increase in the percentage of duplicate genes over that found in the base or noninbred population.

The calculation of the relationship coefficient between full brothers and sisters is similar to the above example, except that there are two common ancestors in such a case, and there are two pathways of gene flow. The calculation of the coefficient of relationship is as follows:

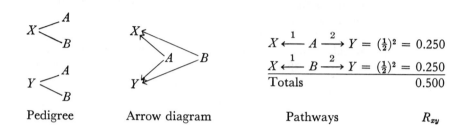

| Pedigree | Arrow diagram | Pathways | R_{xy} |

$$X \xleftarrow{1} A \xrightarrow{2} Y = (\tfrac{1}{2})^2 = 0.250$$
$$X \xleftarrow{1} B \xrightarrow{2} Y = (\tfrac{1}{2})^2 = 0.250$$

| Totals | 0.500 |

The relationship coefficient for individuals X and Y in this example is 0.50.

13.3.2 Relationship Coefficient for the Sire and Dam of Real Prince Domino 33rd

The following is the arrow diagram made from the pedigree of this Hereford bull, as shown in Fig. 13.3

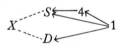

Real Prince Domino (S), his sire, is inbred, so we must first calculate his inbreeding coefficient, which would be 0.250, since he was from a mating of Prince Domino (1) to one of his own daughters (4) Belle Domino 115th. The next step is to list the number of paths through the common ancestor (1) connecting the sire S and the dam D. These would be

$$S \xleftarrow{1} 4 \xleftarrow{2} 1 \xrightarrow{3} D = (\tfrac{1}{2})^3 = 0.1250$$
$$S \xleftarrow{1} \phantom{4 \xleftarrow{2}} 1 \xrightarrow{2} D = (\tfrac{1}{2})^2 = 0.2500$$

| Total | 0.3750 |

The remaining calculations would be

$$R_{sd} = \frac{0.3750}{\sqrt{(1 + 0.25) \times (1 + 0)}} = \frac{0.3750}{\sqrt{1.25}} = \frac{0.3750}{1.1180} = 0.3354$$

or the relationship between S and D would be 33.5 percent.

13.4 DIRECT RELATIONSHIPS

Many times it is of interest to know something about the relationship between an individual and some outstanding ancestor in the pedigree. This is of particular value when linebreeding has been practiced, although the same procedure may be used for calculating the degree of relationship to any particular ancestor. The formula used is

$$R_{xa} = \Sigma \left(\tfrac{1}{2}\right)^n \sqrt{\frac{1 + F_a}{1 + F_x}}$$

This formula is correct only when the relationship is direct. It there is a combination of direct and collateral relationships, the general formula must be used. In this formula, of course, when neither of the individuals involved is inbred, the figures under the square root sign will be equal to one and can be disregarded. For example, the following method is used to calculate the relationship between a sire (S) and his daughter (D).

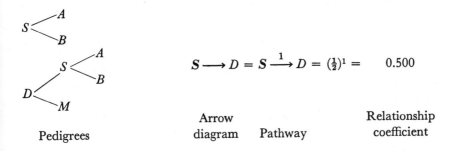

| Pedigrees | Arrow diagram | Pathway | Relationship coefficient |

$$S \longrightarrow D = S \xrightarrow{1} D = (\tfrac{1}{2})^1 = \quad 0.500$$

For further illustration, let us calculate the relationship coefficient between Real Prince Domino 33rd (X) and Prince Domino (1). We can use the arrow diagram shown previously. The first step is to calculate the inbreeding of Prince Domino (1), but in this pedigree he shows no inbreeding. Next, we calculate the inbreeding coefficient of Real Prince Domino 33rd, which is

$$X \leftarrow S \xleftarrow{1} 4 \xleftarrow{2} 1 \xrightarrow{3} D \longrightarrow X = (\tfrac{1}{2})^3 = 0.1250$$

$$X \leftarrow S \xleftarrow{1} 1 \xrightarrow{2} D \longrightarrow X = (\tfrac{1}{2})^2 = 0.2500$$

Total	
	0.3750

The inbreeding coefficient is $\tfrac{1}{2} \times 0.3750$, or 0.1875.

The third step is to list all paths connecting Real Prince Domino 33rd(X) with Prince Domino (1), which are

$$X \xrightarrow{1} S \xrightarrow{2} 4 \xrightarrow{3} 1 = (\tfrac{1}{2})^3 = 0.1250$$

$$X \xrightarrow{1} S \xrightarrow{2} 1 = (\tfrac{1}{2})^2 = 0.2500$$

$$X \xrightarrow{1} D \xrightarrow{2} 1 = (\tfrac{1}{2})^2 = 0.2500$$

Total	
	0.6250

The figure 0.6250 expresses the probable proportion of genes individual X inherited from ancestor 1. The final relationship coefficient is

$$R_{X_1} = 0.6250 = \sqrt{\frac{1+0}{1+0.1875}} = 0.6250 \sqrt{\frac{1}{1.1875}} = 0.6250 \times \sqrt{0.8421}$$

$$= 0.6250 \times 0.9177 = 0.5736$$

This coefficient represents the probability that individuals X and 1 will have duplicate genes which were unlike in the base or noninbred population.

13.5 COVARIANCE TABLES FOR MEASURING INBREEDING COEFFICIENTS

In some herds, such as experimental herds where inbreeding is being practiced, covariance charts for all animals within the inbred line may be calculated in the early stages of the development of the inbred line. Once these are established, it is fairly simple and rapid to calculate inbreeding coefficients for different individuals [3].

To illustrate the calculation of covariance charts, let us use the following pedigree for individual Z, which is given in the form of an arrow diagram. The first step is to list all animals in the order of their birth and also the parents of each individual, as is done in Table 13.1.

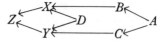

Note that this pedigree does not list the parents of individuals A or D, because they are not inbred. One parent (individual A) is shown for both B and C. Both parents are shown for individuals X and Y, because they are involved in the inbreeding of individual Z. The component for any covariance cell in Table 13.1 between any two individuals is one-half of the components of the parents of that individual. For instance, the covariance for A and B is $(1 + 0)/2 = 0.50$. Other components are calculated as follows:

For A and D it is 0, because the parents are not related.

For B and Y, it is $\dfrac{\text{cov}(BD) + \text{cov}(BC)}{2}$ or $\dfrac{0 + 0.25}{2} = 0.125$

For B and Z, it is $\dfrac{\text{cov}(BX) + \text{cov}(BY)}{2}$ or $\dfrac{0.50 + 0.125}{2} = 0.3125.$

For X and Y, it is $\dfrac{\text{cov}(XD) + \text{cov}(XC)}{2}$ or $\dfrac{0.50 + 0.125}{2} = 0.3125$

For X and Z, it is $\dfrac{\text{cov}(XX) + \text{cov}(XY)}{2}$ or $\dfrac{1.00 + 0.3125}{2} = 0.6563.$

All diagonal cells in the covariance chart are calculated from $1 + F_x$, where F_x is equal to one-half of the covariance between the parents of an individual. Thus, the inbreeding coefficient of individual Z would be one-half of the covariance between individuals X and Y, or one-half of 0.3125, or 0.15625. The diagonal component for individual Z then becomes $1.0 + 0.15625$.

If no inbreeding is involved, the coefficient of relationship may be read directly from the charts or the off-diagonal cells. If inbreeding is involved, corrections must be made for inbreeding by using the formula

$$R_{xy} = \frac{\sum (\tfrac{1}{2})^n (1 + F_a)}{\sqrt{1 + F_x} \cdot \sqrt{1 + F_y}}$$

Table 13.1 Covariance chart constructed for individuals in the pedigree of individual Z

Sire		A	A		D	D	X
Dam					B	C	Y
Individuals	A	B	C	D	X	Y	Z
A	1.000	0.500	0.500	0	0.250	0.250	0.250
B		1.000	0.250	0	0.500	0.125	0.313
C			1.000	0	0.125	0.500	0.313
D				1.000	0.500	0.500	0.500
X					1.000	0.313	0.656
Y						1.000	0.656
Z							1.156

Note: The inbreeding coefficient for individual Z, for example, is one-half the coefficient of relationship of the parents (0.313), or 0.1565. Also, all animals should be included in the covariance chart and not just those contributing to the F_x of an individual.

For example, the relationship coefficient for Y and Z would have to be corrected for the inbreeding of Z as follows:

$$R_{YZ} = \frac{0.6563}{\sqrt{1.0 + 0} \cdot \sqrt{1 + 0.1563}} = \frac{0.6563}{1.075} = 0.610$$

For further illustration, let us use data from an inbred line of Duroc swine maintained at the Missouri Agricultural Experiment Station. Boar 303 and sows 43, 47, 49, 136, and 144 will be used in this example. None of the original stock was inbred or related, except sows 43, 47, and 49, which were littermates. The covariance chart for offspring of the five sows bred to boar 303 is given in Table 13.2. The inbreeding of any offspring from male 1 mated to either females 2, 3, 4, or 5 could be calculated by taking one-half of the covariance component from the off-diagonal cells. The relation coefficients and inbreeding coefficients among descendants of the original stock would be easily calculated by extending the covariance chart. In doing this, however, the covariance components for any two individuals should be corrected for inbreeding as described previously.

Table 13.2 Covariance chart for calculating inbreeding coefficients in swine

Sire	24	10	10	10	241	251	303	303	303	303	303
Dam	75	36	36	36	346	540	43	47	49	136	144
Individuals	303♂	43♀	47♀	49♀	136♀	144♀	1♂	2♀	3♀	4♀	5♀
303♂	1.000	0	0	0	0	0	0.500	0.500	0.500	0.500	0.500
43♀		1.000	0.500	0.500	0	0	0.500	0.250	0.250	0	0
47♀			1.000	0.500	0	0	0.250	0.500	0.250	0	0
49♀				1.000	0	0	0.250	0.250	0.500	0	0
136♀					1.000	0	0	0	0	0.500	0
144♀						1.000	0	0	0	0	0.500
1♂							1.000	0.375	0.375	0.250	0.250
2♀								1.000	0.375	0.250	0.250
3♀									1.000	0.250	0.250
4♀										1.000	0.250
5♀											1.000

Note: The inbreeding of the offspring out of any one male and female in this chart would be one-half of the covariance component. For example, offspring from a mating between male 1 and female 2 would be one-half of 0.375, or 0.1875.

13.6 OTHER METHODS OF ESTIMATING HOMOZYGOSITY IN A POPULATION

Blood types in animals have received some attention from research workers as a possible means of estimating the degree of homozygosity within certain inbred lines. The *B* blood group system is of considerable interest in this respect, because a very large number of alleles is known to be present at this locus on the chromosomes [5] .If a high degree of inbreeding is involved, there should be a reduction in the number of alleles in this group found in an inbred line.

Blood antigen studies were made at the Colorado Experiment Station in the formative years of the development of inbred lines of beef cattle [6]. It was found that the incidence of certain antigens was high in some lines but almost totally absent in others. The rate of disappearance of the antigens from the inbred lines was faster than the predicted rate, based on the loss of heterozygosity estimated by means of Wright's inbreeding formula.

In a study with poultry [2], it was found that the number of alleles at the *B* locus was reduced to two when inbreeding was 66 percent or more. On the other hand, up to eight alleles were present in noninbred populations.

It is possible that inherited characteristics other than blood types may be used to estimate the degree of homozygosity of individuals within a line or strain, but little research has been done along this line. It has been suggested that nipple numbers in swine might give a rough estimate of the homozygosity within an inbred line, although this trait is apparently affected by several pairs of genes, some of which have an additive and some a non-additive action [1]. It seems very unlikely that environment would have much influence on nipple numbers in swine, and so the variations observed must be largely due to heredity. That individuals within some inbred lines do vary less in nipple numbers than those in other lines was shown by a coefficient of variation of 5.70 percent in an inbred Landrace line as compared to 8.22 percent in another line of inbred Polands. No study was made of changes in variation in nipple numbers as inbreeding increased.

REFERENCES

1. Allen, A. D., L. F. Tribble, and J. F. Lasley. "Inheritance of Nipple Number in Swine and the Relationship to Performance," MoAESRB 694, 1959.

2. Briles, W. E., C. P. Allen, and T. W. Millen, The *B* Blood Group System in Chickens; I; Heterozygosity in Closed Populations," G, 42: 631, 1957.

3. Emik, L. O., and C. E. Terrill. "Systematic Procedures for Calculating Inbreeding Coefficients," JH, 40: 51, 1949.

4. Lush, J. L. *Animal Breeding Plans* (3rd Ed.) (Ames: Iowa State College Press, 1945).

5. Neimann-Sorensen, A. *Blood Groups of Cattle*, A/S, (Carl Fr. Mortensen, Copenhagen, 1958).

6. Stonaker, H. H. "Breeding for Beef," ColoAESB 501-S, 1958.

7. Wright, S. "Coefficients of Inbreeding and Relationship," AN, 56: 330 1922.

QUESTIONS AND PROBLEMS

1. Explain what is meant by an inbreeding coefficient.

2. Explain what is meant by a relationship coefficient.

3. Of what use are inbreeding and relationship coefficients?

4. Compute the inbreeding coefficients for individuals X and Y in the following pedigrees.

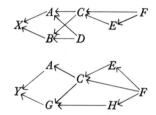

5. Compute the relationship coefficients for individuals X and Y in the above pedigrees.

6. What is the relationship between individuals Y and F in the second pedigree?

14

Linebreeding

Linebreeding is a system of mating in which the relationship of an individual, or individuals, is kept as close as possible to some ancestor in the pedigree. The ancestor is usually a male rather than a female, because a male generally produces many more descendants than a female, and this allows a greater opportunity to prove his merit by means of a progeny test.

14.1 ILLUSTRATION OF LINEBREEDING

Pedigree 1 in Fig. 14.1 illustrates systematic linebreeding. This is a hypothetical pedigree, with letters and numbers used in place of names to simplify the example. Only one ancestor in this pedigree is responsible for the relationship between the sire and the dam of individual X. This is ancestor 5. We have purposely placed him in this pedigree as the only great-grandfather of individual X. The arrow diagram shows that individual X traces by four separate lines to his ancestor 5. This is why it is called linebreeding. If linebreeding or inbreeding had not been practiced, there would have been only

Fig. 14.1 Pedigrees and arrow diagrams of the pedigrees which illustrate some different systems of linebreeding

No. of pedigree	Pedigree	Arrow diagram	Inbreeding coefficient	% inher. from 5*	Relationship coefficient†
1.	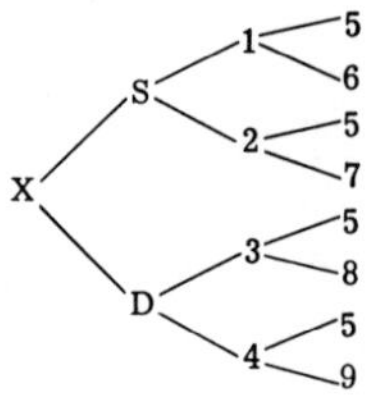	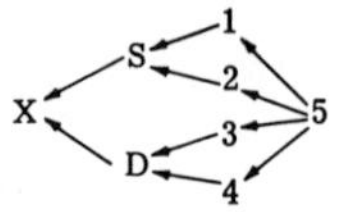	0.125	50	0.47
2.	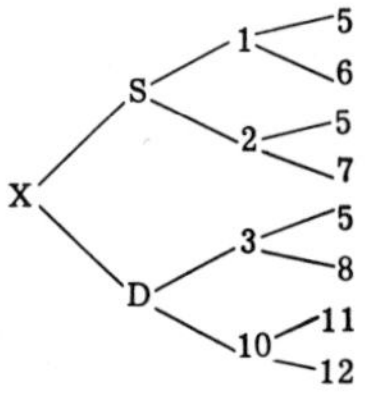	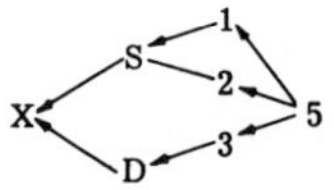	0.0625	37.5	0.36
3.	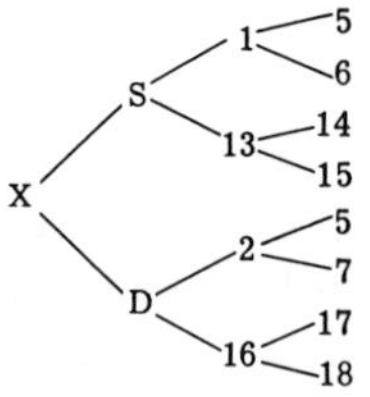	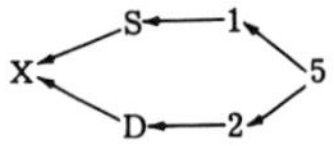	0.031	25.0	0.06
4.	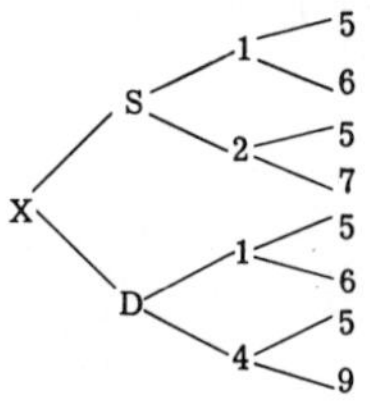	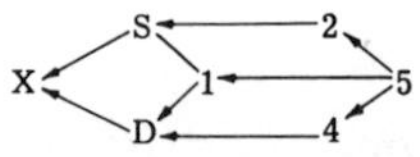	0.219	50.0	0.45
5.	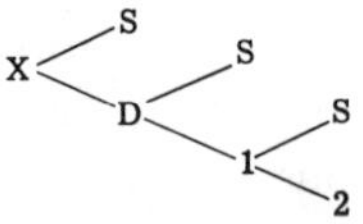	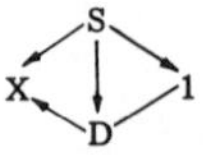	0.375	87.5	0.78

*Probable percentage of inheritance received from ancestor 5.

†Probable duplicate genes because of their relationship as compared to the base or noninbred population. This is between individual X and ancestor 5 in the first four pedigrees and to ancestor S in the fifth pedigree.

one line involved. In this pedigree, individual X probably received about 50 percent of his inheritance from ancestor 5. This is approximately as much as an individual usually receives from a parent. If linebreeding had not been practiced, individual X would have received only about 12.5 percent of his inheritance from individual 5.

An actual pedigree of Real Prince Domino 13th, a Hereford bull, together with an arrow diagram of his pedigree, is given in Figs. 14.2 and14.3. Since Real Prince Domino 13th traces by several paths to Prince Domino, we would say that he is linebred to Prince Domino. He also traces by two paths to Beau Aster, but apparently no great attempt has been made to keep the relationship high to this ancestor, whereas it is obvious that a deliberate attempt has been made to do so as far as Prince Domino is concerned.

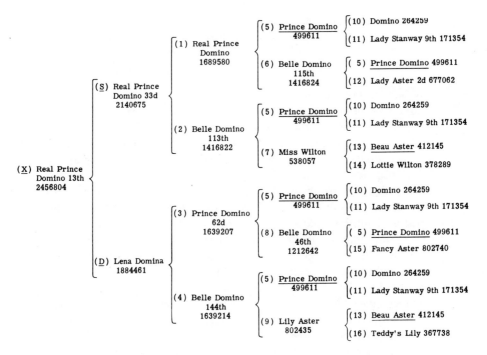

Fig. 14.2 Pedigree of Real Prince Domino 13th, showing linebreeding to Prince Domino.

Several different procedures for linebreeding may be followed, but they must be planned in advance. Some of these systems are illustrated in Fig. 14.1. In pedigree 1, half-sib matings have been made in the pedigrees of the sire (S) and the dam (D), with four individuals being used in that mating that are half sibs. In pedigree 2, half-sib matings were used to produce

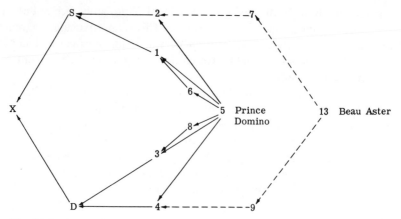

Fig. 14.3 Arrow diagram of the pedigree of Real Prince Domino 13th showing intense linebreeding to Prince Domino and very mild linebreeding to Beau Aster.

the sire but not the dam, but the sire of the dam (D) is a half brother of the parents of the sire of individual X. In pedigree 3, the sire and dam of X are related only because their sires were half brothers. In these examples, the relationship between X and 5 becomes less and less as the number of paths connecting them decreases. Individual X is not highly inbred in any of these three pedigrees. Even in pedigree 1, where it seems to be high, it is only 0.125.

Pedigree 4 in Fig. 14.1 illustrates linebreeding to ancestor 5 through a single son, individual 1. The relationship of X and its ancestor 5 in this pedigree is about the same as it was in pedigree 1. For this reason, there would be an advantage of only one son of ancestor 5's being required in this pedigree, whereas in pedigree 1, two sons were required. The disadvantage of the system illustrated in pedigree 4 is that the inbreeding coefficient of individual X would be about 0.22 ,or considerably higher than in pedigree 1. Since the inbreeding coefficient would be higher, the offspring might be less vigorous, and more recessive genetic defects might appear in the offspring in this pedigree than in pedigree 1.

Pedigree 5 in Fig. 14.1 is a system of linebreeding in which a sire is mated to two successive generations of his own daughters. Such a system would certainly result in a high relationship between individual X and ancestor S, but the inbreeding coefficient would be high, which might be undesirable. This system illustrates a situation in which linebreeding could be practiced while the ancestor to which linebreeding was directed was still alive, whereas the other systems would more likely be used after the ancestor was dead.

Linebreeding to a single ancestor cannot be carried on for many generations without recurrent linebreeding through a descendant, as illustrated in pedigree 4. The arrow diagram of the extended pedigree of Domino given in Fig. 14.4 shows the system used by Gudgell and Simpson, early American breeders of Hereford cattle, in which there was linebreeding to several different ancestors, with a blending of the inheritance of these to produce this particular individual.

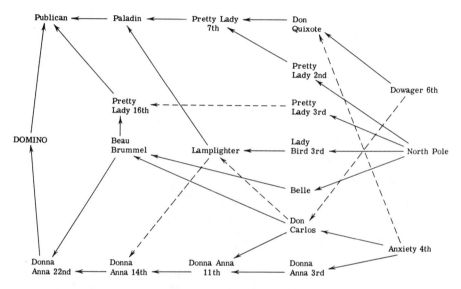

Fig. 14.4 Arrow diagram of the pedigree of Domino showing the system of linebreeding as used by Gudgell and Simpson.

14.2 COMPARISON BETWEEN ORDINARY INBREEDING AND LINEBREEDING

Inbreeding is a system of mating in which offspring are produced by parents that are more closely related than the mean of the population from which they come. This definition also applies to linebreeding, because linebreeding is a special form of inbreeding. However, in linebreeding, as defined earlier in this chapter, the relationship of an individual is kept close to a particular ancestor.

Inbreeding other than linebreeding is a system of mating in which related parents are mated with no particular attempt to increase the relationship of the offspring to any one particular ancestor in the pedigree. This concept is illustrated in Fig. 14.5. A comparison of this pedigree with pedigree 1 in Fig. 14.1 illustrates the basic difference between linebreeding and ordinary inbreeding. The arrow diagram of the pedigree in Fig. 14.5 shows that four different ancestors are responsible for the inbreeding of individual X. The inbreeding coefficient of X in Fig. 14.5 is also 0.125 as in pedigree 1 in Fig. 14.1, but the inheritance received by individual X from any one ancestor is only about 25 percent. Thus, in the mating system illustrated in Fig. 14.5 no attempt was made to concentrate the inheritance of any one ancestor.

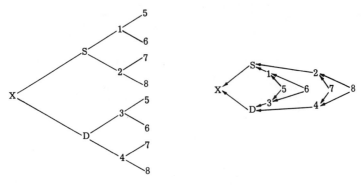

Fig. 14.5 This hypothetical pedigree and its arrow diagram illustrate a form of inbreeding that is not line-breeding because four different common ancestors (5, 6, 7, and 8) are responsible for the relationship between the sire (S) and the dam (D). Individual (X) has an inbreeding coefficient of 0.125, and the probable amount of inheritance received by X from any one of the common ancestors is 25 per cent.

The genetic effects of linebreeding are the same as those of inbreeding explained in Chapter 13. Thus, linebreeding tends to make pairs of genes carried in the heterozygous condition in the ancestor more homozygous in the linebred offspring. In addition, linebreeding increases the probability that the linebred offspring will possess the same genes as the ancestor to which linebreeding is directed. If the ancestor possessed many desirable genes, the offspring are more likely to possess these same desirable genes. If he possessed detrimental recessive genes in the heterozygous state, the offspring are more likely to possess them. Linebreeding will bring some of these recessive genes together in the homozygous state in some of the offspring. This serves to emphasize the importance of linebreeding to an ancestor that is superior genetically and carries a minimum of detrimental recessive genes. Accurate and adequate progeny tests help to identify such individuals.

When ordinary inbreeding is practiced, the relationship between the sire and dam often depends upon several common ancestors. It is evident that it is more difficult to distinguish three or four genetically superior ancestors than it is to distinguish one. A larger number of ancestors increases the probability of a wider variety of homozygous detrimental recessive genes appearing in the offspring when inbreeding is practiced.

14.3 WHY BREEDERS FAVOR LINEBREEDING OVER INBREEDING

Linebreeding is often looked upon with favor by breeders, but inbreeding is not, probably because linebreeding is usually not so intense, and when it is used, the inheritance of truly outstanding individuals is concentrated in the pedigree. Perhaps, also, the name of some great show animal in the pedigree has caused breeders to remember the superior individuals produced by linebreeding and to overlook others that were less superior.

Although when linebreeding is practiced, inbreeding is usually not intense, it can be if a linebreeding system such as that illustrated in pedigree 5 in Fig. 14.1 is used, where a male is mated to his own daughters. The first generation of such a mating would result in an inbreeding coefficient of 0.25, which is intense enough to cause a decline in the vigor of the offspring produced and an appearance of genetic defects. Two instances in which Prince Domino was mated to his own daughters are shown in the pedigree of Real Prince Domino 13th in Fig. 14.2.

As a general rule, a sire is not mated to his own daughters when linebreeding is practiced, but half-sib matings are made among the offspring of this particular sire. Perhaps this is done because the breeder wants to avoid intense inbreeding or because he has not yet recognized the outstanding merit of the ancestor in question. On the other hand, there may be no great need for linebreeding at that stage, since the offspring of a parent are related to that parent by about 50 percent.

Linebreeding is often used after the death of the ancestor or when it is no longer available for breeding purposes. Then the half-sib matings would be the closest possible matings in a linebreeding program. Full-sib matings would be possible in litter-bearing animals such as swine, but this would be no more effective, at least at the beginning of the linebreeding program, than half-sib matings in increasing the relationship to one particular ancestor in the pedigree. The disadvantage of full-sib matings would be that the inbreeding coefficient of the first generation would be 0.25, as compared to 0.125 in the half-sib matings, and there would be a possibility of the offspring getting detrimental recessive genes from at least two common ancestors instead of from one ancestor.

14.4 LINEBREEDING IN THE PAST

Linebreeding was used extensively in the development of the British breeds of cattle, such as the Angus, Herefords, and Shorthorns. This system was also used by the King Ranch in Texas in the development of the Santa Gertrudis breed of cattle. Mr. Gentry of Sedalia, Missouri, who was a successful breeder of Berkshire hogs in the early 1900's, made use of this system of breeding in producing many show-ring winners.

Linebreeding and inbreeding were used to a great extent by early breeders of Hereford cattle in England. Benjamin Tomkins and his son in the 1700's used such a system of mating to improve their cattle, as did William and John Hewer to obtain the desired color and type. Lord Berwick, another notable early breeder of Herefords in England, improved his herd by purchasing and using the bull, Sir David (349). He was probably the greatest show bull of his day and was from an accidental mating of a cow to her own sire.

Early breeders of Herefords in the United States also used linebreeding and inbreeding to improve their cattle. One of the most famous of the herds which followed this course was that founded by Gudgell and Simpson, at Independence, Missouri. They imported foundation cattle from England, and among the imports were the bulls, Anxiety 4th and North Pole. They tried outbreeding at first but apparently were disappointed in the results, for they later turned to linebreeding. They concentrated and blended the inheritance of Anxiety 4th and North Pole to develop one of the most outstanding herds in the United States. The pedigree of Domino in Fig. 14.4 shows the system of breeding used by Gudgell and Simpson. Actually, this pedigree shows that Domino was linebred to three different ancestors, Anxiety 4th, North Pole, and Dowager 6th, receiving a probable 20 to 24 percent of his inheritance from each of them. Even though it would appear that Domino was very highly inbred, the inbreeding coefficient calculated from this pedigree is only about 0.17.

14.5 WHEN TO LINEBREED

Linebreeding should be used only in a purebred population of a high degree of excellence. Its use is indicated when some truly outstanding individual has been identified and proved by an adequate progeny test. If such an individual is still living and available for breeding purposes, it seems more desirable to use him on females that are not related to him, since all of his offspring from such a mating will receive about 50 percent of their inheritance from him anyway. If he is mated to his own daughters, the relationship

Fig. 14.6 A linebred Charolais bull, LCR Sam of Perfecto 24th. His sire and dam were half-sibs, both being sired by Sam 951. (Courtesy of The Litton Charolais Ranch.)

coefficient between him and the offspring will be about 0.78. The inbreeding coefficient for the offspring will be 0.25 and high enough to increase the probability of a decline in vigor and the appearance of genetic defects.

Linebreeding is probably most useful when the outstanding individual is dead or not available for breeding purposes. One of the systems of line-breeding illustrated in Fig. 14.1 could be used with the main objective of concentrating the genes of that ancestor in the herd or flock. This requires a planned, systematic mating system. If practiced for a long period of time, linebreeding sooner or later must be directed through one or more of the ancestor's sons or other descendants, as illustrated in pedigree 4 in Fig. 14.1.

Linebreeding to a particular outstanding male may be practiced by a breeder who does not own the male or cannot purchase him. If he can pur-chase one or more high-quality sons of the admired sire, this would allow the practice of linebreeding to increase the relationship between the young pro-duced and that particular sire.

A word of caution about linebreeding is appropriate. It will be most successful when used by breeders who recognize its potentialities and its limitations and by those who have made a detailed study of the principles of breeding. Even in these instances, success will depend upon the breeder's ability to find and recognize individuals of outstanding merit. Nevertheless, we probably could make more use of this system than we have been making in the past few years.

QUESTIONS AND PROBLEMS

1. Define *linebreeding*.

2. Explain the main difference between inbreeding and linebreeding.

3. What are the main genetic effects of linebreeding?

4. Explain why breeders often look upon linebreeding with favor but avoid inbreeding.

5. Outline a system of linebreeding in which you would linebreed to a male you do not own or is now dead. Explain why you might use this system in preference to others that could be used.

6. What are some limitations for the use of linebreeding?

7. When should linebreeding be used?

8. In your opinion, was Prince Domino a carrier of the dwarf gene? Explain why you gave this answer.

15

Outcrossing
and Crossbreeding

Crossbreeding is the mating of animals of two or more different breeds. Outcrossing is the mating of unrelated animals within a breed. Although crossbreeding is more extreme than outcrossing, the genetic effects of both are similar.

15.1 GENOTYPIC AND PHENOTYPIC EFFECTS OF OUTCROSSING AND CROSSBREEDING

The genetic effects of outcrossing and crossbreeding are exactly the opposite of those of inbreeding. Whereas inbreeding tends to make more pairs of genes homozygous, outcrossing and crossbreeding tend to increase heterozygosity in the genes for which the parents possess different alleles. For example, if breed one is homozygous dominant (frequency of the dominant gene is 1.0) and breed two is homozygous recessive for a particular gene pair (frequency of the recessive gene is 1.0), all of the offspring from a cross of animals of these two breeds will be heterozygous. For breeds and lines that are homozygous for different alleles of a pair, the maximum amount of heterozygosity is attained in the F_1 generation. With the segregation of the genes in later generations, heterozygosity diminishes, as shown in the following example.

Generation	Genotypes of:		Percentage of heterozygous individuals
	Breed one	Breed two	
P_1	AA	aa	0
F_1	All Aa		100
F_2	1 AA⎫ 2 Aa⎬ 1 aa⎭		50

These statements also apply when more than one pair of genes affect a particular trait, so that varying degrees of heterozygosity are expressed. Thus, we speak of an animal being more, or less, heterozygous than another for a particular trait.

Outcross or crossbred animals are less likely to breed true than are inbred animals. This is due to heterozygosity and means that they are less likely to transmit the same genes to all of their offspring. This may be illustrated by the following example. A male of genotype *AABBCCDD* produces spermatozoa all of which carry genes *ABCD* to their offspring. A male of genotype *AaBbCcDd*, however, may produce 16 different kinds of spermatozoa with respect to gene combinations and will not breed as true as the individual homozygous for these four gene pairs.

Groups of outcross or crossbred animals in the F_1 generation are likely to be uniform in traits related to physical fitness, especially if the parents are homozygous in opposite ways for the two alleles of a particular pair.

15.2 HYBRID VIGOR OR HETEROSIS

Heterosis, or hybrid vigor, is the name given to the increased vigor of the offspring over that of the parents when unrelated individuals are mated. Hybrid vigor includes more than just hardiness. It includes greater viability, a faster growth rate, greater milk-producing ability and fertility. This phenomenon has been known for many years. The best-known example in animals is the mule, which is noted for its ability to withstand hot weather and hard work. It is the F_1 hybrid resulting from a cross between the male ass and the mare. The offspring of the reciprocal cross, called the hinny or jennet, are also hardy, but fewer of such crosses are made because of the scarcity of female asses and because the hybrids are usually smaller that the mule.

The production of hybrid seed corn by developing inbred lines and then crossing them is probably the most important attempt man has made to take advantage of hybrid vigor. Almost all of the corn grown in the corn belt today is grown from hybrid seed, because the yield is greater and the corn is of better quality. In recent years, swine producers have also used cross-breeding to induce hybrid vigor in the commercial production of hogs. This will be discussed in detail in a later chapter.

15.3 HOW TO ESTIMATE HETEROSIS FOR ECONOMIC TRAITS

Heterosis is expressed by some traits but not by others. Thus, it is desirable to determine how the degree of heterosis may be estimated for each trait in order to know, in a general way, what traits may be improved by out-breeding and crossbreeding. Heterosis cannot be estimated accurately for a single mating because nongenetic factors may cause a great deal of variation in a particular trait in a single mating. It can be estimated more accurately by comparing groups of crossbred and purebred animals. Animal breeders are not in complete agreement on how this should be done. Some feel that the best measure is the amount that the F_1 exceeds the higher parent. Others feel that heterosis is best measured by comparing the mean of the F_1 offspring with that of the purebred parents, by the following formula:

$$\text{Percent heterosis} = \frac{\text{mean of } F_1 \text{ offspring} - \text{mean of parent breeds}}{\text{mean of parent breeds}} \times 100$$

For example, suppose that the average litter size at weaning in swine is 7.0 for Breed A, 8.0 for Breed B, and 8.5 in the F_1 litters. The average litter size in the parent breeds would be 7.0 + 8.0/2 or 7.5 pigs. The amount of heterosis would be 8.5 minus 7.5 or 1 divided by 7.5 times 100 or 13.33 percent.

Estimating the amount of heterosis by comparing the mean of the F_1 offspring with that of the parents seems to be a reasonable method from the genetic standpoint. As was pointed out earlier, one of the characteristics of additive gene action is that the mean of the F_1 individuals coincides exactly with the mean of the parents if environmental variations are not considered. Thus, this kind of gene action is not responsible for heterosis. When non-additive gene action is important, on the other hand, the mean of the F_1 does not coincide with that of the parents but is either above or below it. In some instances, it may even be higher than the high or lower than the low parent.

15.4   GENETIC EXPLANATION OF HETEROSIS

Heterosis is caused by heterozygosity involving genes with nonadditive effects. Nonadditive gene action includes dominance, overdominance, and epistasis. The effects of these three types of action will be discussed separately.

15.4.1   Dominance

The decline in vigor due to inbreeding indicates very strongly that there are many recessive genes in farm animals that have deleterious effects on the vigor of the animal, varying from the lethal to the slightly detrimental. Since there are so many different pairs of genes influencing the expression of quantitative traits, some pure breeds and some inbred lines could be homozygous recessive for some pairs of genes. Many of our breeds in the United States could be homozygous for many genes, or at least the frequency of one allele may be much higher than that of the other, since they have descended in many cases from a relatively few imported animals. It has been reported that over one-half of the American Shorthorns trace to one bull in their pedigree [2]. For example, where several pairs of genes control one trait, one breed could be homozygous dominant for several pairs and homozygous recessive for another ($AABBCCdd$), while the second could be, respectively, homozygous recessive and homozygous dominant for those pairs ($aaBBCCDD$). When individuals of the two breeds were crossed, the F_1 ($AaBBCCDd$) would be superior to both parents in that particular trait, having at least one dominant gene in each pair.

Data presented in Table 15.1 illustrate further how dominance is a cause of heterosis. Note that in this example, the mean of the F_1 population exceeds the mean of the two parents. Note also that in the F_2 there is the beginning of a regression toward the mean of the P_1.

Since dominance is responsible for heterosis, it should be theoretically possible to capture this superiority in a single line by making individuals homozygous dominant for all pairs of genes. For instance, in the example in Table 15.1, a few of the individuals in the F_2 were $AABB$. If animals of this genotype were mated *inter se*, their offspring would all have this same genotype. However, these homozygous dominants would be difficult to distinguish, for they would resemble the heterozygotes in phenotype. In addition, it is very likely that there are many more than two pairs of genes involved, and these would further complicate efforts to establish homozygous dominance. Thus, even though it is theoretically possible to get a strain that is homozygous dominant for several genes, it is not practical or probable.

Table 15.1   Example* showing how dominance may be responsible for the expression of heterosis

Assumptions:
  That environment has no influence on the expression of genes.
  That animals of genotypes *AAbb, Aabb, aaBB,* and *aaBb* gain
  1.80 pounds per day, those of genotypes *AABB* and *AaBb* gain
  2.20 pounds per day, whereas those of genotype *aabb* gain
  1.60 pounds per day.

Generation	Genotypes	Phenotypes daily weight gain in pounds	Average daily weight gain in pounds
P_1	AAbb	1.80	1.80
	aaBB	1.80	
F_1	AaBb	2.20	2.20
F_2	1 AABB	2.20	2.01
	2 AABb	2.20	
	1 AAbb	1.80	
	2 AaBB	2.20	
	4 AaBb	2.20	
	2 Aabb	1.80	
	1 aaBB	1.80	
	2 aaBb	1.80	
	1 aabb	1.60	

*Example includes some epistasis as well as dominance.

15.4.2 *Overdominance*

This type of gene action is also responsible for heterosis. This is illustrated in Table 15.2, using two different pairs of genes that affect the same trait. In actual practice, several pairs of genes with overdominant action may affect the same trait, but effects of the different pairs may not be equal some having a greater effect than others.

With this kind of gene action, it would never be possible to fix heterosis-in a single pure strain, because the gene action is entirely dependent upon heterozygosity. The only way to take advantage of such kind of gene action would be to first form inbred lines and make them homozygous by inbreeding. Then the lines would have to be tested in crosses to find which lines combine best and induce the most heterosis in their offspring. Once this series of test crosses was made, the best-combining original parents would have to be crossed again to produce heterozygous individuals. (This is exactly what is done in producing hybrid seed corn.) The fact that this heterosis could not be fixed is obvious in Table 15.2. If F_1 heterozygotes mated *inter se*, already in the F_2, by segregation and recombination of genes, the means would regress, and only four out of every 16 individuals would retain the induced heterozygosity and heterosis. Further mating among the individuals of succeeding generations would only result in further disruption of the heterozygosity in the population, an over-all decline in heterosis, and a regression of the mean of the population to the mean of the original parents.

Table 15.2 Example showing how overdominance may be responsible for the expression of heterosis

Assumptions:

That environment has no influence on the expression of rate of gain.

That each pair of heterozygous genes has an equal effect on rate of gain.

That animals with both pairs of genes homozygous gain at the rate of 1.60 pounds per day; those with one pair homozygous and one pair heterozygous, at the rate of 1.90 pounds per day; and those with both pairs heterozygous, at the rate of 2.20 pounds per day.

Generation	Genotypes	Phenotypes daily weight gain in pounds	Mean daily weight gain in pounds
P_1	$A^1A^1B^1B^1$	1.60	1.60
	$A^2A^2B^2B^2$	1.60	
F_1	$A^1A^2B^1B^2$	2.20	2.20
F_2	1 $A^1A^1B^1B^1$	1.60	1.90
	2 $A^1A^1B^1B^2$	1.90	
	1 $A^1A^1B^2B^2$	1.60	
	2 $A^1A^2B^1B^1$	1.90	
	4 $A^1A^2B^1B^2$	2.20	
	2 $A^1A^2B^2B^2$	1.90	
	1 $A^2A^2B^1B^1$	1.60	
	2 $A^2A^2B^1B^2$	1.90	
	1 $A^2A^2B^2B^2$	1.60	

15.4.3 Epistasis

There are many different kinds of epistatic gene action, but their effects on the quantitative traits are difficult to measure accurately because of their complexity. The illustration of epistasis in Table 15.3 shows very clearly that this kind of gene action also can be responsible for heterosis or hybrid vigor.

In dominance and overdominance, the heterotic effect is due to the interaction of genes that are alleles, even though several pairs of alleles may affect the same trait. In epistasis, the interaction is between pairs of genes that are not alleles. It is theoretically possible to fix heterotic effects that are due to epistasis in a single pure line, as is the case with dominance, but this would also be extremely difficult and may not be probable or practical.

From the foregoing discussion, it becomes very apparent that it is difficult, if not impossible, to fix heterosis; that is, to attempt to maintain heterosis by mating those individuals having the highest degree of heterosis. The most practical procedure of making use of heterosis seems to be the formation of distinct lines or breeds and then crossing them to find those which give the greatest hybrid vigor. Such a procedure will work regardless of the kind of nonadditive gene action that is responsible for the heterosis.

Table 15.3 Example* showing how epistasis may be responsible for the expression of heterosis

Assumptions:
  That environment has no influence on the expression of rate of gain.
  That individuals of genotypes combining one or more A and B genes, gain at the rate of 2.00 pounds per day, whereas all other genotypes gain at the rate of 1.60 pounds per day.

Generation	Genotypes	Phenotypes daily weight gain in pounds	Average daily weight gain in pounds
P_1	AABB	2.00	1.80
	aabb	1.60	
F_1	AaBb	2.00	2.00
F_2	1 AABB	2.00	1.83
	2 AABb	2.00	
	1 AAbb	1.60	
	2 AaBB	2.00	
	4 AaBb	2.00	
	2 Aabb	1.60	
	1 aaBB	1.60	
	2 aaBb	1.60	
	1 aabb	1.60	

*Includes complete dominance as well as epistasis.

15.4.4 Additive Gene Action

This kind of gene action is not responsible for heterosis, since the average of the F_1 would coincide with the average of the two parents. This is illustrated in Table 15.4.

Table 15.4 Example showing how additive gene action is not responsible for the expression of heterosis

Assumptions:

That environment does not influence the expression of rate of gain.

That the residual genotype of *aabb* gains at the rate of 1.60 pounds per day, and the plus or contributing genes *A* or *B* each add 0.20 pound to the daily rate of gain.

Generation	Genotypes	Phenotypes daily weight gain in pounds	Average daily weight gain in pounds
P_1	AABB	2.40	2.00
	aabb	1.60	
F_1	AaBb	2.00	2.00
F_2	1 AABB	2.40	2.00
	2 AABb	2.20	
	1 AAbb	2.00	
	2 AaBB	2.20	
	4 AaBb	2.00	
	2 Aabb	1.80	
	1 aaBB	2.00	
	2 aaBb	1.80	
	1 aabb	1.60	

15.5   WHAT DETERMINES THE DEGREE OF HETEROSIS?

Not all traits in farm animals are affected to the same degree by heterosis. Those traits that are expressed early in life, such as survival and growth rate to weaning, seem to be affected the most, whereas feed-lot performance, as measured by rate and efficiency of gain after weaning, is only moderately affected. In general, heterosis seems to have very little influence on carcass quality in farm animals. It may be pointed out again that those traits that show the greatest degree of heterosis are the same ones that show the greatest adverse effects when inbreeding is practiced.

Traits that are highly heritable seem to be affected very little by heterosis, whereas those that are lowly heritable are affected to a greater degree. A good example of this phenomenon is fertility or litter size in swine. This trait is only about 15 to 17 percent heritable but is affected greatly by hybrid vigor.

Evidence also indicates that the degree of heterosis depends upon the degree of genetic diversity of the parents that are crossed. Thus, a higher degree of heterosis should be obtained when different breeds are crossed than when lines within the same breed are crossed. Furthermore, crossing breeds having greater differences in their genetic backgrounds should give more heterosis than crossing breeds having similar genetic backgrounds. This is reasonable from a theoretical standpoint. Unrelated parents are less likely than related parents to be homozygous for the same pairs of genes. This is because the original stocks of the unrelated parents differed in their gene complements; or, if the unrelated parents are descendants of the same original stock, segregation and recombination of genes have caused their relationship to become more remote. For example, if individuals of genotype *aaBBCCdd* were crossed, the F_1 would not be superior to the parents because of heterosis, regardless of the kind of nonadditive gene action involved. On the other hand, if individuals of that genotype were crossed with individuals of genotype *AAbbccDD*, heterosis would be expressed in the F_1 if nonadditive gene action were important. In summary, then, we can say that heterosis depends upon nonadditive gene action and upon one parent being homozygous for one allele in which the other parent is homozygous for the other.

15.6 PHYSIOLOGICAL BASIS OF HETEROSIS

Very few investigations have been made to determine the physiological basis of heterosis in farm animals. Gregory and Dickerson [1] made comparisons of inbred and topcross pigs (those from the mating of inbred boars to outbred sows) on the same level of feed intake. They found that topcross pigs required 10 percent less feed per unit of gain, made 10 percent faster gains, and possessed less fat and more muscle and bone at a market weight of approximately 200 pounds. These advantages occurred even though there was no significant difference in the ability of inbred and topcross pigs to digest their food. This led to the conclusion that the topcross pigs had a more efficient metabolic system, which could have been due to a lower maintenance requirement, or less energy loss during the growing-fattening period.

15.7 PRACTICAL USE OF OUTCROSSING

Outcrossing is the system of mating used most widely by purebreeders in the production of purebreds at the present time. It is practiced to offset the adverse effects accompanying inbreeding. Also, many breeders feel they can purchase from other breeders males of higher quality than they can produce in their own herds. One contributing factor to the widespread use of outbred sires has been the extreme importance that has been placed on show-ring winnings. The breeders feel that they must obtain a certain amount of advertising for their own herds by purchasing progeny of grand champion sires. This does increase the chances of bringing more desirable show-ring type into the herd, but the chief advantage is that a well-known name thereby appears on one side of the pedigree of the young stock they have for sale. This has resulted in more attention being paid to names than to merit in many instances.

15.8 OTHER SYSTEMS OF OUTCROSSING

Other systems of outcrossing are grading and topcrossing. In grading, pure-bred males of superior merit are mated to grade females or females of low quality. This system has been used considerably in the past, especially in the West, where purebred Hereford bulls were used on Longhorn and grade cows. It has also been used extensively where tested purebred dairy bulls have been used by artificial insemination in grade herds.

One of the features of grading is that the greatest improvement is usually made with the first cross, with less and less improvement made in later generations as the level of quality in the herd increases. Some of the improvement made in the first cross is due to heterosis; in later generations the level of heterosis tends to regress. The factor that could be responsible for continued improvement when grading is practiced is the introduction of desirable additive genes with plus effects into a herd that originally lacked them. Then, selection for higher quality in succeeding generations maintains the genes in the breeding stock. This should be especially true for traits that are highly heritable. Intensive selection must be practiced, however, if these traits are to be continuously improved.

Until recently, topcrossing was practiced in the production of purebred animals. Topcrossing refers to the last male in the top side of the pedigree and derives its name from this fact. An example is "Scotch-topped Short-horns," which means that the dam was a regular Shorthorn but the sire was of straight Scotch breeding. More recently topcrossing has been used to refer to the mating of inbred males with unrelated females. If the females belong to the same breed as the inbred male, this is, of course, a form of outcrossing. However, if the females to which the inbred male is mated belong to a different breed this is a form of crossbreeding. Usually, no attempt is made to distinguish between topcrossing that is outcrossing or crossbreeding.

15.9 PRACTICAL USES OF CROSSBREEDING

Crossbreeding is a system of mating often used by the producer of market animals. One reason for using it is to take advantage of heterosis that cannot be fixed within a line or breed.

Crossbreeding is also used to combine the desirable traits of two or more breeds in their crossbred offspring. For example, the Brahmans have been crossed with the British breeds in the tropics or semitropics to combine the heat and disease resistance of the Brahmans with the superior meat and performance qualities of the British breeds. In Canada, Herefords have been crossed with Highlander cattle in an attempt to introduce more winter hardiness into the F_1 crossbred calves.

In farm animals, a breed that is excellent in maternal traits but average or below in postweaning performance may be crossed with another breed that has only average maternal ability but is excellent in postweaning performance. For example, the majority of broiler chickens on the market today, and there are millions of them, are produced by crossing the white Cornish male, which is excellent in growth and meatiness, with white Plymouth Rock females, which are good layers and only fair in growth and meatiness. The reciprocal cross would be undesirable, because the Cornish hens would be poor in egg production and would increase the cost of chick production.

Crossbreeding has been used in recent years to establish a broad genetic base for the development of a new breed. The initial crossbreeding is then followed by inbreeding and selection for the characteristics desired in the new breed. An example where this system was used to develop a new breed is the Santa Gertrudis breed of cattle developed by the King Ranch in Texas from a Shorthorn × Brahman cross. Other examples are the development of new breeds of swine such as the Minnesota 1, 2 and 3, and the miniature pig. New breeds could be superior to older breeds if properly developed to combine the desirable genes of two or more older breeds, because no single old breed possesses all of the possible favorable or desirable genes.

Many breeders object to the use of crossbreeding for various reasons. One reputed disadvantage of crossbreeding is that the offspring lack uniformity of coat color. This objection is more likely to be valid where three or more breeds are used in a crossbreeding program. Some breeders have overcome this difficulty, however, by developing new breeds of the same color for crossing. Other producers do not consider variations in coat color a disadvantage if there is uniformity in other economic traits. A group of crossbred animals, especially F_1 crossbreds, may be as uniform, or more so, for some of the economic traits related to vigor and physical fitness, such as litter size at weaning, as purebred or inbred animals. The mating of crossbreds with crossbreds will eventually result in considerable variation because of the segregation and recombination of genes in the crossbred progeny. A lack of uniformity in unimportant traits should not be considered a fundamental disadvantage of the crossbreeding system of mating.

15.10   SYSTEMS OF CROSSBREEDING

The growing urgency of efficiency in livestock production has focused attention on crossbreeding in recent years. Crossbreeding systems, as they apply to each species of farm animals, will be discussed in later chapters dealing with each species. Certain kinds of breed crosses may be used that are applicable to all species, and these will now be discussed.

Crossbreeding has two important genetic advantages. As stated previously, it tends to average the phenotypic merit of all parent breeds used in the cross, and results in hybrid vigor for some traits. Since the livestock producer is largely interested in total over-all efficiency of production, he may be interested in more than hybrid vigor alone. What he desires is to combine the various breeds in such a way that the efficiency of total production is maximum. To do this he must use breeds for crossing that are superior in highly heritable traits affected mostly by additive gene action and those that give superior combining ability as shown by the expression of more hybrid vigor. Thus, the goal of the purebred producer should be to improve the highly heritable traits within the pure breeds by finding and mating the best to the best. Then, the improved breeds may be crossed to take advantage of hybrid vigor in traits that express it. Again, let us keep in mind that certain inbred lines may show greater combining ability when crossed than others, and since these inbred lines tend to be homozygous, the superior combining ability should be recurrent in subsequent crosses of these lines. Also, there seems to be a tendency for more hybrid vigor to be expressed, in traits showing hybrid vigor, when the breeds are more diverse genetically, or are farther apart in their relationship.

15.10.1   Two-Breed Crosses

This system of crossbreeding has been used in several species of animals for many years. Purebred animals of two different breeds are used for crossing and the purity of the parent breeds is not altered. Only the F_1 offspring are crossbreds and show hybrid vigor. Hybrid vigor is not expressed in the parents because they are not crossbred.

The two-breed cross requires considerable thought and planning to give optimum results. Let us suppose that we have two breeds that we wish to cross. One breed excels in fertility and mothering ability, whereas the other excels in postweaning growth rate and carcass quantity and quality. The question arises as to which breed should supply the females and which the males in the crossbreeding program. Obviously, the breed excelling in fertility and mothering ability should be used as the females, since the females govern, to a great extent, the number of young that are born and survive to weaning as well as the weight of the young at weaning. The sire, on the other hand, influences the young only through the genes he transmits to his offspring through the spermatozoa.

The use of the two-breed cross means that eventually the purebred females used for breeding must be replaced. This can be done by purchasing purebred females or by breeding the best females in the herd to a male of their own breed to produce purebred replacement females. The production of purebred replacement females within a herd for their use in crossbreeding has the advantage of the females, being acclimated to conditions on that particular farm or ranch, and they can be selected from the best-performing dams. A disadvantage is that when purebred animals are being produced within a herd, crossbred offspring cannot be produced from all matings.

15.10.2 The Backcross or Crisscross System of Crossbreeding

This system of crossbreeding logically follows the two-breed cross. Crossbred F_1 females are kept for breeding and are mated to nonrelated males from one or the other of the two original pure breeds used in the two-breed cross. For example, suppose breeds A and B are crossed. Crossbred females (AB) are retained for breeding and are mated to a male from breed A. Later, crossbred females from these matings are retained for breeding and are mated to a male of breed B. Thus, males from breeds A and B are used alternately on successive generations of crossbred females. This means that a heifer sired by a male from one breed will be mated to a male from the other breed.

As shown in Table 15.5, the first backcross progeny will possess about 75 percent of the genes of one breed and 25 percent of the other. Later cycles of crisscrossing or backcrossing will eventually carry about two-thirds of the genes of one breed and about one-third of the other, or *vice versa*.

An advantage of the two-breed backcross is that the dams as well as the offspring are crossbreds and will possess hybrid vigor for traits expressing

Table 15.5 The calculated percentage of inheritance from each of two breeds in successive generations of crossbred females and progeny

Number of each generation	Percentage of each breed in the dams	Breed of sire	Percentage of each breed in the progeny
1	breed 1	breed 2	50% breed 1
			50% breed 2
2	50% breed 1	breed 1	75% breed 1
	50% breed 2		25% breed 2
3	75% breed 1	breed 2	37.5% breed 1
	25% breed 2		62.5% breed 2
4	37.5% breed 1	breed 1	68.7% breed 1
	62.5% breed 2		31.3% breed 2
5	68.7% breed 1	breed 2	34.4% breed 1
	31.3% breed 2		65.6% breed 2
6	34.4% breed 1	breed 1	67.2% breed 1
	65.6% breed 2		32.8% breed 2
7	67.2% breed 1	breed 2	33.6% breed 1
	32.8% breed 2		66.4% breed 2
8	33.6% breed 1	breed 1	66.8% breed 1
	66.4% breed 2		33.2% breed 2
9	66.8% breed 1	breed 2	33.4% breed 1
	33.2% breed 2		66.6% breed 2
10	33.4% breed 1	breed 1	66.7% breed 1
	66.6% breed 2		33.3% breed 2

Note: After the third generation of crossing, the amount of inheritance each breed contributes to its progeny stabilizes, with about two-thirds coming from one breed and one-third from the other.

hybrid vigor. This should give more hybrid vigor than expressed in the two-breed cross, where only the calves show hybrid vigor. Repeated cycles of backcrossing or crisscrossing should result in some loss of hybrid vigor. As shown in Table 15.5, the optimum hybrid vigor should be expressed in the animals that possess 50 percent of the inheritance of each breed, but not all of the hybrid vigor is lost because of the approximate 67 to 33 percent ratio of the inheritance from the two breeds in later generations.

15.10.3 Three-Breed Crosses

Crosses consisting of three breeds may involve a three-breed rotation cross, where males from each of the three breeds are used in succession on crossbred females.

The three-breed rotational cross may also logically follow the two-breed cross. In this system of crossbreeding, F_1 females from the two-breed cross are mated to males of a third breed. Each new generation of females thereafter would be mated to males of one of the three breeds in rotation. Once the rotation of sires is completed, it can begin over again. For this reason it is called the three-breed sire rotation cross because purebred males from three different breeds are used in rotation on crossbred females. This is illustrated in Table 15.6.

Table 15.6 The calculated percentage of inheritance from each of three breeds in successive generations of the crossbred females and progeny when the three-breed rotational cross is used

Number of each generation	Percentage of each breed in the dams	Breed of sire	Percentage of each breed in the progeny
1	breed 1	breed 2	50% breed 1 50% breed 3
2	50% breed 1 50% breed 2	breed 3	25% breed 1 25% breed 2 50% breed 3
3	25% breed 1 25% breed 2 50% breed 3	breed 1	62.5% breed 1 12.5% breed 2 25.0% breed 3
4	62.5% breed 1 12.5% breed 2 25.0% breed 3	breed 2	31.2% breed 1 56.3% breed 2 12.5% breed 3
5	31.2% breed 1 56.3% breed 2 12.5% breed 3	breed 3	15.6% breed 1 28.1% breed 2 56.3% breed 3
6	15.6% breed 1 28.1% breed 2 56.3% breed 3	breed 1	57.8% breed 1 14.1% breed 2 28.1% breed 3
7	57.8% breed 1 14.1% breed 2 28.1% breed 3	breed 2	28.9% breed 1 57.1% breed 2 14.0% breed 3
8	28.9% breed 1 57.1% breed 2 14.0% breed 3	breed 3	14.5% breed 1 28.5% breed 2 57.0% breed 3

Note: Decimals are rounded off in some instances to make the total inheritance from the three breeds equal to 100 percent.

A maximum amount of hybrid vigor should be obtained in the three-breed cross because it takes advantage of hybrid vigor in both the offspring and their dams. Later generations in the three-breed rotation cross may lose some hybrid vigor as compared to the original three-breed cross. As shown in Table 15.6, after the fourth generation of crossing one breed never contributes more than 56 to 57 percent of the inheritance as compared to 50 percent from one breed in the first generation of the three-breed cross. This assumes, of course, that purebred sires are used that are not related to the crossbred females.

The commercial producer may use a four-breed rotational cross if he so desires. The results should be similar to those obtained with the three-breed rotational crossing system.

The commercial producer may also use a three-breed terminal sire (or static) crossing system. This means that females of a particular two-breed cross are always used as the females and are mated to males of a third breed. All progeny produced are marketed, and when replacement females are needed they are purchased. This has the advantage of maintaining maximum hybrid vigor found in the original three-breed cross. It also makes it possible to use certain superior F_1 females from the cross of two breeds that possess high fertility and mothering ability.

The three-breed static cross may also have the advantage that a sire line may be developed with special emphasis and selection toward superiority in postweaning traits such as rate and efficiency of gains and carcass quality. Males from such a line would probably be purebreds and could be used for the production of commercial crossbred calves only, unless they demonstrated genetic superiority and combining ability through the production of superior crossbred progeny. They could then be used for breeding in the pure sire line for the production of sires for future crossing purposes. Sires from such a line are the only ones used in the three-breed static crossing system. This is why such a sire line is often referred to as a *terminal sire line*. Females used in the three-breed static cross could be crossbreds (F_1) of two or more breeds selected for crossing because of their superior milk producing ability. Replacement females to use in the three-breed static cross could be purchased as needed from the producer who has specialized in breeding such females. The female line could be produced by the crossing of the same pure lines each generation, or they could be produced by the two-breed backcross or crisscross system discussed previously.

Certain disadvantages may be encountered when the commercial producer must purchase replacement females for his herd. The producer has the problem of finding such females when needed, and then these females may not be acclimated to the conditions on his farm or in his herd. It is not known how important the problem of acclimation might be. Also, the purchaser may have to be content with purchasing only average females rather than having the opportunity of purchasing superior females selected from superior parents.

15.10.4 Crossbred Males for Breeding

In the past, animal breeders have avoided using crossbred or non-purebred males for breeding purposes. More recently, however, an increased interest has been shown in their use, especially in the commercial production of swine and cattle.

Hybrid boars have been used by many commercial swine producers with considerable success. The term "hybrid" in swine is often a misnomer since most boars sold under this name are really from new breeds which have been developed (synthesized) from the combination of two or more older established breeds. The new synthetic breed is developed by crossing two or more older breeds, interbreeding the cross, and then selecting for superiority in performance. Only highly superior breeding stock from the standpoint of performance is used to establish these synthetic breeds.

The use of crossbred bulls has gained in popularity with the importation of a number of exotic breeds of cattle such as the Simmental and the Limousin into Canada and the United States. The use of crossbred bulls containing some exotic breeding is favored because few purebreds of each of these breeds are available for breeding purposes.

Crossbred males that are superior for highly heritable traits such as rate and efficiency of gain and carcass quantity and quality should have an important place in commercial animal production. They should also transmit a portion of this superiority for traits determined by additive gene action to their progeny almost as efficiently as purebred males. However, they should be more heterozygous than purebred males, on the average, and would not be so likely to breed true for heterotic traits determined by nonadditive gene action. If crossbred males are bred to females representing a breed different from their ancestry, considerable heterosis could still be realized in their progeny for those traits normally expressing it.

The most value of using crossbred breeding stock may be realized by using crossbred females. Hybrid vigor appears to be most important for fertility and certain maternal traits that affect the young from conception to weaning. It is probable, however, that crossbred males would be more fertile on the average than purebred males, and they should be more vigorous sexually, reaching puberty at an earlier age. This could be an important advantage in practical animal production. Future research will give a better evaluation of the advantages and disadvantages of crossbred males for commercial animal production.

REFERENCES

1. Gregory, K. E., and G. E. Dickerson. "Influence of Heterosis and Plane of Nutrition on Rate and Economy of Gains, Digestion and Carcass Composition of Pigs," MoAESRB, 493, 1952.

2. Wing, J. M., and C. J. Wilcox. "Inbreeding—Friend or Foe?" Contribution from the Purebred Dairy Cattle Association Research Committee, Box 150, Peterborough, New Hampshire, 1960.

QUESTIONS AND PROBLEMS

1. Define outcrossing and crossbreeding.

2. What are the genotypic effects of outcrossing and crossbreeding?

3. What are the phenotypic effects of outcrossing and crossbreeding?

4. How do the genotypic and phenotypic effects of outcrossing and crossbreeding differ from those of inbreeding?

5. Why does the proportion of heterozygosity decrease in the F_2 generation as compared to the F_1 generation?

6. In general, the farther apart the breeds are in relationship, the more hybrid vigor should be expressed for those traits that show hybrid vigor. Why?

7. Why does the purebred breeder prefer outcrossing to inbreeding and linebreeding?

8. Define heterosis and give examples where it has been used to advantage in livestock production and farming operations.

9. In swine, assume that the average litter size at weaning in Breed A is 7.0, whereas in breed B it is 8.0. When the two breeds are crossed, reciprocally, average litter size is 9.0. What is the percentage of heterosis in this example?

10. How much heterosis is expressed for lowly heritable traits, as a general rule? For highly heritable traits? Why?

11. What types of phenotypic expression of genes are responsible for the expression of hybrid vigor? What is the importance of this?

12. What traits in farm animals appear to be affected most by hybrid vigor?

13. What seems to determine the degree or amount of heterosis obtained through crossbreeding?

14. What procedure should be followed in using heterosis to the greatest possible extent in farm animals?

15. What appears to be the physiological cause, or causes, of heterosis?

16. What are the main uses of crossbreeding?

17. Define grading and topcrossing.

18. Discuss the advantages and disadvantages of the two-breed cross.

19. Discuss the advantages and disadvantages of the three-breed cross.

20. Discuss the advantages of the three-breed rotational cross, the three-breed static cross, and the four-breed rotational cross.

21. Which crossbreeding system (two-breed or three-breed cross) should give the most hybrid vigor? Why?

22. It has often stated that the three-breed rotational crossbreeding system will eventually result in poor performance after several generations. Is this true? Explain.

23. Why is it important to use superior performing breeds in a crossbreeding system?

24. In your opinion, what is the future of crossbreeding in the commercial production of livestock?

16

Summary
of Animal Breeding
Principles

In the preceding chapters we discussed in detail the principles involved in animal breeding. Before showing how they may be applied to the improvement of the various species of farm animals we shall summarize and restate these principles.

16.1 SAMPLING NATURE OF INHERITANCE

The inheritance of individuals is transmitted from parents to offspring by means of genes located on the various chromosomes. There are hundreds and probably thousands of genes, and they occur in pairs except those on the nonhomologous parts of the X and the Y chromosomes. Of each pair of genes, the individual receives one from the mother and one from the father; and to each of his offspring the individual will give one member of the pair. Thus, for every pair of genes, each offspring receives a sample one-half of its inheritance from each parent. Since there are many pairs of genes, and some are in the homozygous and some in the heterozygous state in the individual, the samples may be quite different for any two.

16.2 ROLE OF THE ANIMAL BREEDER

Differences in genes, or variations in individuals due to differences in genes, are the raw material with which the animal breeder must work. The animal breeder cannot create new genes, but must work with the genetic variation that already exists in his animals.

Genetic variation is the result of mutations of genes which have occurred during the production of many, many new generations of animals. These mutations, along with selection, have made some animals more suitable for a particular purpose or a particular environment.

The animal breeder today is more interested than ever in producing animals that are highly efficient for a particular purpose. The dairy-cattle breeder is interested in developing a herd that gives the largest amount of milk per animal for the smallest amount of feed consumed. The beef producer is interested in the efficiency of beef production, but recently he has been giving increased attention to carcass quality. In other words, the problem now is not to develop new and better breeds, but to improve existing ones or to combine them in a way as to take advantage of heterosis. Of course, some new breeds have been developed from crossbred foundations, and new ones are being introduced from other countries, but their superiority to the more popular older breeds has still to be proved. There seems to be some potential for the development of new and superior breeds from a crossbred foundation if the two or more parent breeds possess different desirable genes.

The role of the animal breeder is to identify those animals that possess desirable genes or combinations of genes and to concentrate in his herd as many of these genes as possible. In attempting to find superior animals, he is often confused by environmental effects and by the different modes of gene action and interaction. He should compare prospective breeding animals in as like environments as possible, preferably in one similar to that in which their offspring would be raised, and he should compare animals near the age at which he would market the offspring. He must choose superior individuals on the basis of information in pedigrees, the individuality of the prospective breeding animals, and on information on collateral relatives and progeny when their records are available. His breeding program will be effective if the traits to be selected for have been measured accurately and if they are highly heritable, indicating that additive gene action is the cause of most of the genetic variation. If nonadditive gene action is the most important genetic influence on the traits he plans to select for, his breeding program will have to include the development of inbred lines first. Then these will have to be tested in crosses to identify those with the superior combining ability in order to take advantage of nicking effects or heterosis.

16.3 INFORMATION NEEDED TO FORMULATE EFFECTIVE MATING AND SELECTION SYSTEMS

When developing a breeding program the breeder must first decide what traits are the most important to select for from the economic standpoint. His decision will depend upon the species of farm animals with which he is working, the feeding program he intends to use, the kind of product he intends to market, and the sale price of that product. In most cases, the breeder will limit the number of traits selected for and will include these traits in an index. The amount of weight he gives each trait in the index will depend upon the heritability of that trait, its relative economic value, and the genetic correlation of that trait with others of economic importance. Example index formulas will be given for each class or species of farm animals in later chapters.

Next he must acquaint himself with the methods that have been devised to make accurate measurements and records; thereby, the breeder can distinguish more satisfactorily between genetic and environmental effects both in prospective breeding animals and in generations during the breeding program. Actual measurements of such traits as weight, milk production, or percentage of lean cuts should be made and not estimated. The use of correction factors to adjust records of all animals in the herd to a comparable age, to the same-age-of-dam basis, for sex, and other variables whenever applicable, enables the breeder to make comparisons more accurately. His accuracy in choosing genetically superior animals for breeding purposes and in evaluating the progress of his breeding program will be increased if he keeps detailed written records.

Another fact the breeder needs to know in planning his program is which kind of gene action, additive or nonadditive, has the greater influence on each of the important economic traits. Additive gene action is indicated when the heritability of the trait is high, as measured by the resemblance between parents and their offspring, and when the crossing of breeds results in an average of the F_1 that closely approximates the average of the parents. Additive gene action is also indicated, but not yet proven, when sex differences for a trait are large. When additive gene action has more influence, mass selection (mating the best to the best) will be effective. Nonadditive gene action is indicated when the heritability of a trait is low, when inbreeding has had detrimental effects, and when outbreeding or crossbreeding has had beneficial effects. Nonadditive gene action is also indicated when the average of the F_1 individuals differs from the average of the two parental groups (heterosis). When nonadditive gene action has more influence on a trait, the greatest improvement in performance will come from the crossing of strains or lines known to have good nicking or combining ability. The breeder may want to breed and select for several different traits of which some are affected by additive and some by nonadditive gene action. The recommended procedure here would be to form pure lines or breeds by selection for improvement in those traits that are highly heritable; then, to cross these lines or breeds to improve those traits that show heterosis.

The breeder should also know whether genetic correlations are important among the different traits selected for and whether the correlations are positive or negative. Some information is available for the important traits, and these will be discussed in the chapters dealing with each of the species of farm animals.

The breeder can make more effective plans if he also can determine whether or not genetic-environmental interactions influence the traits he wishes to select for. As discussed earlier, today's breeder has little information about this factor, but further research should provide more.

16.4 WHEN TO USE INBREEDING AND LINEBREEDING

Inbreeding and linebreeding are practiced to produce seed stock. Intensive inbreeding is done as a general rule, with the intention of using inbred animals for crossing purposes.

If this is not the breeder's main objective, intensive inbreeding might not be desirable, for the main phenotypic effect of inbreeding and linebreeding is a decline in the performance traits that are affected greatly by nonadditive gene action. Much of this decline must be due to the fact that detrimental recessive genes are revealed by increased homozygosity. In general, the traits affected most by inbreeding are those associated with physical fitness. The decline in performance in these traits seems to be greater in some inbred lines than in others, but in general, the decline occurs in spite of selection against it.

The producer of seed stock, himself, must be prepared to accept a certain amount of decline in performance of his stock and must expect the appearance of some inherited defects. He must decide whether or not the expected increase in prepotency or uniformity of genetic composition will overbalance the decline in performance.

If inbreeding is practiced for some other purpose than the production of seed stock to be used in crosses, such as the production of purebreds for show-ring purposes, the degree of inbreeding should be held to a minimum and should be increased slowly, with intensive culling and selection of breeding animals that reproduce that line. These methods of breeding should not be used by the commercial livestock producer, who is an animal multiplier and not a producer of seed stock. His purpose is to produce the greatest amount of salable product per breeding animal, and a lowered productivity of the breeding animals will not accomplish this.

Inbreeding and linebreeding should not be used in herds made up of average or mediocre breeding stock, for several undesirable recessive genes may be present and frequently will be brought together and appear phenotypically; this may result in the discarding of the whole inbred line sooner or later.

The building of superior inbred lines of livestock is a slow, time-consuming, and methodical process and probably should be undertaken only by the breeder who has the knowledge, the time, and the necessary capital to continue the process to its completion.

16.5 WHEN TO USE OUTCROSSING AND CROSSBREEDING

Outcrossing is the form of mating most often used by present-day purebred breeders in the production of purebreds. It will probably continue to be widely used, because by mating females in his herd or flock to nonrelated males the breeder avoids the effects of inbreeding. In outcrossing, the breeder must attempt to purchase the best genetic material he can find at a price he can afford to pay, and he must always try to find a male that is superior to the females in his herd. Recently, breeders have been selecting males on the basis of both type and performance rather than on type alone. The use of superior males proved by performance tests should improve the over-all performance of the herd and produce superior seed stock to sell to the commercial livestock producers.

The purebred breeder should purchase older males that have been progeny-tested and proved of superior genetic worth. Often the older male of outstanding breeding ability is sold because he has so many daughters in the herd that the owner would be forced to use inbreeding if he were retained in the herd. Superior genotypes of this kind should be used in some-one's herd as long as the animal remains fertile, in order to improve the overall production of a herd and possibly a breed.

Crossbreeding is the mating system that should be used by the livestock multiplier or the commercial livestock producer. The class of livestock may be important, however; crossbreeding is useful in swine, but may not be in dairy cattle. Crossbreeding effects are the opposite of inbreeding and linebreeding effects because traits associated with physical fitness are improved, but the breeding worth of the individuals may be lessened to a certain degree because they are het erozygous and do not breed as uniformly true as inbred animals.

Some livestock producers become so enthusiastic about crossbreeding that they may forget that some traits in farm animals do not show heterosis. Thus, it is important for the breeder to be familiar with which traits show heterosis and which ones do not. These will be discussed in detail for each species of farm animals in later chapters.

In conclusion, it is well to emphasize that, for the greatest improvement in livestock breeding, the breeder must not only be familiar with the raw material and the tools available to mold this raw material, but he must also know how to use both to the greatest advantage.

17

Systems of Breeding and Selection in Swine

Swine are known as "mortgage lifters," and they are reared in large numbers on farms in the Corn Belt region. The past few years have seen a definite change from the lard-type to the meat-type hog because of the surplus of lard on the market caused by the substitution of vegetable oils for lard for cooking purposes and of detergents for soaps made from lard. Another factor has been the consumer demand for leaner pork. This demand is also a result of growing calorie-consciousness; slim and trim figures have become more desirable and seem to be conducive to longer life and better health.

Swine breeders no longer select breeding animals on type alone. Results of swine breeding research have impressed on breeders the importance of selection for performance as well as for type. The fact that breeders are now more interested in systematic selection and breeding practices is demonstrated by the number of swine-evaluation stations that have been established in the United States during the past few years.

Swine-breeding research has yielded many practical results that are now being used by swine breeders. Much more research work still needs to be done, however, especially in the determination of genetic correlations between traits, in possible genetic-environmental interactions, and in improved methods of selecting and breeding for superior performance. This chapter will be devoted to a discussion of selection and breeding methods now recommended on the basis of research in swine breeding.

17.1 TRAITS OF ECONOMIC IMPORTANCE AND HOW TO MEASURE THEM

Only those traits that are the most important from the economic standpoint should be considered in a breeding program. Selection for many different traits, of which some are of minor importance, such as color patterns, will result in less selection pressure being applied for those traits that are most important. For this reason, the discussion here will be limited to only the most important traits.

17.1.1 Total Litter Weight at Weaning

Litter weight at weaning is a measurement of net merit for preweaning performance in swine. It gives a measurement of the fertility of the sow, because the heavier litters at weaning are usually the larger litters. It also is an indication of the milking and mothering ability of the sow and the vigor and growth rate of the pigs. Litter size and weight at weaning are determined by the number of pigs born per litter and the ability of these pigs to survive to weaning. To wean a litter of 10 pigs, sows must farrow at least 10, and in most cases one or two more. On the other hand, large litters at birth are not advantageous if the sow is a poor mother and crushes

Fig. 17.1 Grand Champion barrow at the International Livestock Show in 1919. Consumers now demand more lean and less fat in pork products. (Courtesy of the University of Missouri.)

Fig. 17.2 Large litters at weaning are necessary for a profit in pork production. This is a litter in which eighteen pigs were born and sixteen weaned with a total 56-day litter weight of 687 pounds. (Courtesy American Landrace Association.)

many of her pigs during their first few days of life, or if the pigs are born so weak that they fail to survive to weaning. Much of this trouble can be prevented by proper feeding and management of the sows and litters, but gene actions of various types are known to affect these traits.

To measure litter size and weight at weaning accurately, it is necessary to identify each litter, and preferably each pig, at birth, by using an ear-marking system or by tattooing a number in the pig's ear. Many ear-marking systems have been proposed by breed associations and by experiment stations, all of which are satisfactory. The ear-notching system shown in Fig. 17.3 may be used to identify several thousand individual pigs by litter number and individual number.

In the past, most pigs have been weaned at 56 days, but since the advent of sow milk replacers and higher quality creep rations, many pigs are being weaned at from three to five weeks of age. The age at weaning will depend upon the wishes of the swine breeder, his facilities, and the farrowing system he is following. For selection purposes within a herd, however, all pigs should be weaned at the same age, so that litters may be compared on the same age basis. Often it is not possible to weigh all litters as they reach an exact age. Correction factors have been calculated to adjust litter weights to a 56-day basis. The one given in Table 17.1 may be used for this purpose. To use these

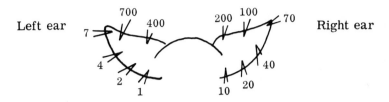

Position value of ear notches

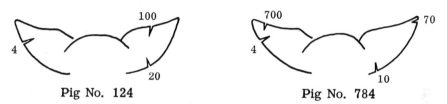

Pig No. 124   Pig No. 784

Fig. 17.3 This system of ear marking makes it possible
to give each pig a specific number. Thousands of pigs can
be marked by this system without duplication. (From
Missouri Agricultural Experiment Station Research
Bulletin 587.)

correction factors, let us suppose that pigs were weighed at 61 days of age
instead of 56. The weight is adjusted by multiplying the weight of the litter
at 61 days by the correction factor (0.8779) given in Table 17.1.

17.1.2 Weight at 154 Days of Age

The weight of each pig at 154 days of age is another important economic
trait. A weight taken at this time is actually a total measurement of all gains
made from birth. If 56-day weights for each pig are available, the rate of
gain from weaning to 154 days of age can be calculated by dividing the weight
gained by 98. Since it is not always possible to weigh each pig or each litter
at exactly 154 days of age, correction factors have been devised to adjust
the weights to this standard age basis [32]. The following has been suggested
for this purpose:

$$\text{Adjusted weight} = \frac{\text{actual weight} + 154}{\text{actual age} + 45}(199) - 154$$

Other breeders prefer to record the average daily gain from shortly after
weaning to a market weight of 200 pounds. As a general rule, no correction

Table 17.1 Correction factors for adjusting litter weights of pigs to a 56-day basis

Age when weighed (days)	Multiply total litter weight by this factor to adjust to 56 days
50	1.1801
51	1.1471
52	1.1154
53	1.0849
54	1.0555
55	1.0272
56	1.0000
57	0.9738
58	0.9485
59	0.9241
60	0.9006
61	0.8779
62	0.8560
63	0.8359

is made for the age of the pig, but the numbers of pounds gained divided by the total days on feed gives an average daily gain figure.

Postweaning gains are easy to calculate for each pig, whether a standard 154-day weight or the average daily gain from weaning to market weight is used. For accuracy of records, it is to be emphasized that pigs fed in groups or litters must bear individual identification numbers.

17.1.3 Economy of Gains from Weaning to Market Weight

The economy of gains is also of considerable importance in pork production. The amount of feed required per pound of gain by individuals is difficult to calculate, however, because it requires individual feeding, which is expensive and impractical. Individual feeding is usually practiced only for boars, because a boar is the most important single individual in the herd; he supplies one-half of the inheritance for many litters during a season, whereas each sow supplies the inheritance for only one. In addition, fewer boars are required for breeding purposes, and only the top herd-sire prospects need to be individually fed.

Estimates of the feed efficiency for a pig may be obtained by feeding an entire litter together in one pen or by feeding two barrows (or boars) and two females together from weaning to market weight. This gives a figure for each litter and thereby an estimate for each member of the litter. Feeding pigs in this manner is less expensive than individual feeding, but still is not practical enough to be used by many breeders. It is being used to a considerable extent by swine-evaluation stations now in operation in the United States.

All pigs within a herd should be compared on as nearly the same basis as possible for efficiency of gains by starting them on feed at about the same weight and taking them off the gain test at the same final weight. This is necessary because a pig requires more feed per pound of gain between 100 and 200 pounds than between 50 and 100 pounds. For this reason, the beginning and ending weight for all of the pigs is held as constant as possible in measuring the economy of gains.

Table 17.2 The variation in performance of Duroc boars fed in individual pens from weaning to 200 pounds live weight

No. of boar	Avg. daily gain (lb)	Feed consumed per day (lb)	Lb of feed per pound gain
104	2.03	7.51	3.70
105	1.96	7.54	3.85
148	2.14	7.94	3.70
157	2.05	7.51	3.67
207	1.89	7.49	3.97
209	2.03	7.67	3.78
209–1	1.76	6.96	3.96
236	1.81	7.06	3.89
238	1.94	6.99	3.61
276	1.86	6.67	3.59
355	2.17	7.48	3.43
500	1.66	6.69	4.03

Pigs to be compared for efficiency of gains must all be fed the same ration. Less pounds of feed per unit of gain will be required with concentrated, high-energy rations than with bulky, less-concentrated rations. Different herds do not all get the same kind of rations, and for this reason, it is difficult to compare their economy-of-gain figures where management and feeding practices may vary widely also, making comparisons difficult. Another consideration is that, unless self-feeders are adjusted very carefully, feed wastage may be considerable, and the resultant economy-of-gain figures may be inaccurate.

17.1.4 Type and Conformation Score at Market Weight

Different views may be expressed on the value of recording type and conformation scores of pigs at market weight. Until the last few years, most of the emphasis in swine production and selection was on show-ring type. More recently, however, type and conformation have been emphasized less, and preference has been given to the percentage of lean cuts whenever possible. Nevertheless, there is a certain amount of correlation between meatiness and scores for meat type. Thus, this trait should be given consideration by purebred breeders.

Many complicated score cards have been used in the past, but probably they are of no more value than a simple scoring system in which the animals are scored A, B, C, D, and E, with A being the most desirable and E the least desirable. Scores of this kind are more valuable when a committee of three or more judges do the scoring and an average is recorded for each pig.

17.1.5 Carcass Desirability

More and more attention is now being paid by breeders to carcass desirability in swine, mainly because of the demand for more lean and less fat in pork products. Formerly, carcass desirability could be determined only after slaughter, but in recent years attempts have been made to measure carcass quality in the live animal before slaughter. One of the most useful methods developed is the measurement of backfat thickness in the live animal by using a scalpel to cut the skin and then thrusting a ruler (or a ruled probe) into the fat until it is stopped by contact with the muscle fibers of the body [19]. Measurement of backfat thickness in the live hog by this method is as accurate in most instances as measurement of backfat taken in the carcass, and has been a valuable criterion in selecting breeding stock.

Other methods of estimating the amount of lean and fat in the live animal are determination of the level of creatinine in the blood and urine, blood lipids, blood volume, and changes in the leucocyte numbers at the end of the fattening period. None of these has proved to be accurate enough for selection purposes, but continued studies may develop methods of greater value. Machines utilizing the principle of high-frequency sound, which give a different reading in contact with fat than in contact with lean, are being used under practical conditions.

Carcass data on littermates of breeding animals have been used for selection purposes. This method of selection has been used in Denmark for many years in the development of bacon qualities in the Landrace breed. It is now being used in many other countries, and in the United States, many of the breed associations have initiated the Meat Hog Certification programs in which a barrow and a gilt from a litter are used to obtain carcass data to certify the litter. Certification of the litters is dependent upon the length of body, backfat thickness, and area of the loin-eye muscle at certain weights.

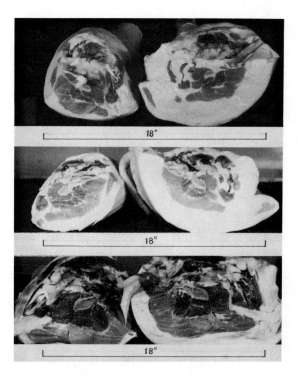

Fig. 17.4 Definite differences in meatiness in swine
show up after slaughter. These were from the same herd
in the same year. (Courtesy of the University of
Missouri.)

Litters that qualify must also meet certain age standards at slaughter and must be from a production-registry litter.

Carcass data obtained in swine-evaluation stations include body-length measurements, backfat thickness, area of the loin-eye muscle and the percentage of lean cuts. These criteria are used to identify those strains or individuals that possess inherent meat type and other desirable performance qualities so they may be used to produce their kind in an attempt to raise the genetic worth of the entire population.

17.2 HERITABILITY OF ECONOMIC TRAITS IN SWINE

Many heritability estimates have been calculated for the different traits, including conformation, performance characters, and carcass quality. A summary of these heritability estimates is presented in Table 17.3.

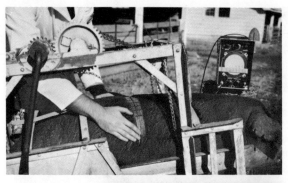

Fig. 17.5   Loin eye area in the live hog may be estimated through the use of a Sonoray machine. (Courtesy of the University of Missouri.)

Litter size and weight at weaning are lowly heritable, with an average estimate from many studies of approximately 20 percent. The weight of pigs at 154 days is approximately 30 percent heritable, whereas growth rate from weaning to 180 to 200 pounds is 30 percent heritable. Efficiency of gain is about 30 percent heritable and is high enough to indicate that this trait is affected by additive gene action and that some progress could be made in selection for its improvement. Undoubtedly, if more accurate methods of measuring economy of gain could be developed in which feed wastage was less, the heritability of this trait would be higher.

Carcass traits and conformation items are highly heritable and they should be improved through selection.

17.3   SELECTION INDEXES FOR SWINE

Breeders are seldom interested in selection for the improvement of a single trait. Usually they are interested in improving several traits of the greatest

Table 17.3   Heritability estimates for certain traits in swine*

	Approx. average
Items of conformation:	
Length of body	59
Length of legs	65
Number of vertebrae	74
Number of nipples	59
Type	38
Performance characters:	
Number of pigs farrowed	15
Number of pigs weaned	12
Weight of litter at weaning	17
Weight of pig at approximately 5–6 months	30
Growth rate (weaning to 180–200 lb.)	29
Economy of gain	31
Items of carcasses:	
Length	59
Loin-eye area	48
Thickness of backfat	49
Thickness of belly	52
Percent of ham (based on carcass wt.)	58
Percent of shoulder (based on carcass wt.)	47
Percent of fat cuts (based on carcass wt.)	63
Percent of lean cuts (based on carcass wt.)	31
Carcass score	46

*Mostly after Craft [9].

economic importance. A selection index may be used for this purpose. The following index [14] may be used when selecting for sow performance:

$$\text{Index} = 2\left(N_b + 2N_w + \frac{2T_w}{30}\right)$$

where

N_b = number of pigs born.
N_w = number of pigs weaned.
T_w = total litter weight at weaning.

For example, if a sow farrowed 10 pigs and weaned eight with a total litter weight of 320 pounds at weaning, her index would be

$$2\left(10 + 2 \times 8 + \frac{640}{30}\right), \quad \text{or} \quad 94.6$$

Another index for selecting pigs on the farm [4] is

$$\text{Index} = -0.5X_1 + 7X_2 - 0.02X_3 + 0.5X_4$$

where

X_1 = number of pigs farrowed in the litter.
X_2 = number of pigs per litter at 154 days.
X_3 = litter weight at 154 days.
X_4 = individual pig weight at 154 days.

Swine-testing stations all over the United States are using indexes for comparing pigs on feeding tests. The index used varies with different stations. The following is an example of one that could be used:

$$Index = 500 + 30G - 100F - E$$

where

G is the average daily gain in pounds.
F is backfat thickness in inches.
E is feed efficiency in pounds of feed required
to produce 100 pounds of gain.

The index for a pig that gained 2.00 pounds per day on 280 pounds of feed per 100 pounds of gain and had 1.20 inches of backfat at 200 pounds would be $500 + 30(2.00) - 100(1.20) - 280$, or 160.

The index for each individual in a herd should be calculated and those animals with the highest index kept for breeding purposes. If type or conformation were given consideration, those individuals of the poorest type could be culled and the remainder compared on the basis of the index.

17.4   SELECTION EXPERIMENTS WITH SWINE

The amount of progress expected from selection for the various economic traits may be calculated from the selection differentials, heritability estimates for these traits, and the generation interval. The real proof of progress in selection for these traits, however, depends upon results obtained in actual selection experiments.

A cooperative study was made of the amount and effectiveness of selection practiced during the development of inbred lines of swine at several cooperating experiment stations in the Regional Swine Breeding Laboratory [14]. Data used to measure the amount of selection were obtained on 4521 litters from 38 lines during the period from 1932 to 1948. After adjusting for the expected detrimental effects of increased inbreeding, the average results indicated that selection practiced in the development of mildly inbred lines failed to improve measurably the genetic merit of the lines. One conclusion was that the effectiveness of artificial selection might have been reduced because most of the selection practiced for litter size actually was automatic.

That is, this much selection would have occurred if a random sample of the pigs weaned had been chosen for breeding purposes.

A selection experiment was conducted at the Illinois Agricultural Experiment Station for rate of gain [24, 1]. The experiment involved selection for fast rate of gain in one line of Hampshire pigs and for slow rate of gain in another. At the end of eight to nine generations, there was a difference of 61.8 pounds in the 180-day weights of pigs in the two lines. The results indicated that selection for rate of gain was effective.

The Washington Agricultural Experiment Station conducted an experiment to determine if selection for growth rate is as effective when swine are fed on a low plane of nutrition as when they are full-fed [16]. A crossbred foundation stock of Danish Landrace × Chester White pigs was used, and two lines were formed. One line was full-fed from weaning to 150 pounds, then limited-fed from 150 pounds to parturition, and then full-fed again during lactation. The other line received 70 percent as much food as the full-fed group for each of these periods. Selections were made on the basis of an index including litter size at farrowing and weaning and average daily gain from weaning to 150 lb. Selection for rate of gain was effective in both lines, the progress made being very close to that expected from the selection differentials and the heritability estimates. When pigs from the two lines were exchanged after the sixth generation of selection, the pigs from the limited-fed line made faster gains than those selected for fast gains on a full feed, whereas when both lines were fed on a limited ration, the pigs selected for rapid gains on the limited ration made the faster gains. It was concluded that the results of this experiment supported the contention that breeding animals should be produced and selected in the same environment in which their progeny will be produced.

A selection experiment was conducted at the Alabama Agricultural Experiment Station [13] for efficiency of gains in Duroc swine. Two different lines were established from the same foundation stock, with selection for high feed requirements in one line and low feed requirements in another. This was the only trait considered in the selection experiment. All pigs from each litter were full-fed a mixed ration in individual pens with concrete floors from 72 days of age to a live weight of 225 pounds. At the end of five generations of selection, 67 pigs from the efficient line had required an average of 352 pounds of feed per 100 pounds of gain, whereas 22 pigs from the inefficient line had required an average of 377 pounds, or a difference of 25 pounds.

An experiment designed to study the effects of selection for a single character, backfat thickness in swine, was started at Beltsville in 1954 in Yorkshires and in Durocs. The experiment was designed to select for a line low in fatness and one high in fatness, with a third group as an unselected control [22]. Preliminary results indicate that selection for thinner backfat is effective, although final results have not been published.

Results of five generations of selection for low backfat thickness at approximately 175 pounds of body weight in swine have been reported by the Missouri Agricultural Experiment Station [17]. In this experiment backfat thickness as measured in the live hog by backfat probes was reduced by 20 percent. At the end of five generations of selection backfat thickness in boars had been reduced to 0.85 inch in a spring line and 0.87 inch in a fall line. In gilts, the average backfat thickness had been reduced to 0.90 inch in the spring line and 0.89 inch in the fall line.

Results from several swine-testing stations in the United States over a period of several years show that, in general, efficiency of feed utilization, percentage of ham and loin and loin eye area have increased, and backfat thickness has decreased by 12 to 20 percent, but there has been little or no change in body length and average daily gains (Topel, 1968) [34].

Selection for small body size at 140 days of age in swine over a period of 11 years resulted in a 29 percent decrease in this trait. Heritability estimates indicated that additional progress should be made in selection for this trait. The average 140-day weight at the end of 11 years was 38.6 lb (Dettmers, 1965) [10]. The object of this experiment was to develop a small pig for medical research.

17.5 SELECTION FOR SWINE IMPROVEMENT IN DENMARK

Denmark has exported most of its surplus of pigs in the form of bacon to Great Britain for more than 50 years [25], and the Danish breeding policy has been concentrated toward producing a type of pig to meet this demand for quality. In the 1890's, the Danes started the development of a superior breeding system; it has been used ever since without a change of the general principles, although some changes and improvements have been made occasionally in feeding and handling the pigs.

Special State Recognized Breeding Centers were established for both Landrace and Large Whites in order to supply commercial producers with purebred breeding stock for crossing and to improve the Landrace by selection for carcass quality. The first breeding centers were established in 1895. By 1938, the number of Landrace breeding centers had increased to approximately 250; in 1956 there were 254. The number of Large White centers reached 33 in 1933, but then decreased rapidly, leaving only two breeding centers for this breed in 1956.

It soon became apparent to the Danes that an external examination of pigs was not sufficient to bring about improvement in carcass quality, so this was supplemented by a slaughter test. The first permanent pig progeny-testing station in Denmark, and the first of its kind in the world, was built for use in 1907. By 1926, five such testing stations were in operation. In 1950, these five stations were replaced by three new identical stations constructed for individual feeding of all pigs under test.

The breeding centers are supervised by special committees and breeders must submit groups of progeny (four littermates) of every approved boar and sow to one of the testing stations. The rate of gain and feed efficiency of each pig from a weight of 20 to 90 kg. live weight is recorded. The pigs are then slaughtered at cooperative bacon factories, where the carcasses are examined the day after slaughter. This information is used as a basis for selecting breeding stock and to help breeders improve their herds.

Although the Danes attach great importance to fertility [33], less selection has been practiced for milk-producing ability and fertility of sows than for carcass quality. To enable a breeding animal to be approved or registered in the National Herd Book, however, certain minimum requirements for fertility and number of teats must be met. A committee visits the pig-breeding centers in its district twice each year to score and approve the animals approved for breeding. To be approved, the full pedigree for three generations must be known, with full information about each pig in the pedigree. The animal itself must be of good conformation (exterior), typical for the breed, and of good constitution. Each individual must have at least 12 teats, but at the present time most of them have 14, and some 16 to 18. Each dam in the pedigree must, as an average for all litters, have given birth to at least 10 and weaned eight pigs. In the pedigree, there must be testing results from the Progeny Testing Stations for the animals, and the records must not be inferior on important points to the average of all litters tested.

Since 1907, litter size at weaning in Danish Landrace swine has increased slightly more than one pig per litter. A large proportion of this improvement may have been due to improved feeding and management methods, but some may have been due to genetics. Since 1926 daily gains in the feed lot in Danish Landrace swine has increased by about 10 percent, feed efficiency by 10 to 15 percent, and body length by 8 to 10 percent; backfat thickness has decreased about 30 percent, and the percentage of slaughter hogs producing grade A carcasses is now about 93 percent as compared to 40 percent in 1926–27.

In general, results of selection experiments with swine indicate that reported heritability estimates for the various traits do accurately predict the amount of progress to expect in selection. The higher the heritability of a trait, the more progress one could expect to make in selection.

Fig. 17.6 Excellent type Landrace sow with a good
underline. (Courtesy American Landrace Association.)

17.6 GENETIC CORRELATIONS AMONG TRAITS IN SWINE

Two methods may be used to estimate genetic correlations between traits. One of these is to use statistical procedures to estimate the probability that many of the same genes affect two traits. The other is to conduct experiments where selection is practiced for only one trait and then determine the correlated response of other traits as progress is made in selection.

Considerable evidence indicates that genes that promote rapid growth relative to body size cause more efficient utilization of the food consumed. This has been demonstrated by statistical studies [11] and by selection experiments. In the Illinois selection experiment [1] for rapid and slow rate of gain, there was not only an average of 61.9 pounds difference in 180-day weights in the 8th to 9th generations, but the slow-line pigs required 3.64 pounds of feed per pound of gain and the rapid-line pigs required only 2.76 pounds of feed. In the Alabama selection experiment for efficiency of gains [13], pigs from the efficient line made daily gains of 1.40 pounds per day in the fifth generation of selection, while pigs from the inefficient line gained 1.24 pounds per day. In the Washington selection experiment for fast growth rate on a full feed and on a limited feed, an increase in rate of gain through selection, especially in pigs fed a limited ration, was accompanied by more efficient gains. The correlation between rate and efficiency of gain is not perfect, however. Results from several swine-testing stations in the United States show that feed efficiency has improved in recent years even though there has been little or no improvement in average rate of gain in growing-fattening pigs.

Some evidence for a negative genetic relationship between food utilization and maternal influence has been reported [13, 12]. However, rate of gain and milk production seemed to be more nearly independent of each other genetically. A comparison of large-type with small-type swine indicated that the small type, which was thought to mature earlier, was poorer in milk production [38]. This would suggest a possible negative relationship between early maturity and milk production.

A positive genetic correlation has also been reported between rate of gain and degree of fatness in pigs [13]. However, swine-testing results clearly show that a decrease in backfat thickness is associated with a larger percentage of lean without a decrease in the postweaning growth rate. A United States Department of Agriculture experiment [23] in which selection was practiced in opposite directions for backfat thickness in Duroc and Yorkshire breeds of swine indicated that daily gains were decreased only slightly in the Yorkshire high and low fat lines, but gains were not affected in comparable Duroc lines. Results of a Missouri experiment showed that daily gains did not decrease as backfat thickness was reduced through selection.

Research results in recent years suggest that increased leanness in market hogs may be related to an increased incidence of pale, soft, watery, or exudative pork, called PSE for short. PSE pork results in poorer quality and greater cooking losses, and occurs at higher incidences in some breeds than in others. Although certain environmental factors are known to be related to PSE, breed differences suggest that inheritance may also be involved. Additional evidence has shown recently that when market-sized swine are stressed by fighting or routine handling, some of them may go into a shocklike syndrome often associated with their death. Deaths from this type of syndrome have occurred on farms where feed and management practices are above average and where the producer has bred for more lean and less fat in his market animals (Topel, et al., 1968) [34]. This condition is known as the porcine stress syndrome, or PSS, and it may be related to the occurrence of PSE.

17.7 INBREEDING RESULTS IN SWINE

Experiment stations in the Regional Swine Breeding Laboratory [8] have been developing inbred lines and studying inbreeding effects. One hundred twelve inbred lines were started at the various stations, some from purebred and some from crossbred foundations. In all cases, an attempt was made to obtain as good stock as possible from which to form the inbred lines. Many of the original lines were dropped after a year or two because of poor fertility or performance. Other lines were retained, and in some cases, two inbred lines were crossed to form a new inbred line. The results of these studies have shown that, although some inbred lines performed fairly well, there was a general decline in performance, with some traits being affected more than others. In addition, a few inherited defects due to recessive genes, such as hemophilia [8], were revealed and caused these lines to be discarded.

A summary of inbreeding effects on performance traits is given in Table 17.4. The data were obtained on 538 litters from four different stations, and

Table 17.4   Change in performance for each 10 percent inbreeding from intraseason comparison of linecrosses and parental inbred lines*

	Inbreeding of:	
	Litters†	Dams‡
Litter size at birth	−0.20	−0.17
Litter size at 21 days	−0.35	−0.31
Litter size at 56 days	−0.38	−0.25
Litter size at 154 days	−0.44	−0.28
Pig weight at birth (lb)	0.02	−0.06
Pig weight at 21 days (lb)	0.08	−0.11
Pig weight at 56 days (lb)	0.03	−0.06
Pig weight at 154 days (lb)	−3.44	−0.13

*Reference 14.
†Litter size adjusted to zero difference in age of dam.
‡Dams adjusted to zero difference in litter inbreeding data.

the inbreeding ranged up to an average of 41.7 percent for some of the lines. Inbreeding of the pigs affects their performance directly because of their genetic constitution, whereas the inbreeding of the dams affects the pigs through the maternal environment provided them from conception to weaning. It will be noted that the genetic constitution of the pigs, or their own inbreeding, caused a decrease in litter size at birth, 21, 56, and 154 days, with the effects becoming progressively less as the pigs grew older. This indicates that the vigor of the pigs was adversely affected by inbreeding, and the death loss before and after birth increased as the degree of inbreeding increased. The inbreeding of the pigs had little or no effect on their growth rate up to 56 days of age, but at 154 days there was 3.44 pounds less weight per pig for each 10 percent increase in inbreeding. Inbreeding seemed to affect rate of gain less than it affected survival rate.

As shown in Table 17.4, inbreeding also affected the performance of the sow. Increased inbreeding of the sow resulted in a reduction in litter size and, to a lesser extent, the weights of the pigs. Since litter size up to the time of birth is determined by the ovulation rate and embryonic death losses, the results show that these factors were affected adversely by inbreeding. Maternal influences on pig weight after birth and up to 154 days of age are a reflection of milking and mothering ability of the sows. Inbred sows were inferior to noninbred sows in this respect.

Inbreeding also delays the onset of sexual maturity in gilts [31] and in boars [18]. Reports from farms indicate that inbred boars do not perform as satisfactorily as noninbred boars because of a lack of mating desire, or libido. This causes a delay in the time of farrowing and may result in considerable economic loss, because pigs should be marketed at a definite time to command highest prices. For this reason, few inbred boars are sold for breeding purposes, and usually breeders prefer to sell line-cross boars. Inbred gilts generally produce fewer eggs during estrus and farrow smaller litters than those that are not inbred.

Inbreeding does not seem to have an adverse effect on economy of gains in swine [8]. In fact, in some lines, economy of gain seemed to improve somewhat when inbreeding accompanied by selection for greater efficiency was practiced.

Few data are available on the influence of inbreeding on carcass quality, but this influence seems to be very small or nonexistent.

17.8 PERFORMANCE OF CROSSES OF INBRED LINES OF SWINE

The main purpose of developing inbred lines is to make the lines homozygous; these are then crossed to determine which combine well for commercial production. The actual performance of the inbred lines does not seem to be a good indication of their combining, or nicking, ability. Therefore, the only sure way to find if lines will "nick" is to cross them and observe the performance of the crossbred offspring.

Data summarized in Table 17.5 show the performance of inbred lines as pure lines and when combined in crosses. The inbreeding of each of the three lines ranged between 25 and 30 percent. Their performance was poor as inbred lines, but when they were combined in a three-line cross using crossbred Landrace × Poland sows as dams and Duroc boars as sires, performance was excellent. Thus, the performance of pigs depends not only upon the kind of genes they possess, but upon the ability of the genes to work together or complement each other when combined properly.

Detailed studies of the performance of line crosses as compared to

Table 17.5   Illustration of the nicking effect observed from crossing three inbred lines of swine*

Characteristics	Inbred Durocst	Inbred Polandst	Inbred Landracet	3-line cross	Avg. of 3-linest	3-line cross as % of avg.
Number of litters	47	76	65	60		
Litter size at birth	7.59	7.61	8.67	9.94	7.87	126.3
Litter size at 56 days	5.26	4.57	6.15	8.39	5.31	158.0
Litter size at 154 days	4.77	4.16	5.45	8.06	4.79	168.2
Litter weight at birth (lb)	23.77	26.48	26.59	32.33	25.15	128.6
Litter weight at 56 days (lb)	170.78	168.03	224.32	326.12	183.48	177.7
Litter weight at 154 days (lb)	765.18	694.51	870.18	1544.58	773.76	199.6
Wt. per pig at birth (lb)	3.13	3.48	3.07	3.25	3.20	101.6
Wt. per pig at 56 days (lb)	32.47	36.77	36.47	38.87	34.55	112.5
Wt. per pig at 154 days (lb)	160.42	166.95	159.67	191.64	161.87	118.4

*Reference 26.
†Dams' records adjusted to a gilt basis but not adjusted for inbreeding of sow and litter.
‡Calculated by giving one-half weight to Durocs and one-fourth weight each to Polands and Landrace. Thus, the calculated 3-line average has the same proportion of genes of each line as was present in the 3-line cross.

inbreds has been made at a number of experimental stations in the Regional Swine Breeding Laboratory [9]. The results of these studies show that economic traits most adversely affected by inbreeding are those that show the greatest response when lines are crossed. Information on several of these experiments is presented in Table 17.6. Among the traits most responsive to crossing are litter size and weight at 56 days of age and postweaning rate of gain. Efficiency of gains and carcass traits were improved very little by crossing.

Line-cross pigs possessed more fat than inbred pigs. Similar results were also obtained [28] when Landrace × Poland F_1 pigs were compared with the average of the inbred parent lines for backfat probes in live hogs at 200 pounds. In this study, crossbred pigs exceeded the average of the two parental lines by 6.5 percent in backfat thickness measured only in inbred gilts of the two lines and gilts of the crosses of these two lines.

Sows from two-line crosses of the same breed have been bred to boars

Table 17.6 Various traits in two-line crosses expressed as a percentage of the average of the two inbred parental lines*

Trait	Two-line crosses as a percentage of the parental inbred lines†
Litter size at birth	108.5
Litter size at 56 days	120.8
Weight per pig at 56 days	112.3
Weight per litter at 56 days	139.8
Postweaning rate of gain	114.8
Postweaning efficiency of gain	105.4
Carcass:	
Dressing percentage	100.9
Percent lean cuts	100.0
Backfat thickness	104.7
Percent fat cuts	101.6

*References 5, 7, 11, 21, 26, 27.
†Most data adjusted for age of dam where appropriate, but not for inbreeding. Part of the increase over the parental inbred lines, therefore, is heterosis, and part is recovery from inbreeding effects.

from a third line within that same breed and comparisons made between the performance of the two-line- and three-line-cross pigs [7]. In this comparison at the Oklahoma Station, the pigs of the three-line cross had the benefit of having noninbred mothers. The two-line-cross sows definitely exceeded inbred sows in numbers of pigs farrowed and weaned. They also exceeded noninbred sows of the same breed.

Crosses of lines from different breeds usually show more hybrid vigor for litter size and growth rate than crosses within the same breed. This suggests that for best results in crosses, it is very important to cross those lines that are as far apart as possible in their genetic origin and relationship [30].

Studies were made at the Purdue Station in which crosses of various inbred lines were compared with purebred noninbred controls and with crossbreds of the conventional purebreds. These results showed that linecrosses of inbreds were superior to the controls, especially in the number and weight of pigs raised to 154 days.

17.9 TOPCROSSING

Topcrossing refers to the use of inbred boars on sows of various breeds that may or may not be inbred. It is more practical to use inbred boars for crossing purposes than it is to use inbred sows and gilts. The boar affects its offspring only in a genetic way, whereas the inbred sows and gilts must provide much of the environment for their pigs during both pregnancy and lactation. Since inbreeding causes a decline in vigor of the sows as well as of the pigs, inbred sows would be at a disadvantage from a practical standpoint as compared to noninbred sows because of their inferior fertility and mothering ability. In addition, it would be more costly to furnish inbred sows to a farmer than it would inbred boars because of the larger number required and the greater expense of producing them.

The Wisconsin Station [15] made a comparison of the progeny of inbred and noninbred boars used in two-sire herds on Wisconsin farms. In this study, it was possible to compare records of test litters on the same farms in the same seasons. The data included records of 38 boars from several inbred lines and pigs from 680 litters produced on 44 farms. It was found that litters by inbred boars from four of the inbred lines definitely were superior to those of litters by noninbred boars, as evidenced by the performance of their litters. It was concluded that inbred lines differ in their ability to combine with noninbred stock. Thus, for topcrossing purposes, some lines are superior and others are not, and considerable testing is necessary to find those inbred lines with superior crossing ability.

In the same study [15] 200 gilts from litters by inbred boars were compared with 238 gilts by noninbred boars on the same farms. Topcross gilts from inbred boars produced an average of one more pig per litter, and their litters weighed an average of 37 pounds more at weaning time.

17.10 CROSSBREEDING RESULTS WITH NONINBRED STOCK

Many experiments have been conducted to determine the merits of crossbreeding for the production of market hogs. Much has been written about the pros and cons of crossbreeding. Commercial hog producers have used it, as evidenced by the fact that approximately 85 percent of the hogs in the United States marketed commercially are crossbreds.

Results of several studies of crossbreeding in swine are summarized in Table 17.7. In some of these studies, where only two-breed crosses were involved, both parental breeds were not available to compare with the performance of the F_1. Since several studies were averaged, however, the performance of crossbred pigs as compared to purebred pigs from one of the breeds, especially that of the dam, should give a fair estimate of the amount of heterosis involved in a two-breed cross.

Heterosis from a two-breed cross represents increased vigor only in the crossbred pigs, since the dams in each case are purebreds. Results summarized in Table 17.7 show that litter size at farrowing was slightly smaller in pure-

Table 17.7 Illustration of the amount of hybrid vigor* in crosses of noninbred breeds of swine†

Traits	2-breed cross as a percent of purebreds	3-breed cross as a percent of 2-breed cross	3-breed cross as a percent of purebreds
Litter size at birth	99	108	107
Litter size at 56 days	119	123	142
Weight per pig at 56 days	107	100	107
Weight per litter at 56 days	128	123	151
Postweaning rate of gain	107	100	107
Postweaning efficiency of gains	99	101	100

*Hybrid vigor is estimated by subtracting 100 from each of the above figures.
†References 6, 20, 27, 29, 36.

bred sows bred to boars of another breed than in purebred sows bred to boars of the same breed. The difference is small and may be due only to sampling error.

Litter size at birth is determined by the number of eggs produced by the sow at the time of ovulation, by the number of eggs that are fertilized, and by the number of embryos and fetuses that survive to birth. In these experiments, since sows producing pigs were from the same breed, and within each experiment they were maintained under similar environmental conditions, ovulation rate should be the same regardless of whether sows produced crossbred or purebred pigs. If true, fertilization rate and the degree of embryonic death loss would be the two factors responsible for differences in litter size at birth between purebred and two-breed cross pigs. We can only conclude from the data presented that embryonic death loss seemed to be as great in crossbred as in purebred pigs. It should be pointed out, however, that a lower rate of fertilization occurs when sows are mated to purebred boars of another breed than when they are mated to boars of the same breed. This could reduce litter size at farrowing, even if embryonic death losses were greater in purebred pigs. We have no experimental evidence to offer that would suggest that this is true, however.

Litters of crossbred pigs from the two-breed cross averaged 28 percent heavier at weaning than litters of purebred pigs. Most of this improvement was due to less mortality in the crossbred pigs between birth and weaning, but some of the improvement was due to their slightly heavier weaning weights. Crossbred pigs also made slightly faster gains from weaning to market weight, but there was little or no difference in the amount of feed required per unit of gain.

At the Iowa Station [27], purebred sows were double-mated, so that some of them produced both purebred and crossbred pigs in the same litters. By making matings in this way, variations due to maternal environment could be controlled more accurately. The crossbred pigs from such matings were more vigorous at birth than purebred pigs in the same litters, and a

Fig. 17.7 Two-breed cross pigs out of a purebred sow. Only the offspring benefit in this case. (Courtesy of the University of Missouri.)

Fig. 17.8 Pigs from a three-breed cross out of a crossbred sow. The three-breed cross gives added hybrid vigor in the sow not found in the two-breed cross. (Courtesy of the University of Missouri.)

larger percentage of those farrowed survived to weaning. Crossbred pigs also averaged about four pounds heavier than purebreds at weaning and made slightly faster and more efficient gains from weaning to market weight.

The performance of crossbred sows as compared to that of purebreds is shown in Table 17.7. This comparison is made by expressing the performance of the three-breed cross as a percentage of that of the two-breed cross. This gives an estimate of the advantage of crossbred over purebred sows.

Crossbred sows produced larger litters at farrowing than purebred sows, with the advantage averaging about eight percent. The greatest advantage of the crossbred sows was in their ability to raise more pigs to weaning age. Crossbred sows exceeded purebred sows by 23 percent in this particular trait. Most of the advantage of crossbred sows is evident by the time the pigs are weaned. Crossbred pigs from crossbred mothers gained no faster nor made more economical gains than crossbred pigs from purebred mothers.

A comparison of the performance of the three-breed crosses with that of purebreds gives an estimate of the combined hybrid vigor in both sows and pigs. This averaged 51 percent for litter weight at weaning and is large enough to be of great value in commercial pork production.

17.11 CONCLUSIONS REGARDING CROSSBREEDING IN SWINE

The preceding discussion has shown rather clearly that the chief advantage of crossbreeding in swine production lies in the resulting increase in the size and weight of the litter at weaning and, in some instances, slightly faster rate of gain from weaning to market weight. Most of the advantage due to crossbreeding is due to the increased vigor of the pigs and, to a certain extent, the vigor of the crossbred sows.

Crossbreeding has certain disadvantages that should be pointed out. Quite often, crossbreeding results in the production of pigs that vary widely in coat color. Some people consider this a disadvantage, whereas others do not. It is possible, however, by watching the color of the breeds or lines used for crossbreeding, to control coat color so that it is uniform in the crossbred offspring. This may be done simply by using breeding animals all of one color or by using boars from a line or breed that is dominant in color to that of the sows used for breeding purposes. In the latter case, the boar must be homozygous dominant for all pigs to be of one color.

Crossbreeding alone will not cause much improvement in traits such as economy of gain and carcass quality. In other words, crossing parents from two families that are too fat will not give offspring of meat type. Or crossing two strains that are not efficient in their food utilization will not give efficient offspring. These traits must be present in the two strains or breeds that are crossed if they are desired in the crossbred offspring.

17.12 SYSTEMS OF CROSSBREEDING

To be successful, the commerical hog producer must follow a definite and systematic crossbreeding program. Several crossbreeding plans may be used for commercial pork production (see Chapter 15).

One plan is to use either purebred or very high-grade sows and mate them to a purebred boar of another breed. This is referred to as a single cross. The sire in this system of crossbreeding should come from breeds and herds which are superior in conformation and performance and especially noted for production of meat-type pigs. Sows should also be selected on a similar basis, but with special attention to fertility and nursing ability, along with performance and carcass quality. The chief disadvantage of this plan is that if it is followed in its entirety, all breeding stock must be discarded sooner or later, and the breeder must start over again. Another disadvantage of such a system of crossbreeding is that the swine producer does not take advantage of the hybrid vigor in the crossbred sows, because only purebred sows are used.

Another system of crossbreeding is referred to as "backcrossing" or "crisscrossing." In this system, a single cross between two breeds is made first, and from their offspring crossbred gilts are mated back to a boar of one or the other of the two original breeds. From the offspring of this mating, the crossbred sows are mated with a boar from the other of the two original breeds. This plan will result in about two-thirds of the inheritance of the pigs coming from the breed of the boar used last and one-third from the other breed. When this system is followed on a long time basis, it should be possible to retain some of the hybrid vigor originally obtained in both sows and pigs, but some of this is lost after the first few generations.

A third and different system of crossbreeding is used quite widely by commercial swine producers. This is the three-breed or four-breed system of crossbreeding, in which purebred boars from the different breeds are used in rotation on selected crossbred sows. In this system of breeding, an attempt is made to retain the advantage from hybrid vigor that was attained when the first crossbred sows were used for breeding purposes. The optimum amount of hybrid vigor is attained with the three-breed cross, and after that in the four-breed or five-breed cross one merely attempts to retain that level in later generations. Contrary to the beliefs of some persons, using a three-breed or four-breed boar rotation as described above does not cause a great decline in the level of heterosis after several generations if purebred boars are always used.

The three-breed static cross may also be used in the commercial production of swine. Two-breed crossbred gilts such as the cross between the Yorkshire and Hampshire breeds are excellent mothers, weaning large litters of healthy pigs and possessing good carcass quality and quantity. These crossbred gilts are bred to a boar of a third breed such as the Duroc, which possesses superior post-weaning performance and good carcass traits. All of the pigs from this three-breed cross are marketed. When new breeding stock is needed, both sows and boars are purchased. Such a system of crossbreeding should give maximum heterosis, since both sows and pigs are crossbreds.

Regardless of the method of crossbreeding that is followed, it is extremely important to use as good breeding stock as possible, placing special emphasis on carcass quality and performance. Purebred boars should be used that are from strains of known superior quality, and gilts should also be selected on this basis.

Breeds to use in rotation will depend a great deal on what is available to the commercial swine producer. In general, the principle to apply is to use a boar from a breed that is especially strong in points in which the previous boar or the gilts were the weakest. Not enough information is available to indicate for certain which breeds "nick" best in crosses. Actually, since breeds are not highly homozygous, there may not be any breeds that "nick" better than others. Perhaps certain strains within each breed, however, do have better combining ability than others. If this fact is established, it should be exploited to the fullest extent.

17.13 KINDS OF GENE ACTION AFFECTING SWINE TRAITS

Having discussed the heritability of traits and inbreeding and crossbreeding effects on these traits in swine, we should be in a position to make some estimate of the kinds of gene action that affect the important economic traits in swine. Each trait may be affected by several kinds of gene action, but the proportional influence of some may be greater than others. Data are summarized in Table 17.8 to show these effects.

Litter size and weight at weaning seem to be affected greatly by nonadditive gene action, which includes dominance, overdominance, and epistasis. The evidence for this is that the heritability of these traits is low and effects of inbreeding and crossbreeding have considerable influence on these traits. Little progress could be made in selection for these traits by mating the best to the best; the most improvement would come from crossing distinct lines from different breeds that genetically are as unlike as possible,

Table 17.8 Kinds of gene action affecting important economic traits in swine

Trait	Heritability	Effects of: Inbreeding	Crossbreeding	Proportion of genetic variation due to different types of gene action Nonadditive	Additive
Litter size and weight at weaning	low	large	large	large	small
Rate of gain, weaning to market wt.	moderate	moderate	moderate	moderate	moderate
Economy of gain, weaning to market wt.	moderate to high	small	small	very small	large
Conformation	high	moderate	moderate	moderate	large
Carcass quality	very high	very small	very small	very small	large

to take advantage of heterosis. To improve these traits, crossing, and not selection within a pure line, strain, or breed, should be practiced.

Rate of gain from weaning to market weight is about 30 percent heritable and is affected only moderately by inbreeding and crossbreeding. This suggests that both nonadditive and additive gene effects are moderate, and that some progress should be made by selecting for improved rate of gain through selection on the basis of individuality and families within a pure line. Some improvement could also be made in this trait by crossing lines of known superior combining ability.

Economy of gain from weaning to market weight is affected little by inbreeding and crossbreeding, and the heritability of this trait is about 40 percent. Therefore, the evidence indicates that economy of gain is affected mostly by additive gene action, and selection for this trait within a pure line should be moderately effective. Possibly this trait would be more highly heritable if environmental variables such as feed wastage could be controlled more effectively. Results of the Danish Pig Testing Stations [33] indicate that considerable progress has been made in selecting for this trait, with feed requirements per pound of live weight gain being reduced from 3.44 to 2.97 in the past 30 years.

Heritability of carcass quality in swine seems to be high and inbreeding and crossbreeding effects very small. Thus, the additive type of gene action seems to be very important for this trait, and selection on the basis of family is the method indicated. Selection on individuality cannot be practiced, because the trait cannot be measured until after death, with the exception of backfat, which can be measured in the live animal. That selection for improved carcass quality is effective is borne out by the results in Denmark, where considerable progress has been made through selection on the basis of family [33]. The meat-hog certification program of the various breed associations in the United States is based on the high heritability of most carcass traits, and progress is being made in finding within the breeds those strains of inherent superior meat type, and the numbers of meat-type hogs is being increased by this selection procedure.

Items associated with conformation are, in general, highly heritable, but little has been reported concerning inbreeding and crossbreeding effects. It would seem, however, that any decline in vigor, such as observed on inbreeding, would result in less desirable conformation. This is particularly true of sound feet and legs and bloom of the coat, which is associated with vigor. Crossbreeding, on the other hand, should improve some items of conformation because of increased vigor, and might be more favorable to the development of sound feet and legs, especially if these traits are influenced by recessive genes. The high heritability of most items of conformation, however, indicates that considerable progress should be made in selection for these traits within pure lines, strains, and breeds.

Since different kinds of gene action are of more importance for some economic traits than for others, and since each requires a different kind of selection and breeding method to make the most improvement, how can we make the best possible use of this knowledge in over-all swine improvement? This is an important question, because we do not select for one trait alone in swine, but are interested in several. The answer is that obviously we must apply as much selection pressure as possible in our pure breeds for those traits that are highly heritable and are affected by additive gene action. This should result in the improvement of such traits in our pure strains and breeds. Then, to take advantage of heterosis in traits such as litter size and weight at weaning, we must cross those lines that are superior in the other traits such as carcass quality, economy of gain, and, to a lesser extent, rate of gain. Obviously, if crossing has very little or no effect on these traits, we can expect the crossbreds only to equal the average of the parent lines or breeds for these particular traits. This is an important point often overlooked by breeders, for they do not understand that crossbreeding does not improve all traits.

Because some strains and breeds seem to be considerably superior to others in degree of fertility, more satisfactory results should be obtained when lines or breeds are crossed which are of high fertility. Heterosis effects, in addition to the average effects of the parents, should give much more satisfactory results for this than the crossing of two lines or strains that are decidedly inferior in fertility and prolificacy.

17.14   PROGRAMS FOR SWINE PRODUCTION

The foregoing discussion of breeding and selection systems for swine makes it very obvious that improvements can be made in swine through attention to breeding methods, but improvement will be much slower in some traits than others, and different systems of breeding and selection may be required for each. The kind and amount of selection to apply will depend upon the purpose for which the animals are produced or the objective the breeder has in mind. Attention to disease control, management, and proper nutrition geared to the needs of the animals at various periods during life are necessary for a successful pork-production enterprise. We shall outline only those factors related to the improvement of swine through breeding methods.

17.14.1   Production of Purebred Swine

The first step in swine production of any kind is to set up a system of record-keeping for each pig and sow in the herd. Examples of records are shown in Fig. 17.9. At birth, each pig in the litter should be given an individual number, and the date recorded.

Litter size and weight at weaning may be obtained merely by weighing each pig or the entire litter on the 56th day following birth. Or litters may be weighed as near to 56 days of age as possible and corrected to a 56-day basis by means of correction factors given earlier in this chapter. These weights should be recorded in a permanent record book so that they can later be used for selecting breeding stock from the largest and heaviest litters.

Records should be obtained on the daily rate of gain from shortly after weaning to a market weight of near 200 to 220 pounds. Efficiency of gain made during this period would also be desirable but seldom can be obtained for each individual pig, because it is not practical to feed them individually. If this cannot be done, it would be desirable to feed pigs by litters and obtain the amount of feed required per 100 pounds of gain by the entire litter. Sometimes this is a problem because of the necessity of feeding boars and gilts together, which may result in some of the gilts being bred before the final weight is obtained. Boar and gilt pigs may be fed separately if the breeder desires, and if trouble is experienced with the boars ranting toward the end of the feeding period, final weights should be taken so that records can be made before this occurs.

If room is not available to feed entire litters, samples of two to four pigs per litter from several litters by the same sire may be fed together. Although the economy of gain by litters will not be obtained by such a

RECORD SHEET FOR SWINE LITTERS

Sow number _____   Sire number_____

Date sow born _____   Litter of sow _____
                                                                        1 st, 2 nd, etc.

Date litter farrowed_____   Litter number _____

Date pigs weaned_____   Date vaccinated _____

No. of good teats on sow at farrowing: R _____ L _____

No. of functional teats at weaning: R _____ L _____

Weight of sow at farrowing _____

Weight of sow at weaning_____

Remarks: _____

_____

_____

INDIVIDUAL PIG DATA

Pig No.	Sex	Weight (lbs.) Birth	Weaning	154 days	B. F. at 200 lbs.	Remarks: defects or abnormalities

Figure 17.9

method, it may be obtained for different sire groups, where more than one sire is used for breeding purposes in a given season.

Several precautions should be taken to insure comparable records for litter or sire groups. Comparable beginning and ending weights for all pigs are necessary, and all pigs should be fed the same ration and handled in the same manner so as to hold environmental variations to a minimum.

Purebred pigs that are being tested should be fed and handled in as nearly as possible the same manner as their offspring will be.

Type scores should be obtained on all pigs at near 200 pounds and recorded for selection purposes.

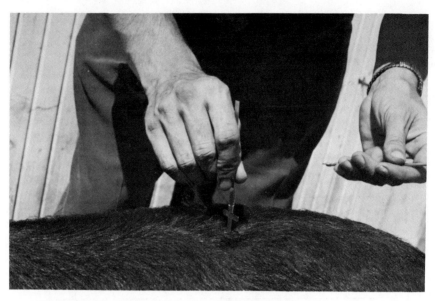

Fig. 17.10 The amount of backfat in the live animal may be measured by using a metal probe or ruler which measures to the nearest tenth of an inch. (Courtesy of the University of Missouri.)

Backfat probes on all pigs taken at three different sites when they weigh about 200 pounds are of considerable use for determining the degree of fatness in breeding groups or individuals. The probes at the three different sites may be averaged for each pig and the average figure used for comparative and selection purposes. Research data indicate that barrows will have from 0.25 to 0.35 inch more backfat than gilts and boars, and this should be given consideration when individuals or litters are compared. This suggests that a boar's barrow offspring will usually have from 0.25 to 0.35 inch more backfat than he possesses; for this reason, it is well to select for the thinnest backfat possible in boars if other traits are also desirable.

Records on carcass data are of considerable value in selection whenever they can be obtained under practical conditions. Carcass data on not less than two pigs per litter should be obtained. Slaughter weights should range from 200 to 220 pounds, and records on length of body, depth of backfat, and the area of the loin-eye muscle should be obtained. When possible to obtain them, the percentages of lean and fat cuts would be very desirable.

The information outlined in the preceding paragraphs should be used to select those breeding animals with superior performance and carcass quality. Traits within a purebred herd that have a heritability estimate of 25 percent or more should be improved by selecting and mating the best individuals to the best. This would include all postweaning traits of economic importance such as rate and efficiency of gains and carcass quality and quantity. One of the selection indexes discussed previously in this chapter could be used if two or more traits are selected for, depending upon the traits the breeder wants to emphasize. The index used should be calculated for all boars and gilts in a given pig crop and those with the highest index kept for breeding.

Although preweaning traits such as litter size and weight at birth and weaning are lowly heritable and little improvement in them would be expected from mating the best to the best, breeding animals should be selected from the largest, healthiest, and heaviest litters at weaning. Litters of this kind indicate that both sows and pigs are probably genetically superior and are healthy and probably free from disease. In general, replacement stock will be selected automatically from the larger litters, because the more pigs weaned per litter, the more pigs available when breeding animals are selected, and the greater the chance they will be selected. This is especially true when gilts are selected for breeding.

Progeny-testing animals selected for breeding requires considerable time and often is not the most efficient method of selection. Selection on the basis of individuality (mass selection) at least for postweaning traits would probably give more progress over a period of years than progeny-testing. However, the selection of replacement stock from litters that are uniformly superior indicates that the dam is genetically superior, as are the littermates that are full sibs. A superior individual with a superior dam and full sibs is more likely to be genetically superior than one from a small or less uniform litter.

17.14.2 Production of Market Hogs

Since commercial pork production requires the greatest efficiency possible, some system of crossbreeding should be followed. But even if this is done, it is still extremely important to use superior stock for breeding purposes. Most breeders will find it desirable to retain females from their own herd for replacement purposes, because they have records on the herd and their relatives and the animals are better adapted to the conditions of the farm where they have been produced. Insofar as practical, the commercial producer should keep the same records as the purebred breeder, and select breeding animals in the same way.

The commercial pork producer relies almost entirely on the purebred breeder for the boars he uses in his herd. He is interested, therefore, in obtaining production-tested boars fed in the same manner as the pigs will be that he intends to produce for the market. Many commercial producers are purchasing top boars from the swine-evaluation stations now operating in various parts of the country. Boars from certified meat-type litters should also be used whenever it is possible to obtain them.

A three-breed rotation system of crossing is now very popular in the United States. This system takes advantage of hybrid vigor in both sows and pigs, and although most of the advantage is gained in the first three crosses, proper emphasis on selection of boars and sows will almost maintain this advantage in later cycles of the rotation. Such a system naturally brings up the question of what breeds and lines should be used in the crossing program. At the present time not enough data are available on different breeds to make sound recommendations, but some general principles may be followed. Regardless of the breed, boars should be selected from strains where the sows are prolific and good mothers and where pigs make rapid and efficient gains and possess good carcass quality. Boars should be selected that are especially strong in those qualities in which the sows are weak. Obviously, the various breeds that have boars of this kind must be available in his area if the commercial pork producer plans to use them for crossing purposes.

Still another point to consider in the choice of breeds to use in a crossbreeding program is how far apart the breeds are genetically. In other words, were they developed in different countries from different genetic material? Research work shows that the greatest improvement from crossing is obtained from crosses of this kind.

Some of the new breeds of swine have been developed to be used for crossing purposes, and producers of seed stock of this kind can make recommendations to individual pork producers as to what breeds to use in a rotation system. The producer of seed stock is very careful about his recommendations, because later sales will depend upon whether or not his customers are satisfiied.

17.15 NEW BREEDS OF SWINE

Several new breeds of swine have been developed in the United States in the past few years, and others are being developed. Several of these are known as new breeds and have been described in U.S.D.A. Farmer's Bulletin No.

1263. Most of them are registered with the Inbred Livestock Registry Association of St. Paul, Minnesota. All of these new breeds were developed from a crossbred foundation of two or more breeds. This has led some to believe that they are just crossbred hogs, which is far from the truth. They are actually new breeds and are much more highly inbred than the old breeds that have existed in the United States for many years. In other words, the new breeds are probably more homozygous than are the older breeds.

Fig. 17.11 Minnesota No. 3 gilt from one of the newer swine breeds. (Courtesy of the Animal Husbandry Department, University of Minnesota.)

The new breeds, besides being inbred, also have other points in their favor. They were developed from crosses of highly fertile breeds of fair-to-good meat type. In addition, many of them were developed by means of performance-testing the breeding stock, and as a result, their production is at a desirable level.

In spite of the fact that the new breeds are inbred, most of them perform very satisfactorily from the standpoint of litter size at weaning and the rate of gain from weaning to market weight. As might be expected, however, there are differences between individual pigs within each breed. Information about the new breeds is given in Table 17.9. Many of the new breeds were developed specifically for crossbreeding purposes, and some of them are used in rotation crossing systems at the present time. Some of the new breeds have become more popular than others because of the more desirable traits they possess and because of their crossing ability. A summary of the production of new breeds as compared to the old breeds in the United States is presented in Table 17.10.

Table 17.9   New breeds of swine in the United States

Breed	Foundation breeds and percentage contribution	Year breed began	Year breed recognized	Percentage inbreeding when recognized
Minnesota No. 1	Tamworth 52% Landrace 48%	1936	1946	32
Minnesota No. 2	Canadian Yorkshire 40% Poland 60%	1942	1948	28
Montana No. 1	Hampshire 45% Landrace 55%	1936	1948	32
Maryland No. 1	Berkshire 37% Landrace 63%	1940	1951	30
Beltsville No. 1	Poland 37% Landrace 63%	1935	1951	35
Beltsville No. 2	Danish Yorkshire 58% Duroc 32% Landrace 5% Hampshire 5%	1940	1952	35
San Pierre	Berkshire ?% Chester White ?%	1950	1953	26
Palouse	Chester White 47% Landrace 53%	1945	1956	9

Table 17.10   Performance of sows from new breeds as compared to those from older breeds*

Characteristics	Purebreds not inbred	New breeds or lines
Number of litters farrowed	4435	3180
Number of litters weaned	4203	2985
Number of pigs farrowed	36965	28463
Number of pigs weaned	25933	20418
Percentage of pigs weaned	70.2	71.7
Number of pigs farrowed per litter	8.33	8.95
Number of pigs weaned per litter	6.17	6.84
Percentage of litters weaned	94.80	93.90

*Data in this table adapted from reference 9, Table 2, page 975.

REFERENCES

1. Baird, D. M., A. V. Nalbandov, and H. W. Norton. "Some Physiological Causes of Genetically Different Rates of Growth in Swine," JAS, 11: 292, 1952.

2. Bell, C. E. Jr. "Hogs Remodeled with Backfat Probes," NHF, 5: 28, 1960.

3. Bereskin, B., et al. "Inbreeding and Swine Productivity Traits," JAS, 27: 339, 1968.

4. Bernard, C. S., A. B. Chapman, and R. H. Grummer. "Selection of Pigs under Farm Conditions: Kinds and Amount Practiced and a Recommended Selection Index," JAS, 13: 389, 1954.

5. Bradford, G. E., A. B. Chapman, and R. H. Grummer. "Effects of Inbreeding, Selection, Linecrossing and Topcrossing in Swine; 2; Linecrossing and Topcrossing," JAS, 17: 441, 1958.

6. Chapel, G. A., and J. I. Cabrera. "Crossbreeding for Swine Production in Puerto Rico," JAUPR, 32: 119, 1949.

7. Chambers, D., and J. A. Whatley, Jr. "Heterosis in Crosses of Inbred Lines of Duroc Swine," JAS, 10:505, 1951.

8. Craft, W. A. "Results of Swine Breeding Research," USDAC No. 916, 1953.

9. Craft, W. A. "Fifty Years of Progress in Swine Breeding," JAS, 17: 960, 1958.

10. Dettmers, A. E., W. E. Rempel, and R. E. Comstock. "Selection for Small Size in Swine," JAS, 24: 216, 1965.

11. Dickerson, G. E., J. L. Lush, and C. C. Culbertson. "Hybrid Vigor in Single Crosses between Inbred Lines of Poland China Swine," JAS, 5: 16, 1946.

12. Dickerson, G. E. "Composition of Hog Carcasses as Influenced by Heritable Differences in Rate and Economy of Gains," IowaAESRB 354, 1947.

13. Dickerson, G. E., and J. C. Grimes. "Effectiveness of Selection for Efficiency of Gain in Duroc Swine," JAS, 6: 266, 1947.

14. Dickerson, G. E., et al. "Evaluation of Selection in Developing Inbred Lines of Swine," MoAESRB, 551, 1954.

15. Durham, R. M., A. B. Chapman, and R. H. Grummer. "Inbred versus Non-inbred Boars Used in Two Sire Herds on Wisconsin Farms," JAS, 11: 134, 1952.

16. Fowler, S. H., and M. E. Ensminger. "Interactions between Genotype and Plane of Nutrition in Selection for Rate of Gain in Swine," JAS, 19: 434, 1960.

17. Gray R. C., L. F. Tribble, B. N. Day, and J. F. Lasley. "Results of Five Generations of Selection for Low Backfat Thickness in Swine," JAS, 27: 331, 1968.

18. Hauser, E. R., G. E. Dickerson, and D. T. Mayer. "Reproductive Development and Performance of Inbred and Crossbred Boars," MoAESRB 503, 1952.

19. Hazel, L. N., and F. A. Kline. "Mechanical Measurement of Fatness and Carcass Value on Live Hogs," JAS, 11: 313, 1952.

20. Hazel, L. N. "Crossbreeding for Commercial Hog Production," ISGDR, April 2, 1958.

21. Hetzer, H. O., O. G. Hankins, and J. H. Zeller. "Performance of Crosses between Six Inbred Lines of Swine," USDAC No. 893, 1951.

22. Hetzer, H. O., and J. H. Zeller. "Selection for High and Low Fatness in Duroc and Yorkshire Swine," JAS, 15: 1215, 1956.

23. Hetzer, H. O., and W. H. Peters. "Selection for High and Low Fatness in Duroc and Yorkshire Swine," JAS, 24: 849, 1965.

24. Krider, J. L., B. W. Fairbanks, W. E. Carroll, and E. Roberts. "Effectiveness of Selecting for Rapid and for Slow Growth Rate in Hampshire Swine," JAS, 5: 2, 1946.

25. Larson, L. H., H. J. Clausen, and J. Jepersen. "Breeding and Feeding of Cattle and Pigs," P6ICAH, 1952.

26. Lasley, J. F., and L. F. Tribble. "The Influence of Breeding Methods on Performance in Swine," MoAESMR, Swine Day, Sept. 5, 1958.

27. Lush, J. L., P. S. Shearer, and C. C. Culbertson. "Crossbreeding Hogs for Pork Production," IowaSAESB 380, 1939.

28. Reddy, V. B., J. F. Lasley, and L. F. Tribble. "Heritabilities and Heterosis of Some Economic Traits in Swine," MoAESRB 689, 1959.

29. Robison, W. L. "Crossbreeding for the Production of Market Hogs," Ohio-AESB 675, 1948.

30. Sierk, C. F., and L. M. Winters. "A Study of Heterosis in Swine," JAS, 10: 104, 1951.

31. Squiers, C. E., G. E. Dickerson, and D. T. Mayer. "Influence of Inbreeding, Age and Growth Rate of Sows on Sexual Maturity, Rate of Ovulation, Fertilization and Embryonic Survival," MoAESRB 494, 1952.

32. Taylor, J. M., and L. N. Hazel. "The Growth Curve of Pigs between 134 and 174 Days of Age," JAS, 14: 1133, 1955.

33. Thomsen, N. R. "Pig Breeding and Progeny Testing in Denmark," in Report of the Meeting on Pig Progeny Testing in Denmark, FAOUN, EAAP, pp. 9–12, 1957.

34. Topel, D. G. *The Pork Industry: Problems and Progress.* (Ames: Iowa State University Press, 1968).

35. Topel, D. T., E. J. Bicknell, K. S. Preston, L. L. Christian and C. Y. Matsushimi. "Porcine Stress Syndrome," MVP, 49: 40, 1968.

36. Winters, L. M., O. M. Kiser, P. S. Jordan, and W. H. Petis. "A Six Year Study of Crossbreeding Swine," MinnAESB 320, 1936.

37. Winters, L. M., P. S. Jordan, R. E. Hodgson, O. M. Kiser, and W. W. Green. "Preliminary Report on Crossing of Inbred Lines of Swine," JAS, 3: 371, 1944.

38. Zeller, J. H. "Swine Types as a Factor in Pork Production," PASAPP, 279, 1940.

QUESTIONS AND PROBLEMS

1. Why has there been a change in the type of hog produced in the United States in recent years?

2. What problems still need to be studied in swine breeding?

3. List the traits of economic importance in swine breeding.

4. Why is total litter weight at weaning an important economic trait?

5. From the practical standpoint, why is it important to use correction factors for adjusting litter weights of pigs to a 56-day basis?

6. Assume that a pig weighed 235 pounds when he was 164 days of age. What would be his adjusted 154-day weight?

7. Why isn't it practical to obtain the economy of gain from weaning to market weight in all pigs in the herd?

8. What precautions should be taken in obtaining economy of gain records on individual pigs in a group on the same farm?

9. Should animals be selected on the basis of type and conformation? Explain.

10. What methods may be used to measure backfat thickness in the live hog? Why is the use of these important?

11. In general, what traits are lowly heritable in swine? Highly heritable?

12. A sow gives birth to a litter of 14 pigs, 12 of which are still alive at weaning with a total litter weight of 425 pounds. Calculate the index for this sow given in the earlier discussion.

13. What traits are usually included in the index used for swine in swine-testing stations in the United States?

14. Would selection for thinner backfat in swine at near market weight be effective? Explain.

15. What country is noted for its program of developing a bacon-type breed by breeding methods? What breed is the most popular in this country?

16. What is meant by genetic correlations among traits in swine?

17. If one selected for more rapid gains in swine, what would probably happen to economy of gains if selection were effective?

18. Do experiments indicate that there is a positive genetic correlation between fast gains and more fat in the carcass?

19. What is meant by PSE and PSS?

20. Assume that you mate a boar to his own daughters. What would you expect to happen to the performance of the pigs?

21. What are some of the effects of increased inbreeding in swine?

22. Theoretically, why would one develop inbred lines of swine? Has this proved practical? Should a commercial producer use crossbred or inbred sows for pigs production? Explain.

23. What is meant by topcrossing? Should a commercial producer use inbred sows for pig production? Explain.

24. Outline the crossbreeding system you would recommend to the commercial swine producer and explain why you recommend this system.

25. What traits would one expect to improve when crossbreeding in swine is practiced?

26. What breeds would you recommend for crossing in the production of commercial swine? Why would you recommend these breeds?

27. What traits should the purebred breeder emphasize in selection for improvement of swine in his herd?

28. Is it important to use superior purebreds in the production of swine by crossbreeding? Explain.

29. Outline a breeding and selection system that you would recommend for the production of purebred swine.

30. Outline a breeding and selection system that you would recommend for the commercial swine producer, naming the breeds to be used, the records that should be kept, and how boars and gilts would be selected for breeding purposes.

18

Systems of Breeding and Selection in Beef Cattle

Information on beef cattle breeding is not as complete as that on swine. One reason for this is that extensive and cooperative efforts toward a comprehensive study of breeding principles in beef cattle were initiated relatively recently. Another reason is that progress is slower in beef-cattle breeding because the interval between generations is considerably longer in cattle than in swine, and cattle are much less fertile than swine. Cattle usually produce only one calf per year, whereas a sow may produce two litters per year, each consisting of eight to ten pigs.

Much work is now being done in beef-cattle breeding, and considerable progress has been made in developing breeding principles for this species. In this chapter, important breeding principles will be presented, and it will be shown how they may be used for improving the performance and carcass quality of this species.

18.1 TRAITS OF ECONOMIC IMPORTANCE AND HOW TO MEASURE THEM

The efficiency of production of beef cattle in the United States depends upon the amount of lean meat yielded per animal at slaughter, its quality and appearance, and the efficiency of production of that meat on a farm or on the range.

Beef cattle are produced under a wide variety of environmental conditions in the United States. Many areas of the West and the Southwest are used only for production. Sometimes only 8 to 10 head of cattle are grazed per section of land on a year-long basis, yet herds may number into hundreds and even thousands. In other areas, such as in the Midwest, herds may be a very small part of a general farming enterprise. Because of the wide variety of conditions under which cattle are produced, methods of management and feeding may vary considerably from one locality to another. However, methods of breeding for the improvement of beef cattle are very similar in all areas.

The improvement of beef cattle through breeding methods requires that accurate and careful records be kept on all animals in the herd. This is done

Fig. 18.1   These yearling steers have been gathered from the range and are headed for the sales pens many miles away. Much of the western range country produces splendid feeder cattle for the fattening pens. (Courtesy of the San Carlos Apache Indian Tribe, San Carlos, Arizona.)

on many farms and ranches at the present time, and special attention is being paid to a few traits of the greatest economic importance. These traits will be discussed in the next few paragraphs.

18.1.1 Fertility

Fertility may be defined in numerous ways, but the definition used here is the percentage of calves raised to weaning age from all mature cows in the herd. Such a definition, of course, includes numerous factors, such as the ability of a cow to rear a calf to weaning and her ability to conceive while raising that particular calf. Of course, fertility in a herd is also dependent on management and nutrition factors and the ability of a calf to survive from birth to weaning.

Fertility as thus defined is one of the most important economic traits in beef cattle in all areas of the United States. In some areas, there is room

Fig. 18.2 "Eye appraisal" for performance is often misleading. The cow on the top weaned a heavy calf each year for eight years. The cow on the side weaned only four average weight calves in the same period of time. Records tell the story.

for much improvement in this particular trait. A survey of the American National Cattlemen's Association in 1956 [14] showed that only 79 percent of beef cows actually bred dropped live calves, and only 62 percent reared calves to weaning age.

The percentage calf crop is an important factor in the efficiency of production, because dry cows eat almost as much as cows nursing calves, and they yield nothing but their added weight for that particular year. If we assume that the cost of keeping a cow for a year is $100, the cost of each calf weaned is strongly dependent on the percentage calf crop weaned, as shown below [48].

Percentage of calf crop weaned	Cost per calf weaned
100	$100
90	$111
80	$125
70	$143
60	$167

Heritability estimates for calving interval in beef cattle show that the degree of heritability [3, 4] and the repeatability are very low for this trait [4, 7], indicating that there is little or no effect of additive genes on calving interval and that most observed variations are due to environmental factors. These low heritability and repeatability estimates probably mean that genes which affect fertility are those with nonadditive effects. Considerable evidence is available that many recessive genes affect fertility in beef cattle [46]. Such conditions are more or less self-limiting, because the homozygous recessive individuals are usually of low fertility (some are sterile) and leave fewer offspring in the herd.

The following points should be given consideration if the percentage calf crop is low and improvement is desired.

1. Thoroughly investigate the management of cows and bulls, making certain that enough bulls are turned in to the breeding pasture each year to assure that every cow has a chance to be bred. When only one bull is used, observe cows as often as possible to make certain that they are not failing to conceive, as evidenced by their coming into heat at regular intervals during the breeding season. A fertility test on the bulls before the beginning of the breeding season will identify many of those of low fertility or those that are sterile.

2. Make certain that the nutrition level, quantity and quality, is adequate and that diseases which may affect fertility are not present in the herd.

3. Cull cows that are hard to settle or those that are dry. This will remove all cows that are poor breeders because of disease, accidents (which may occur at calving time), and heredity. A pregnancy test each fall on all cows will identify those that are not pregnant; these can be marketed when fat without being fed another year.

4. Select breeding stock, both bulls and heifers, from cows that have a record of producing a good calf every year.

18.1.2   Weight of Calves at Weaning

The percentage calf crop and the weight of each calf at weaning, combined, are probably the two most important factors in production. The weaning weight of the calf is of importance, because this represents the pounds of production per cow per year. This trait depends on the milk production of the cow and, to a lesser extent, on the ability of the calf to make fast and efficient gains.

Data presented in Table 18.1 show that the heritability of differences in weaning weights in beef calves is about 25 percent, with a range in 11 different studies from −6 to +64. Thus, this trait is affected to a certain extent by additive gene action but to a larger extent by environmental factors. Careful selection for this trait should result in some improvement over a period of years.

An average of seven different studies shows that weaning weights are about 46 percent repeatable (Table 18.2). This means that the weaning weight of the first calf from a cow is a fairly good indication of the weaning weight of her later calves. Culling heifers or cows that wean light calves will tend

Table 18.1   Heritability estimates in percent for various economic traits in beef cattle*

Trait	Number of studies	Range	Average
Weaning weight	11	−6 to 64	25
Weaning score	8	23 to 53	33
Rate of gain in feed lot	10	26 to 99	57
Efficiency of gain in feed lot	5	17 to 75	36
Slaughter grade	4	38 to 63	47
Carcass items:			
Dressing percent	4	1 to 73	46
Carcass grade	5	16 to 84	48
Thickness of fat	1		38
Area of eye muscle	3		70
Tenderness of lean	2		61

*Averages from many reports.

Table 18.2   Repeatability estimates for economic traits in beef cattle*

Trait	Number of studies	Range	Average
Calving interval	2	−9 to 2	4
Interval from exposure to bull and calving	4	14 to 38	28
Weaning weights	7	40 to 52	46
Weaning scores	2	19 to 21	20

*Average of data from several sources.

to improve the over-all average of the herd in later years, with other factors remaining equal. The fact that repeatability estimates average much higher than the heritability estimates (almost twice, in fact) indicates that the maternal influence of the cow is an important source of variations in weaning weights of calves. This influence is both environmental and genetic. The environmental aspect includes the nutrition of the embryo in the uterus and the influence on the calf after birth, mostly through the milk production of the cow.

Weaning weights of the calves may be used to evaluate the milk production level and mothering ability of cows in a herd, and the differences in the growing ability of the calves. Since many nongenetic factors affecting weaning weights of calves are known to exist, adjustments should be made for these factors when possible. When these adjustments are made, a larger proportion of the remaining variation within a group of calves would be due to inheritance. This is why weaning weight adjustments for age of dam, age of calf, and sex of calf have been developed.

The U.S.D.A. Extension Service has printed guide lines for uniform beef improvement programs. These guide lines are based on research results and will be followed closely in this chapter. Their use gives a uniform program for many areas of the country and world.

A standard weaning age of 205 days is recommended for beef calves. On a practical basis, it is not possible to weigh each calf as it reaches 205 days of age, because this would require that calves be weighed almost every day over a period of 90 or more days. Under practical farm and ranch conditions this would be impossible. A single weight of each calf in the herd taken on the same day and when the average age of all calves is about 205 days is recommended. The weight of each calf can then be adjusted to a 205-day basis by means of an appropriate formula. It is recommended that when a single weight is taken for all calves, it should be an age range of 160 to 250 days.

The following formula may be used to adjust calf weights to a 205-day basis:

$$\text{Adjusted 205-day weight} = \left(\frac{\text{actual weaning weight} - \text{birth weight}}{\text{actual age in days}} \right) \times 205 + \text{birth weight}$$

Records may be adjusted to any age simply by substituting the desired age at weaning for 205 in the above formula.

If birth weights are not obtained on each calf, an average birth weight of 70 1b may be used. If desired, the average birth weight of calves within a particular breed may be used. The adjustment of weaning weights of all calves in a group to a standard age removes much of the environmental variation for this trait.

Brood cows in a beef herd often differ in age, and the age of the cow has an important influence on the weaning weight of her calf. Therefore, weaning weights should be adjusted for age of dam. Many adjustment factors for age of dam have been reported in the literature, but the following adjustments are recommended:

Age of dam in years	Multiply adjusted weaning weight of calf by this factor to adjust for age of dam
2	1.15
3	1.10
4	1.05
5–10	No adjustment
11 and up	1.05

Adjustments of the weaning weights of calves in a particular year to a standard 205-day basis and for age of dam are sufficient to determine which calves within a calf crop should be selected for breeding herd replacements. Calves of a particular sex, however, should be compared with other calves of the same sex.

If the producer prefers, the "weaning weight ratio" for each calf in the herd may be calculated from:

$$\frac{\text{Adjusted weaning weight of calf}}{\text{Average adjusted weaning weight of all calves in the group}} \times 100$$

The calf with an average weaning weight would have a weaning-weight ratio of 100, whereas the heavier calves would have a ratio above 100 and the lighter calves, below 100.

In a beef herd, the production record of the cow depends upon her ability to wean a heavy calf each year. The production record of the sire also depends to a certain extent on the weaning weight of his calves. The breeder often has to make a decision as to which parents to keep and which to cull in order to maintain high performance records in his herd from year to year. Since parents (especially cows) may vary in the sex of the calf they produce, in comparing cows all of their calves should have their weaning weights adjusted to the same sex basis.

Weaning weights of the calves may be adjusted to a bull calf or a steer calf basis, depending upon the sex status of the male calves in a particular herd. For example, in a commercial herd most of the male calves at weaning would be steers, so the weaning weights of the heifer calves should be adjusted upward to a steer basis by multiplying their adjusted weaning weight by 1.15. Weaning weights of any bull calves in the herd may be adjusted downward to a steer basis by multiplying their adjusted weaning weight by 0.95. In a purebred herd, where most male calves at weaning are bull calves, the weaning weights of heifer calves may be adjusted upward to a bull basis by multiplying their weaning weight by 1.10. In a large herd, the producer can calculate his own sex adjustment factors. For example, if all bull calves in a herd average 500 lb at weaning and heifer calves 465 1b, the factor used to adjust heifer calves to a bull-calf basis would be $\frac{500}{465}$, or 1.075.

The most probable producing ability (MPPA) of female farm animals was discussed in Chapter 11. It may be used for ranking beef cows with a different number of records for the 205-day weaning weights of their calves. The MPPA makes it possible to cull cows with a different number of records more accurately. For beef cattle it is recommended that the MPPA be computed on the weaning weight ratio by the following formula:

$$\text{MPPA} = \bar{H} + \frac{NR}{1 + (N-1)R}(\bar{C} - \bar{H})$$

where $\bar{H}$ is 100, or the average weaning weight ratio.

N is the number of calves included in the cow average.

R is 0.40 the repeatability for weaning-weight ratio.

$\bar{C}$ is the average for the weaning-weight ratio for all calves the cow has produced.

The calculation of the MPPA may be simplified by using data presented in Table 18.3, which shows the repeatability of varying numbers of records from 1 to 10 where the repeatability of one record is 0.40. To illustrate the calculation of the MPPA, let us use the example of a cow with seven records (calves) with an average weaning weight ratio ($\bar{C}$) of her calves of 115. Her MPPA would be $100 + 0.82(115 - 100)$, or 112. The figure 0.82 represents the portion of the formula

$$\frac{NR}{1 + (N-1)R},$$

where N is 7 and R is 0.4. The MPPA of this cow is considerably above average.

Table 18.3 Repeatability of one or more records when the repeatability of one record is 0.40

Number of records (N)	Repeatability of records $\left(\dfrac{NR}{1+(N-1)R}\right)$
1	0.40
2	0.57
3	0.67
4	0.73
5	0.77
6	0.80
7	0.82
8	0.84
9	0.86
10	0.87

The MPPA when calculated for all cows in a herd gives a more accurate method of culling cows of different ages having a different number of records.

The following points are suggested to improve the weaning weights of calves through attention to breeding methods:

1. Identify each cow and her calf by a tattoo, brand, or some other means.
2. Keep continuous, accurate records on each calf and correct weaning weights for age of dam, sex of calf, and age of calf.
3. Keep replacement heifers, insofar as possible, from those calves with the heavier weaning weights.
4. Cull cows, especially younger ones, that produce a light calf at weaning.
5. Select herd bulls of superior type and performance from a herd where complete and accurate records are available and from cows which have demonstrated their ability to produce a heavy calf at weaning year after year.

18.1.3 Rate and Efficiency of Gain in the Feed Lot

The ability to make fast and efficient gains in the feed lot is an important trait in beef-cattle production. This has been recognized for many years by beef-cattle feeders. It has also been noted that there is considerable variation among different steers in their ability to make fast and efficient gains when placed on a full feed in the feed lot.

Many cattlemen have contended that it is possible to select the faster gaining and more efficient animals by paying attention to conformation. Many studies, however, show rather clearly that this is not the case and that selection for performance on the basis of conformation is ineffective.

Experimental research work at the U.S. Range Livestock Experiment

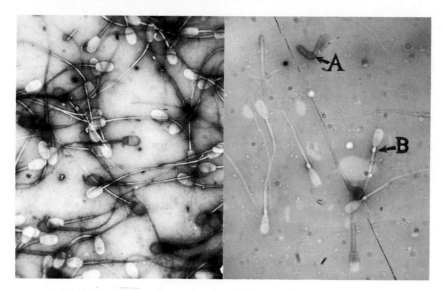

Fig. 18.3 Right, abnormal sperm of the bull. (A), Sperm with a coiled tail and (B), sperm with a protoplasmic droplet near the base of the head. Left, normal sperm of the ram. The stained cells were dead and the nonstained cells were alive when the slides were made. (Courtesy of the University of Missouri.)

Station, Miles City, Montana, several years ago indicated that the ability to make fast and efficient gains in the feed lot was more prevalent in the progeny of some bulls than in the progeny of others [21]. A later study of data from the same station [39], in which results were presented for eight steers each from 13 different sires fed during 1948 and 1949, showed striking differences among progeny of different sires. When the progeny of the best bulls and those of the poorest were compared, there was a difference of 211 pounds in the final feed-lot weight, 0.43 pound per day in average daily gains, and $45 in the returns per head above feed costs. Thus, for eight steers, one bull sired progeny that yielded $360 more above feed costs than those of another bull.

Many later studies indicate that the ability to make fast and efficient gains in the feed lot is highly heritable (Table 18.1). Therefore, bull calves that make fast gains should sire calves that make fast gains under the same conditions. This has led to the feeding of bull calves under standard periods and conditions; this practice is known as performance-testing. Heifers usually are not full-fed on test because of the time and expense involved and because it is possible that a highly fitted condition might interfere with their breeding efficiency. Cattlemen in some areas try to obtain performance-tested bulls

Fig. 18.4 Both of these bulls were fed the same ration in the same barn for the same length of time. The bull at the top, however, gained 3.22 pounds per day as compared to 1.79 for the bull at the bottom. Much of this difference was probably due to inheritance. (Courtesy of the University of Missouri.)

and pay a premium for those with superior records. It should be mentioned here, however, that performance-testing of bulls does not make them any better genetically. It merely gives a bull a chance to show whether or not he possesses the inherent ability to make rapid and efficient gains. Before purchasing a performance-tested bull, one should make sure that he has a good record.

A strong correlation exists between the ability to make rapid gains and the ability to make efficient gains in the feed lot. A summary of six studies [44] showed an average correlation coefficient of 0.45 and a range from 0.06 in one study to 0.69 in another. This degree of correlation suggests that bulls selected for rapid gains are likely to make efficient gains, although the correlation is not perfect. If further research work indicates that the correlation is high enough, individual feeding may not be required, so that bulls can be fed in groups and only the rate of gain during a certain period measured. This would be much simpler and considerably less expensive than individual feeding.

One important question is whether animals of superior inherent gaining ability can be identified as accurately by feeding them a limited ration as by feeding them all they will eat. In one study [54], the heritability of gain on a limited ration in dry lot was 34 percent, and in the same steers later on a limited ration on pasture, the heritability of gains was 43 percent. Many of the same genes seemed to be responsible for fast gains during both periods. In another study [26], it was found that animals making the fastest gains on the range also tended to make the fastest gains in the feed lot.

The heritability estimates for gains made on a limited ration on dry lot or on pasture were high enough to indicate that considerable progress could be made by selecting the faster-gaining cattle on such feeding regimes. These heritability estimates, however, are considerably lower than estimates for cattle on a full feed. Although more data are needed to answer this question, it seems that the most progress would be made by full-feeding cattle for a minimum of 140 days and then selecting those which made the fastest gains. Since such a practice is often impractical on the farm, the next best plan would be to select the animals that made fast gains on either pasture or a limited ration.

On the basis of evidence at hand, the following recommendations should be made on the performance-testing of animals:

1. Feed only bulls that show promise from the standpoint of their own individuality and that of each parent.

2. Bulls to be compared should not vary in age by more than one or two months.

3. Bulls to be compared should have a similar pretest environment. That is, do not compare bulls that have been creep-fed with those that have not. Other things being equal, thinner calves should make faster gains on feed.

4. Use average of two weights taken at least two days apart for the initial and for the final test weights. Bulls should be shrunk on weigh day for 6 to 12 hours. Differences in fill can account for as much as 40 to 60 pounds of apparent gain.

5. Use the same feed for all bulls and see that they have equal opportunity to get their share of feed. This is usually done by feeding in individual stalls two or more times per day. Many prefer to feed a complete mixed ration of chopped hay and grain rather than to feed hay and grain, free-choice. This controls the ratio of hay to concentrate, for, if fed free choice, some animals would eat more roughage than others and would make less rapid gains. Some prefer to feed a pelleted ration, [36], which has resulted in very rapid gains with very little feed wastage.

6. Animals should be fed for at least 140 days, and at the end of the test the daily rate of gain and the feed required per 100 pounds of gain should be calculated. A final type-score should also be taken. These three measurements can be used in an index to select the top bulls for breeding purposes.

7. Most heifers are not full-fed as are bulls, but it is possible and of considerable value to obtain gain tests on them by measuring gains made on pasture during the spring and summer months.

18.1.4 Weight at One Year of Age

Yearling weight is a trait that is of value for selection of both heifer and bull replacements for the breeding herd. Yearling weight is particularly valuable for heifers grown under practical conditions on the farm or ranch where they are not full-fed. Too much fat in heifers during the growing period often causes lowered fertility and milk production later in their productive life.

Adjusted yearling weights taken at 365 days and 550 days may be used. The 365-day weight would probably be more applicable to bull selection when the calves are placed on a full feed at weaning or shortly thereafter. If placed on feed at weaning, the calves should be given a period of approximately 20 days where the amount of concentrated feed is gradually increased each day until they are on a full feed. This would give about 140 days on a full feed by the time the calves are one year of age. The adjusted 365-day weight may be computed as follows:

$$\begin{array}{l}\text{Adjusted} \\ \text{365-day} \\ \text{weight}\end{array} = \left(\frac{\begin{array}{c}\text{actual} \\ \text{final wt}\end{array} - \begin{array}{c}\text{actual} \\ \text{weaning wt}\end{array}}{\begin{array}{c}\text{number of days} \\ \text{between weights}\end{array}}\right) \times 160 + \begin{array}{l}\text{205-day weaning wt} \\ \text{adjusted for age of} \\ \text{dam}\end{array}$$

Yearling weight is medium to highly heritable and includes all gains from birth to one year of age.

Adjusted 550-day weights would be of more value in selecting heifer replacements for the herd. Weights should be taken at the end of the first grazing season when the heifers are yearlings. The recommended formula for calculating adjusted 550-day weights would be:

$$\begin{array}{l}\text{Adjusted} \\ \text{550-day} \\ \text{weight}\end{array} = \left(\frac{\begin{array}{c}\text{actual} \\ \text{final wt}\end{array} - \begin{array}{c}\text{actual} \\ \text{weaning wt}\end{array}}{\begin{array}{c}\text{number of days} \\ \text{between weights}\end{array}}\right) \times 345 + \begin{array}{l}\text{205-day weaning wt} \\ \text{adjusted for age of} \\ \text{dam}\end{array}$$

Either 365-day or 550-day weight ratios may be calculated by dividing the adjusted yearling weight for an individual by the average yearling weight for all individuals in the group multiplied by 100. Individuals with the largest yearling weight ratios should then be selected for breeding purposes.

18.1.5 Type and Size

By type is meant the body form and structure supposed to be ideal for the purpose for which the animal is produced. It includes something more than just size and scale; type is estimated by visual appraisal and cannot be measured by a ruler or tape.

Type has been used very widely in the past for the selection of breeding stock. Perhaps the main reasons for this are that a visual appraisal is rather simple to make and that the price per pound of feeder and slaughter animals is dependent to a considerable extent on this particular trait. In spite of the popularity of type, however, it has been shown time and time again to be inadequate as a criterion for selecting animals that are best suited for the feed lot and for the packer and retailer. A more adequate and accurate measurement of performance in the feed lot is to record the rate and efficiency of gain as discussed previously. Carcass quality can best be measured after the animal is slaughtered. This phase of production will be discussed in a later section.

Various experiment stations have used type scores and slaughter grade scores based on visual appraisal in research studies in the past several years. One of the methods of scoring used is presented in Table 18.4. Experimental data [44] indicate that type-score and grades are subject to considerable error, in that the agreement among different judges for the same animal is difficult to achieve, and the repeatability of the score of the same judges on the same animal at different times is low. Undoubtedly, the condition of the animal at the time of scoring is a factor, for fat imparts a pretty color, and even an excellent animal appears mediocre when it is thin and in poor condition.

Type-scores are often taken at weaning time and, for bulls, at the end of the feeding period when they have been on test. Many experiment station workers have given steers a score for slaughter grade that is often similar to scores given to bulls at the end of the feeding trials.

Table 18.4   Beef-cattle scoring form

Fancy +	15
Fancy	14
Fancy −	13
Choice +	12
Choice	11
Choice −	10
Good +	9
Good	8
Good −	7
Medium +	6
Medium	5
Medium −	4
Common +	3
Common	2
Common −	1

Fig. 18.5 These bulls gained at the same rate on an individual feeding test, but the one at the top had a much lower conformation score. If both type and fast gains are desired, they must both be selected for since they seem to be inherited independently. (Courtesy of the University of Missouri.)

Data summarized in Table 18.1 show that weaning score is about 33 percent heritable, whereas slaughter grade is about 47 percent heritable. Both are highly enough heritable so that they should be improved through selection. Since type is of considerable economic importance, it is well to use this trait in the improvement of beef cattle through breeding. Scores at weaning and at the end of the feeding period should be taken on as many of the animals in the herd as possible.

The tendency of producers recently toward marketing cattle at a younger and lighter age has resulted in the production of cattle that mature earlier and are blockier in type and conformation than those of several years ago. As a result, breeders have emphasized selection for width, depth, and compactness in their breeding animals. The controversy as to which is most desirable, the small, blocky type, the large, rangy type, or the intermediate type, led to research work in which animals of the various types were compared. The American Hereford Association [58] sponsored a test at the Kansas, Oklahoma, and Ohio Experiment Stations to determine what type or size of animals should be the most desirable from the standpoint of rate and economy of gain and finishing ability. In this study, it was found that steers sired by large bulls made faster gains than those sired by medium-size and small bulls. Likewise, steers sired by medium-size bulls gained faster than steers sired by small bulls. These gain advantages were more pronounced during the wintering and grazing phases of the growth period than during the full-feeding phase. Over-all differences among the three groups in the economy of gains were not significant, but when the ration consisted of a large portion of roughage or grass, the small bulls produced progeny that made more costly gains. The results indicated that the medium-size cattle were the most desirable, because they tended to combine the gaining ability of large cattle and the finishing ability of small cattle without a lowered efficiency of gain.

Another study [6] at the Oklahoma Station, compared conventional-type Herefords with comprest-type. The mating of comprest bulls to comprest cows resulted in a lower reproductive performance and a lower calving percentage, apparently due to one or more forms of hereditary dwarfism. Calves produced by the comprest matings that were not dwarfs either were comprests or resembled calves from conventional Herefords. These large-type calves were definitely superior to their comprest half-sibs in rate and economy of gains in the feed lot, but the comprests attained a finish at an earlier age. Similar results were obtained when comprests were compared with calves from the conventional-type line.

A similar study at the Colorado Station [49] showed that the small, comprest types weighed about 20 percent less at weaning and slaughter than conventional-type cattle, but they had about the same feed efficiency and the same percentage of higher-priced cuts. More calving difficulties in the comprest cows and the occurrence of dwarfism made them less profitable than conventional cattle.

A study at the New Mexico Station [25], in which the performance of large- and small-type cattle was studied over a period of several years, showed that the large cows were superior to the smaller, more compact cows in the percentage calf crop weaned and in longevity.

Past experience indicates that selecting for the small, compact, quick-maturing type of cattle may lead to an increase in the frequency of occurrence of some kind of dwarfism. For this reason, perhaps the intermediate type of animal would be the most desirable in the long run.

How large beef cattle should be for efficient production has become a question of increasing interest in recent years. The mature size of the brood cow as well as the potential mature size of calves in the feed lot should be considered.

Mature cow size is not easily measured. A combination of skeletal size, fat thickness, and several body weights taken at different times after maturity would probably give the most accurate measurement of this trait.

The *maintenance requirement* of cattle appears to be more closely related to body surface area than body weight. The term "maintenance requirement" means the amount of food required to support the animal without its gaining or losing any weight when it is doing no work and yielding no product. In other words, the maintenance requirement means the amount of food necessary for the maintenance of the vital body processes from day to day.

The maintenance requirement seems to be more nearly proportional to the 0.75th power of body weight than to body weight, although there is not complete agreement on this matter. For example, an 1800-pound cow does not require as much food as two 900-pound cows for maintenance but probably requires 20 to 25 percent less food.

Large cows generally wean heavier calves than small cows, although there are individual exceptions. In addition, there is a wide degree of variation in the pounds of calf weaned per year within a group of cows of the same size. However, because large cows tend to have a higher maintenance requirement than small cows, they have to wean heavier calves to equal the efficiency of the small cows. Identifying cows that wean large calves, regardless of their size, so that they and their progeny may be kept in the herd for calf production is probably more important than how large the cows might be.

Smaller cows are more likely to experience calving troubles than large cows, especially when bred to a large bull. This is especially true of heifers when they produce their first calf. It does not take many calves lost because of difficulties encountered at birth to reduce the efficiency of small cows as calf producers. Good management will prevent some of these losses.

It is difficult to define the optimum size in cows for most efficient production. It seems best to avoid extremes in size. If cows are selected for efficiency of calf production, it is possible that size will take care of itself or will not be important.

A potentially large mature size of cattle in the feed lot may be economically desirable. Although calves are fattened at a relative young age long before they reach their mature size, those with a large size potential will tend to gain faster and probably more efficiently. They will also tend to produce more lean and less fat at slaughter and may have to be fed to heavier weights than small cattle to reach a desirable slaughter grade.

The practical importance of large mature size in cows and calves in the feed lot indicates that the following recommendations are of value. Medium-sized cows with superior milk-producing and mothering ability should be used for calf production. These cows should be bred to large bulls of the desired type and conformation except first-calf heifers, which probably should be bred to bulls of a smaller size. Some evidence indicates that small cows from a breed such as the Jersey experience little difficulty at calving time. Possibly selection for ease of calving and culling cows which have experienced difficulty at calving would reduce the incidence of this problem in the cow herd.

18.1.6 Carcass Desirability

The final criterion in judging beef cattle is quality of the meat they produce at slaughter. The previous discussion has dealt with the efficiency of production; the discussion now will deal with the end product, the actual meat produced.

At the present time, breeding animals cannot be selected directly on the basis of their own carcass quality because this can be determined only after death. No good method has yet been developed for the determination of carcass quality in the live animal, even though studies have been directed toward this objective and valuable methods may be obtained in the future. For this reason, selection of breeding animals for carcass quality must be based on progeny or sib tests.

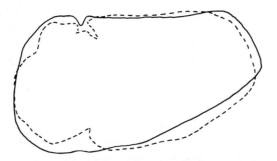

Fig. 18.6 Loin eye area as determined in the live animal by means of the Sonoray (unbroken lines) as compared to the actual loin eye area measured in the carcass after slaughter (broken lines). The total area varied only 0.1 of an inch. (Courtesy of the Animal Husbandry Department, University of Missouri.)

Methods of evaluating carcass quality in steers to meet consumer preference for different kinds of meats are now receiving attention. In addition, emphasis is being placed on the determination of conformation characteristics in the live animal that are related to a higher percentage of the high-priced cuts of meat.

Heritability estimates for certain items of carcass quality are given in Table 18.1. Most of these estimates are high enough to indicate that progress could be made in selecting for these traits and that much of the variation is due to the additive effect of genes.

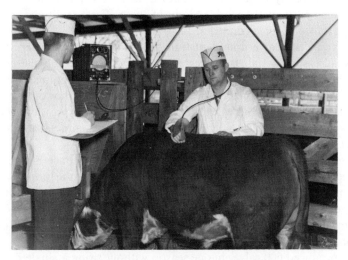

Fig. 18.7 Measuring the loin eye area in the live animal by means of the Sonoray. (Courtesy of the University of Missouri.)

18.1.7 Freedom from Inherited Defects

Another important economic trait in beef cattle is the freedom from certain inherited defects that interfere with performance and fertility. Most of these are probably inherited as recessive traits and will be of no great significance unless the heterozygote is preferred in selection.

That detrimental recessive genes are probably carried by many breeding animals has been shown to be true when progeny tests have been made by mating sires back to their daughters. Progeny-test records on six bulls selected at random showed [34] that each was carrying from one to four recessive genes with undesirable effects. Recessive genes, however, need not cause too much concern, because inbreeding is seldom practiced in commercial beef production. Outbreeding reduces the chances that the recessive alleles will be combined, thus inhibiting the phenotypic expression of these genes.

The most important inherited defect in beef cattle known to date is snorter dwarfism. It has caused an enormous loss to the industry and has concerned almost every breeder either directly or indirectly. Much research to determine the physiological cause of dwarfism has been done. The use of pedigrees and progeny tests in selecting breeding stock have helped to limit it but even so the incidence is still of considerable importance. Evidence seems to indicate strongly that heterozygotes are preferred in selection.

Defects apparently of a recessive nature have been reported for beef cattle in various parts of the world. Many of these affect fertility and have been of importance in some countries.

A defect inherited as a dominant trait should cause little concern to a breeder, because such a trait is expressed by the phenotype and the genes for it may be discarded from the herd merely by culling all animals which show the trait. This kind of inheritance is more complicated, however, when penetrance is not complete and when the genes vary in their expression. Here, the phenotype may not indicate the presence of the genes, and yet the genes for the trait will be transmitted to the offspring.

The use of artificial insemination has heightened the concern over increasing the frequency of detrimental recessive genes in the population. One male may be used to sire several thousand calves in his lifetime. If he is heterozygous for a particular trait, according to the laws of chance, one-half of his offspring will receive the recessive gene from him. The author knew of one case in which a bull, later proved to be heterozygous for the snorter dwarf gene, produced 600 calves in one season without siring a single dwarf. When his sons and daughters were mated to other heterozygotes, or when a certain amount of inbreeding was practiced, many dwarfs were produced.

Fig. 18.8 Two half sisters that have had a long and productive life in the same herd. Longevity should be given attention in selection, especially if it is accompanied by high production.

Thus, hundreds of carriers of a recessive gene were introduced into a herd in one year's time.

18.1.8 Longevity

This is another trait of considerable economic importance to the cattle breeder. The high cost of raising heifers until they produce their first calf makes it imperative that they have a long, productive life span in the herd.

The average generation interval for beef cows (average age at the time their calves are born) was found to be 4.6 years in a Missouri study [8]. The average generation interval for beef bulls in this same study was 3.6 years for one breed and 4.3 years for another.

Little is known about the degree of heritability of the productive life span in beef cattle. Automatic selection has been practiced for this trait, because the cows with the longer productive life will leave more offspring in a herd, making them more likely to leave those selected for breeding purposes. However, bulls may be disposed of before their productive life is completed in order to prevent inbreeding.

Replacement stock should be selected when possible from parents that possess a long, productive life span. Such parents show they possess the inheritance for longevity, and they probably do not possess detrimental dominant, or partially dominant, genes or detrimental recessive genes that express themselves late in life. They also show that they possess genes for good performance.

The opportunity for selection for longevity in beef cattle is greater than in most other classes of livestock. There is especially little opportunity to select for this trait in swine, for a boar and sow are usually discarded after they have produced a few litters, because of their extreme size. In addition, the boar is often sold because he has so many daughters in the herd that he can no longer be used without practicing inbreeding.

18.2 CORRELATION AMONG PERFORMANCE TRAITS IN BEEF CATTLE

Animal breeders are becoming more and more aware of the fact that genes that control the expression of one quantitative trait are quite likely to influence the expression of others. This may be due to the close biochemical and physiological relationships among various life processes. The realization of this has led to the measurement of genotypic, phenotypic, and environmental correlations among traits [19]. Some research on these factors has been done in beef cattle.

18.2.1 Phenotypic Correlations between Traits

Phenotypic correlations between traits are the gross correlations that include both the environmental and the genetic portions of the covariances. Phenotypic correlations for traits are important because they directly affect the size of the selection differentials when several factors are used in a selection index. This is especially true when the correlation is high, whether it is positive or negative.

A summary of many studies in which phenotypic correlations between various traits have been reported is given in Table 18.5. Of particular interest in these correlations is the positive relationship between rate of gain and efficiency of gain and between slaughter grade and some of the other items of carcass quality. A higher slaughter grade is associated with a higher carcass grade, a higher dressing percentage, slightly greater ribeye, and more fat on the outside of the carcass. This would indicate that slaughter grade is one of the better indicators of carcass desirability in the live animal, and attention to this correlation, until a better measure is found, will be of great importance in selecting meat animals.

18.3 GENETIC CORRELATIONS AMONG TRAITS

By genetic correlations is meant the probability that the same gene, or genes, affects two or more traits.

Many of the same genes that affect growth during pregnancy also seem to affect growth rate from birth to weaning [28, 31]. After weaning and during the feed-lot period, many of the same genes affect growth rate. In one study, steers were fed after weaning for 252 days and the period divided into three 84-day periods. The genetic correlations between the various periods ranged from 0.45 to 0.82 [22], indicating that many of the same genes affected the growth rate in the three different periods. Similar conclusions were made in two other studies [42, 54]. Genetic correlations between preweaning and postweaning gains and type scores were large in another study [28]. This seems reasonable, because one would expect the fast-gaining calves to show more desirable beef characteristics than unthrifty, slow-gaining calves.

In beef cattle, there is some evidence that there may be a genetic antagonism between high milk production and fast growth rate [28]. Such an antagonism would mean that an improvement through selection for greater milk production in cows would result in a simultaneous decrease

Table 18.5 Phenotypic correlations between economic traits in beef cattle

	Wean wt.	Wean score	Rate of gain	Efficiency of gain	Slaughter grade	Carcass grade	Dressing percent	Area of rib eye	Thickness of fat
Wean weight	1.00	0.26	0.16	-0.40	-0.05	0.43	—	0.32	0.26
Wean score		1.00	-0.05	-0.04	0.28	0.20	—	—	—
Rate of gain in feed lot			1.00	0.51	0.09	0.32	-0.14	0.36	-0.07
Efficiency of gain in feed lot				1.00	0.15	0.02	-0.24	0.07	-0.03
Slaughter grade					1.00	0.64	0.38	0.29	0.50
Carcass grade						1.00	0.45	0.07	0.54
Dressing percent							1.00	0.36	0.25
Area of rib eye								1.00	0.01
Thickness of fat									1.00

Most of the correlations represent averages from several studies, many of which were summarized in references 7, 43, and 45.

in the rate of gain by the calves produced by these cows. However, more research needs to be done to clarify and prove this point.

18.4 GENOTYPE ENVIRONMENTAL INTERACTIONS

An interaction between the genotype and the environment means that individuals of two or more genotypes may respond differently in one environment than they will in another. Interactions are more likely to occur if the environments are widely different. Beef cattle are grown and fattened under a wide variety of conditions, and when artificial insemination is used the same sire may produce offspring in many different areas. This makes it very important that breeders know if such interactions exist for important economic traits.

A good example of a genotype × environmental interaction in beef cattle is the relative growth rate of Brahmans and British breeds in a hot and a cold climate. The Brahmans and their crosses are often superior to the British breeds and their crosses for performance traits in a hot, tropical climate. This is why the Brahmans and their crosses are preferred to the straight British breeds in many tropical and semitropical regions of the world. The adaptability of the Brahmans to tropical climates is due to their conformation, heat, disease, and insect resistance and possibly other traits.

Genotype × environmental interactions may be important even within a breed for one or more traits. The progeny of two or more bulls may rank differently for performance in one environment than in another.

Research results suggest that genotype × environmental interactions may be important for some traits in beef cattle (Pani, 1970), but reports do not completely agree on this subject. However, until information on this subject is more complete, it is recommended that both the parents and their progeny be produced in as similar an environment as possible (Fig. 18.9).

18.5 EFFECT OF INBREEDING ON PERFORMANCE TRAITS IN BEEF CATTLE

A study of breeding soundness in yearling Hereford bulls at the Colorado station showed that inbreds were significantly inferior to linecrosses in estimated breeding values based on semen evaluations. Inbred bulls were not inferior in breeding soundness to cooperators' outbred bulls, however. Increased inbreeding did not appear to result in increased infertility in the inbred bulls, although there was some suggestion of a retarding effect of

Fig. 18.9 Illustration of genetic X environmental interactions where two breeds (or genotypes) may rank differently in two different environments. (Courtesy of USDA.)

inbreeding on the age of puberty. The data also showed that lower fertility of the inbred bulls as compared to the linecross bulls may have been due to infertility in a few inbred lines rather than among all inbreds. Similar results were observed in another study at the Virginia Station, where fertility was based on actual matings.

At the Colorado Station [49], the weight of calf weaned per 100 cows bred was almost 43 percent greater from crosses of inbred sires on unrelated linecross cows than it was from inbreds. Most of this advantage was due to a 30 percent greater calf crop and 10 percent heavier weights. The linecross calves also outgained and scored higher than the inbreds. These data suggest some effect of inbreeding on these particular traits.

In three studies [5, 27, 35] a decrease in weaning weight varying from 4.8 to 17.5 pounds for each 10 percent increase in inbreeding of the calf was observed. On the other hand, a 10 percent increase in the inbreeding of the dams resulted in a decrease of 11.5 pounds in weaning weights of their calves in one study [5], whereas an increase of 9.50 pounds was found in another [35]. Inbreeding had no effect on type score in the latter study.

The genetic history of the Line 1 Hereford cattle at the United States Range Livestock Experiment Station at Miles City, Montana, gives some information on inbreeding effects in the line from 1936 to 1949. During this time the inbreeding increased slowly and gradually until, in 1948, the average of the herd was 15.95 percent. Record of performance results indicated that the line increased in rate of gain in the feed lot and in weight-for-age without much apparent loss of quality of carcass. (Selection had been made for weight-for-age and rate of gain mostly in the sires, with little selection pressure on the females.) Progress actually made was as great as or even greater than expected as calculated from heritability estimates, selection differentials, and generation intervals.) Apparently, selection for performance was effective in spite of an increase in inbreeding, and may suggest that inbreeding did not have an important effect on weight-for-age or rate of gain in the feed lot in this line.

The effects of inbreeding on beef-cattle performance were studied at the South Dakota Station [13]. Inbreeding of the calves had reached 30 percent and that of the cows 25 percent when the results were reported. Inbreeding effects on the calf and the cow appeared to be more important on weaning traits than on postweaning traits, although significant inbreeding effects were observed in both. It was concluded that breeders could close their herds and use as few as four unrelated bulls initially without encountering severe inbreeding effects.

18.6 TOPCROSSING AND LINECROSSING IN BEEF CATTLE

Topcrossing refers to the mating of bulls from an inbred line with noninbred females not related to them. Although it may be practical to use inbred bulls for breeding providing they are fertile, it may not be practical to use inbred females on the farm or ranch if their fertility and mothering ability has been lowered by inbreeding.

Bulls from two inbred lines were topcrossed on outbred cows in an Ohio experiment [51]. Outbred bulls were also used in the experiment as a control. The results showed that bulls from the two inbred lines sired calves that gained faster on test and made faster gains per day of age than progeny of the outbred bulls. The progeny of the inbred bulls also tended to be more efficient in the conversion of TDN to meat.

Linecrossing may be defined as the crossing of different inbred lines within a breed. Some inbred lines of Hereford cattle have been developed in the United States, and the amount of heterosis shown when these lines were crossed among themselves has been determined.

Results from the crossing of five inbred lines of Hereford cattle developed at the United States Range Livestock Experiment Station, Miles City, Montana have been reported [53]. Matings were made in such a way as to produce all straight-line and crossline combinations of progeny. Heterosis for birth weight was 3.0 percent for bull calves and 3.8 percent for heifer calves. For weaning weight it was 5.1 percent for bulls and 9.4 percent for heifer calves. Postweaning gains showed 2.9 percent heterosis for bulls and 3.7 percent for heifers. Experiments conducted at the Colorado station showed a 10 percent increase in weaning weights and 6.0 percent increase in feed lot gains of linecross over outbred Herefords.

These results show that developing inbred lines within a breed and crossing them will result in hybrid vigor or heterosis. It is not known if enough hybrid vigor results to pay breeders to develop inbred lines for the specific purpose of crossing.

18.7 CROSSBREEDING IN BEEF CATTLE

Most of the early work with crossbreeding in beef cattle was done in the South and Southwest, where Brahmans were crossed with native and British breeds of cattle. Crosses of this kind were produced to combine the heat, insect, and disease resistance of the Brahman with the beefiness and other desirable qualities of the British breeds. In addition, considerable hybrid vigor was also found for traits showing hybrid vigor. This resulted in the superior overall performance of the crossbred individuals.

Several experiments have also been conducted in which the beef breeds (non-Brahman) have been crossed among themselves or with dairy breeds. Many of these experiments, especially those involving beef × dairy crosses, have been initiated, or completed, only recently.

18.7.1 Crosses of European Beef Breeds

The European beef breeds studied more thoroughly in crossbreeding experiments include the Angus, Hereford, Shorthorn, and Charolais. Other European breeds are being introduced into the United States from time to time, but complete experimental data concerning their use in crossbreeding is not available. Table 18.6 includes data from 15 reports on results of crossbreeding with the European beef breeds. Data are not available for each trait in all experiments, so the averages given in this table are those for experiments in which information was given on a particular trait. The data are from crosses of individuals of two different breeds, and therefore only the calves produced are crossbreds and show hybrid vigor.

Table 18.6   Average amount of heterosis for various traits when straight-bred bulls of one breed were mated with straightbred cows of another breed. Only the European breeds were used, and only the calves were crossbred.

Economic trait	Crossbred calves as a percent of purebred calves
Percent calf crop born	105.2
Percent calf crop weaned	107.5
Weaning weight of calf in lb	103.4
Pounds of calf weaned per cow bred	111.2
Postweaning rate of gain	103.3
Postweaning efficiency of gains	101.6
Carcass grade	104.6

Note: Each figure represents averages from 15 different reports. Subtract 100 from each figure to obtain the percent heterosis for each trait.

Fertility.   In beef cattle, fertility may be defined in various ways. From the practical standpoint, it probably should be defined as the percentage of cows exposed to the bull that wean live calves. When defined in this manner, many components of fertility are included, such as conception rate, embryonic and fetal survival, and the number of young that survive from birth to weaning.

Only a few experiments report data giving some indication of the influence of crossbreeding on the rate of conception. The percentage of cows conceiving at first service was 10.1 percent higher when cows of the European breeds produced crossbred than when they produced purebred claves (Wiltbank, *et al.*, 1967) [57]. However, pregnancy tests showed that only about two percent more cows bred to produce crossbred calves were pregnant when checked for pregnancy two to four months after the end of an approximate 90-day breeding season. The prenatal mortality, as determined from the number of cows diagnosed pregnant that failed to produce a calf, varied between 4.3 and 5.0 percent, but was no lower in cows producing crossbred calves than in those producing purebred calves. In a Missouri study, prenatal mortality was higher in purebred cows producing Purebred calves than in those producing crossbred calves. This study included data from five consecutive calf crops involving the Angus, Hereford, and Charolais and their reciprocal crosses.

As shown in Table 18.6, the average calf crop born was 5.2 percent higher when purebred cows produced crossbred than when they produced purebred calves. This margin was increased to an average of 7.5 percent at weaning, indicating that a larger percentage of crossbred than purebred calves survived to weaning. This was probably due to their greater vigor. However, if calving difficulties are increased when a bull from a large breed is mated to cows of a small breed, the percentage of calves surviving to weaning may be less when crossbred calves are produced because of higher death losses at birth.

Weaning weights of crossbred calves from purebred cows averaged 3.4 lb heavier than when they produced purebred calves, as shown in Table 18.6. Since more crossbred calves survived to weaning and were heavier at weaning, about 11.2 percent more pounds of calf were weaned per cow bred than when purebreds were produced.

Crossbred calves gained about three percent faster and made one to two percent more economical gains than purebred calves. They also scored higher for carcass grade at slaughter, probably because they were in slightly higher condition due to their more rapid gains from birth to slaughter.

The three-breed cross when compared with the two-breed cross should give an estimate of the amount of hybrid vigor expressed for maternal traits in crossbred cows. Few data are available on this subject for the European breeds of cattle. Virginia experiments with the European breeds showed that crossbred cows were not superior to purebred cows for weaning and preweaning traits. Preliminary reports from a Nebraska experiment, however, show a seven to eight percent advantage for crossbred cows in the percent calf crop weaned and a five to six percent advantage in weaning weights of the calves. This would result in a 12 to 14 percent advantage for crossbred cows in pounds of calf weaned per cow exposed to the bull. When this advantage for crossbred cows is added to the 10 to 12 percent advantage for crossbred over purebred calves, the three-breed cross would have an advantage of 20 to 26 percent over the production of purebred calves from these same breeds. Crossbred heifers also appear to reach the age of puberty earlier than purebred heifers. This may be an added advantage.

As a general rule, most of the hybrid vigor is expressed in the crossbred calves by the time they are weaned when European breeds are crossed. This is probably because the traits expressing hybrid vigor are related to physical fitness and vigor. Postweaning growth rate is three to four percent faster and one to two percent more efficient in crossbred than in purebred calves. Carcass quantity and quality are affected little by hybrid vigor, with the average for crossbreds being about equal to the average of the pure breeds included in the cross. This average effect can be an advantage, however, since crossing a large, growthy breed with another that is small and early maturing may reduce the amount of fat in the crossbred calf as compared to the small, fatter, early maturing breed and increase the rate and efficiency of gain over the average for the smaller breed.

The crisscross, or backcross, system of crossbreeding has not been thoroughly tested in many experiments. After the F_1 generation of this system of crossbreeding both cows and calves will be crossbred, with the advantage of hybrid vigor being expressed in both cows and calves. Virginia experiments showed that this system of crossbreeding gave results comparable to those of the three-breed cross.

Let us emphasize again that the kind of crossbred animal produced depends upon the breeds used to make up the cross. Some breeds when crossed may give a more desirable slaughter animal than others. The livestock producer should choose those individuals and breeds best suited for his particular need and purpose so that he will produce the largest amount of beef at slaughter per cow in the breeding herd.

18.7.2   British × Brahman Crosses

Since Brahmans belong to the species *Bos indicus* and the European breeds to the species *Bos taurus,* they are quite diverse genetically and therefore should show a large amount of hybrid vigor in those traits that normally express it.

Data presented in Tables 18.7, 18.8, and 18.9 have been compiled from many sources, but most of the experiments were conducted in the southern part of the United States under semitropical conditions, to which the Brahmans and their crosses are better adapted and should have an advantage over the European breeds.

Table 18.7   Performance of F_1 British × Brahman calves as compared to the average of the pure breeds used in the cross

	As a percentage of the average of the pure breeds
Percent calf crop weaned	101.2
Weaning weights of calves in lb	115.5
Lb of calf weaned per cow bred	116.9

Table 18.8   Performance of F_1 British × Brahman cows as compared to the performance of cows of the pure breeds used in the cross

	As a percentage of the average of the pure breeds
Percent calf crop weaned	114.3
Weaning weights of calves in lb	114.6
Lb of calf weaned per cow bred	131.0

Table 18.9 Postweaning performance of Brahman crossbreds as compared to the average of the pure breeds used in the cross

	Crossbreds as a percentage of the average of the pure breeds
Postweaning daily gains	100.3
Carcass weight per day of age	114.3
Postweaning efficiency of gains	104.8
Percent lean in a rib sample	97.3
Percent fat in a rib sample	111.3
Percent bone in rib sample	92.7
Carcass grade	105.3
Tenderness (Warner-Bratzler shear)*	91.8

*Means that crossbreds were more tender than purebred average.

The F_1 cross of the Brahman and European species shows only a small amount of heterosis (1 percent) for percentage calf crop weaned. The reason for this small amount of heterosis is not fully understood. It is possible that the wide genetic diversity of the *Bos indicus* and *Bos taurus* may result in some incompatability of the sperm and egg and/or in the fetus and mother in some matings. Certainly such an incompatability has been observed in the cattle $\times$ American bison cross, as well as in crosses of the sheep and goat.

The F_1 cross of the Brahman and European breeds of cattle show considerable hybrid vigor in gains from birth to weaning, with the cross showing a 15 to 16 percent advantage over the pure breeds making up the cross. An average advantage of 16 to 17 percent in pounds of calf weaned per cow bred is also observed when purebred Brahman or British cows produced crossbred than when they produced purebred calves (Damon, *et al.*, 1959) [11].

Considerable hybrid vigor is noted in F_1 crossbred Brahman $\times$ British cows kept for breeding (Cundiff, 1970) [10]. Data summarized in Table 18.9 show an average advantage of 14.3 percent in the percent calf crop weaned for F_1 cows over the average of purebred cows. They also showed and average advantage of 14.6 percent over purebred cows in pounds of calf weaned. This represents an advantage ranging from 28 to 30 percent for the crossbred cows in the pounds of calf weaned per cow in the breeding herd. It is not known if this advantage of Brahman crossbreds is peculiar to tropical and semitropical regions or if it would also occur in the temperate regions of the world.

18.7.3 Beef $\times$ Dairy Crosses

Certain dairy or dual-purpose breeds of cattle have received increasing attention for possible use in crossbreeding for beef production in the United

Fig. 18.10   A crossbred Brahman X Hereford cow with her steer calf sired by a Charolais bull. The 205-day adjusted weight of this calf was 593 pounds. Crossbred Brahman X British cows are well adapted to tropical and semitropical regions. (Courtesy of Dr. J. W. Turner, Louisiana State University.)

States. Some of the dairy or dual-purpose breeds possess and transmit rapid growth rate, large mature size, more lean and less fat at slaughter, and high milk production as compared to the beef breeds. All of these characteristics are desirable for beef production. It is also possible that dairy × beef crosses may produce a large amount of hybrid vigor in traits normally expressing it, since they should be quite divergent genetically.

Dairy cows mated to beef bulls have produced crossbred calves that have been an important source of beef for many years. Few beef cows have been bred to dairy bulls until recently. Dairy cows or crossbred dairy × beef cows, because of their superior milk production, tend to produce heavier calves at weaning than straightbred beef cows on pasture and on the range. Some concern has been expressed that dairy cows used for beef production may develop more udder troubles than beef cows. This may depend, of course, upon the time in the year when calving occurs. Cows would give more milk when on green, lush pasture than when in dry lot or on dry forage, and this might increase the incidence of caked udders and scours in their calves.

In general, when the Brown Swiss and Holstein breeds have been used in crosses with the beef breeds, the crossbred calves produced have postweaning gains comparable to gains made by crossbred beef calves. They may have a lower carcass grade at slaughter, however, because they possess a larger proportion of lean to fat and may not have enough finish. If induced twinning in cattle should become practical, the high milk production of the dairy cows may make them more suitable for the production of calves than beef cows.

18.7.4 Crosses of Cattle and American Bison

Experimental work at the Canadian Experimental Farm, Manyberries, Alberta has involved the crossing of Herefords (*Bos taurus*) with the American bison (*Bos bison*) [38]. This was done in an attempt to develop a new breed of cattle better adapted to the cold conditions of Canada. This new breed was called the cattalo.

In a six-year study, Hereford cows weaned a higher percentage calf crop than F_1 hybrid, or cattalo, cows, and the calves were heavier at birth. However, F_1 hybrid and cattalo cows weaned heavier calves. Cattalo cows were superior to Hereford cows in their ability to graze on upland winter range in cold weather, but Hereford calves gained faster in the feed lot and had higher carcass grades at slaughter.

18.7.5 Conclusions on Crossbreeding in Beef Cattle

Although there is much more to be learned about the various aspects of crossbreeding in beef cattle, some facts are fairly clear. Evidently, crossbreeding produces the highest weaning weight in calves. It may also increase fertility. Crossbred calves generally show much more vigor than purebreds up to the time of weaning, especially when Brahmans are involved in the crosses. This could be because the Brahmans can transmit their ability to resist heat and other adverse environmental conditions prevalent in the South and Southwest, where these studies have been made.

Crossbred cows also seem to be better mothers than purebred cows from comparisons made to date, but more work is needed before we can state for certain just how superior the crossbred mothers are to the purebreds.

Crossbreds are slightly superior to purebreds, on the average, in postweaning rate of gain, efficiency of gain, and carcass grade. However, most of the advantages of the crossbreds are realized by the time the calves are weaned. Heavier weights at weaning results in heavier weights at the end of the feeding period, even if the daily gains during this period are no faster than in purebreds. If crossbreds do not sell for less per pound at weaning because they are crossbreds, and this is no longer the case, this system of breeding seems to offer definite advantages for more efficient beef production.

The very high heritability estimates for rate of gain in the feed lot and for most carcass traits, together with medium-to-high estimates for efficiency of gains, indicates that, for the present at least, considerable improvement in these traits can be made within the pure breeds. Therefore selection of breeding stock should be made with this goal in mind. The improvement of purebreds would still be very desirable, even if crossbreeding becomes the breeding system of choice in the future. The most efficient production results, even when crossbreeding is practiced, when both parents are superior for the important economic traits.

18.8 KINDS OF GENE ACTION AFFECTING ECONOMIC TRAITS IN BEEF CATTLE

The kinds of gene action of importance in the phenotypic expression of the various economic traits in beef cattle are summarized in Table 18.10. Although much more information on beef-cattle breeding is needed at this time, we already have considerable evidence as to the kind of selection and mating systems that are the most desirable to improve the most important economic traits. The choice of selection and mating system will also depend upon the aims of the breeder and on what traits are to be emphasized to the greatest extent.

Beef cattle are seldom selected on the basis of a single trait, since the

Table 18.10 Kinds of gene action that seem to be the most important in their effect on certain economic traits in beef cattle

Trait	Degree of heritability	Effects of :		Proportion of genetic variation due to different types of gene action	
		Inbreeding	Crossbreeding	Nonadditive	Additive
Weight of calves at weaning	medium	large	medium	medium	medium
Weaning score	medium	low	small	small	medium
Rate of gain in the feed lot	high	probably low	small	small	very large
Efficiency of gains in the feed lot	medium to high	small	small	small	large
Slaughter grade	high	unknown	small	small	large
Carcass items	high	probably small	small	small	large

economic value or net worth of an individual animal may be determined by several traits. A study was made at the Colorado Station [32] to determine the relative economic importance of several traits in determining the net income from a calf-raising and -feeding operation. The traits studied included weaning weight and grade, daily gain in the feed lot, days required to finish to a slaughter grade of low choice, feed required per pound of gain, and cow-maintenance costs. Several selection indexes were constructed to determine which index would be expected to make the maximum genetic progress toward increasing net income per hundred-weight of product marketed. This study indicated that weaning weight alone was an accurate basis for selecting for increased net income in this herd. A simple index, besides weaning weight, with considerable accuracy in this respect was $I = W + 72R$, where W was the weaning weight and R was the rate of gain in the feed lot. Another index, more complicated, was $I = 0.58W + 18.64R - 0.73F - 5.87E$, where W was the weaning weight; R, the daily gain in the feed lot, with a short period between weaning and the beginning of the test, to bring the animals to a full feed; F, the number of days to bring each animal to a low-choice slaughter grade; and E, the amount of feed per pound of gain. To illustrate the calculation of such an index, let us assume the following:

$$W = 400 \text{ lb}$$
$$R = 2.5 \text{ lb}$$
$$F = 200 \text{ days}$$
$$E = 7.5 \text{ lb}$$

The calculated index would be

$$I = 0.58(400) + 18.64(2.50) - 0.73(200) - 5.87(7.50)$$
$$= 232 \qquad + 46.6 \qquad - 146 \qquad - 44.03$$
$$= 88.57$$

This index is given only as an example, to show how several important factors can be considered in the construction of an index so as to aid in selection for maximum genetic progress. It may, or may not, be applicable to the selection of beef cattle in areas other than the region for which it was constructed or for traits other than those selected for here.

18.9 SUGGESTED PROGRAM FOR PUREBRED BEEF CATTLE PRODUCTION

The extension divisions of most states in the United States have outlined improvement programs for their respective states. Therefore, a complete program in every detail will not be given here. Some important points to

BREEDING AND SELECTION IN BEEF CATTLE

consider in a beef-improvement program will be given, however, along with sample record forms in Fig. 18.11 and 18.12.

Breeding stock should be selected as carefully for a commercial program as for the production of purebreds, or seed stock. Therefore, the program previously outlined for the production of purebred beef cattle is recommended.

1. Identify all animals with a tattoo, brand, or some other means.
2. Record the exact birth date of each calf, tattoo at birth, and record the numbers of both the calf and the cow.
3. Obtain a weaning score and weight and correct the weights of calves for age of dam, age of calf, and sex of calf.
4. Retain replacement heifers from those with the heaviest weaning weights and best type-scores.
5. Cull cows after one or two calf crops that consistently wean calves lighter than the average of the herd. The amount of culling will depend upon available replacement stock and upon whether the herd is increasing in numbers or is remaining stationary in size. (Cull on the basis of records, and not on type alone.)
6. Weigh and score heifers again at approximately 18 months of age to obtain information on their rate of gain after weaning and on records of sire groups, as well as information on gaining ability of calves from different cows. Cull those heifers with undesirable type or gains or with obvious undesirable traits.
7. Feed all bull calves that are superior from the standpoint of type and weaning weight. If they cannot be fed individually, feed them as a group, but give all bulls an equal chance at the feed bunk. Feed for at least 150 days, and at the end of the feeding period, calculate the rate and efficiency of gain, and score for type and conformation. Rank the bulls in order for the traits of most economic importance and keep the best for breeding purposes.
8. In purchasing herd bulls, obtain them from a herd where records of the kind mentioned above are kept. Obtain the best bull possible on the basis of type, and rate and efficiency of gain at the end of a feeding test. Select a herd bull from a cow that has a lifetime record of producing a calf each year that is superior in type and weaning weight.

18.10 SUGGESTED PROGRAM FOR COMMERCIAL PRODUCTION OF BEEF CATTLE

Breeding stock should be selected as carefully for a commercial program as for the production of purebreds, or seed stock. Therefore, the program previously outlined for the production of purebred beef cattle is recommended.

Name _____ Address _____ Year _____ Date _____ Breed _____

Calf No. (1)	Sex (2)	Dam (3)	Sire (4)	Age of Dam (5)	Birth Date (6)	Birth Weight (7)	Weaning Date (8)	Weaning Age (Days) (9)	Weaning Weight (10)	210-day Weight (11)	Adj. 210-day Weight (12)	Adj. Daily Gain (13)	Type Grade (14)	Index (15)	Remarks (Creep fed, condition, etc.)

Fig. 18.11 Sample record forms for recording data on calves from birth to weaning.

Name _____ Date _____

Address _____ Year _____ Sex _____ Breed _____

Period Fed _____ Month _____ Day _____ to _____ Month _____ Day _____ No. days on feed _____

Calf No. (1)	Dam (2)	Sire (3)	Initial Weight (4)	Final Weight (5)	Total Gain (6)	Daily Gain (7)	Age at Finish Days (8)	Feed Consumption Total (9)	Feed Consumption Per cwt. (10)	Type Grade (11)	Index (12)	Remarks (Condition etc.)

Fig. 18.12 Sample forms for recording postweaning performance in beef cattle.

A systematic system of crossbreeding should be followed in the commercial production of beef cattle. The system used will depend upon the size of the operation and the climatic and other conditions in which the cattle are to be produced. For example, in a small herd, the producer could use the two-breed backcross or crisscross system very profitably, whereas the larger producer may wish to use the three-breed sire rotation crossing system. However, some producers, large or small, may prefer to use the three-breed static crossing system where both crossbred dams and the purebred sires are purchased. This system of crossing results in optimum hybrid vigor and continues to maintain it.

Breeds used in a crossbreeding system of beef production require careful consideration. Certainly the Brahman crosses contribute much to efficient production of beef in the tropical and semitropical countries. In any region of the world, it is important that crossbred cows used for calf production possess the kind of inheritance that will cause them to be excellent mothers of high milk production. The breed of sire used should be one that possesses and transmits rapid and efficient gains and desirable carcass quality. This would probably mean that the sire would possess a potential heavy mature weight, but such a sire should not be used on first-calf heifers because of the possible loss of the mother and her calf at calving time. This is especially important under range conditions, where cows cannot always be given individual attention at calving time.

The best breeding program possible will not realize its potential unless the herd is properly fed and managed. Proper feed and care allow superior animals to express their full genetic potential. The end result is excellent efficiency of production.

18.11 NEW BREEDS OF BEEF CATTLE

Some new breeds of beef cattle have been developed recently in the South and Southwestern parts of the United States using the Brahman in crosses with other breeds, mostly those of British origin. The Brahmans can withstand the adverse conditions in that part of the country and transmit this ability to their offspring.

Santa Gertrudis

This breed was developed by the King Ranch of Kingsville, Texas, from a cross between Shorthorn cows and Brahman bulls [41]. Its development involved an exploratory phase in which crosses of Brahmans and British breeds were made. This was followed by the multiplication of the progeny

Fig. 18.13 Excellent type Brahman cow. (Courtesy of the Koontz Ranch, Inez, Texas.)

of individuals shown to be superior on the basis of progeny test, accompanied by inbreeding, linebreeding, and selection.

In 1940, 30 years after the first exploratory matings of a Brahman bull with purebred Shorthorn cows, the Santa Gertrudis breed was recognized as a new beef breed. Santa Gertrudis are red in color and are considered to be $\frac{3}{8}$ Brahman and $\frac{5}{8}$ Shorthorn. The breed is very well adapted to the

Fig. 18.14 Purebred Santa Gertrudis cow, age 38 months and weighing 1590 pounds. (Courtesy Santa Gertrudis Breeders International, Kingsville, Texas.)

region where it was developed as well as to other regions of a similar environment. Mature steers and cows average approximately 200 pounds heavier than animals of the British Breeds of the same sex and age.

Charolais

This is a French breed that was developed in the latter part of the 18th century. It was originally used for draft purposes as well as for beef production. The Charolais varies from white to a light-straw color and reaches a very large size at maturity.

Fig. 18.15 Purebred Charolais cow, two years and nine months of age, weight 1734 pounds. (Courtesy of the Litton Charolais Ranch, Chillicothe, Missouri.)

The breed in the United States has been developed from a few individuals imported before laws were passed restricting the importation of breeding stock from countries where foot-and-mouth disease was prevalent. The Charolais breed in the United States may carry genes from other breeds because of the small number of imported purebreds available to breeders in this country. The American International Charolais Breeders Association registers purebred animals and those that are produced by a topcrossing program and possess from $\frac{1}{2}$ to $\frac{31}{32}$ Charolais genetic inheritance. Certificates for animals produced by topcrossing must give the exact percentage of Charolais inheritance as well as that of the other breeds involved. An animal that possesses $\frac{31}{32}$ Charolais inheritance is considered to be a purebred.

Brangus

The Brangus originated from a cross of the Brahman and the Angus. They are black and naturally polled and retain many of the characteristics of the original parent breeds. The International Brangus Breeders Association has its headquarters in Kansas City, Missouri. Registered Brangus are those bred as a breed for several generations or those produced by starting with purebred Angus and purebred Brahmans if regulations of the association are adhered to. Figure 18.17. shows how Brangus may be produced by two different methods by combining the two parent breeds.

Fig. 18.16 Excellent type Brangus bull. (Courtesy of the Bruce Church Ranch, Inc., Yuma, Arizona.)

Charbray

This is an American breed developed by combining the Charolais and Brahman breeds. The American Charbray Breeders Association will register animals that have from $\frac{3}{4}$ to $\frac{7}{8}$ Charolais inheritance, with $\frac{13}{16}$ Charolais and $\frac{3}{16}$ Brahman being the proportion of inheritance that seems most desirable. The breed is horned and is white to light-straw in color.

Beefmasters

This new breed was developed from crosses among Brahmans, Shorthorns, and Herefords. The development of the breed was begun in 1908 by E. C. Lasater of Falfurrias, Texas, and has been continued since 1930 under the direction of his son. No attempt has been made to incorporate a definite percentage of inheritance from the three original breeds, although the genetic composition is probably approximately 50 percent Brahman, 25 percent Hereford, and 25 percent Shorthorn.

Fig. 18.17 Mating systems for the production of Brangus. (Courtesy of *The Cattleman*, Fort Worth, Texas.)

Rigid selection for performance has been practiced within this breed, with special attention to disposition, fertility, weight for age, conformation, hardiness, and milk production. Much less attention has been paid to color and other fine points. Most Beefmasters are horned, but some are polled, and some breeders are selecting for this trait. Although Beefmasters are recognized as a breed, no formal registry association has been formed.

Several breeds of cattle have been imported into the United States and other American countries in recent years. They are often referred to as new breeds, but some of them are only new in that they have been introduced into

Fig. 18.18 The Charbray is a new breed resulting from the crossing of the American Charolais on the Brahman. These are excellent type Charbrays. (Courtesy of the American Charbray Breeders Association.)

these countries only recently. Some of these breeds have existed as a breed for many years in the country of their origin. These breeds include the Limousin, Simmental, Murray Grey, and Maine Anjou among others. Some of these breeds will find a useful place in beef production in their new homes as pure breeds or in a crossbreeding program. They will have to be thoroughly tested and evaluated before their merit is fully known, however.

REFERENCES

1. Baker, A. L., and W. H. Black. "Crossbred Types of Beef Cattle for the Gulf Coast Region," USDAC, 844, 1950.
2. Brinks, J. S., *et al.* "Heterosis in Preweaning and Weaning Traits Among Lines of Hereford Cattle," JAS, 26: 278, 1967.
3. Brown, L. O., R. M. Durham, E. Cobb, and J. H. Knox. "Analysis of the Components of Variance in Calving Intervals in a Range Herd of Beef Cattle," JAS, 13: 511, 1954.
4. Bunch, J. M. "Influence of Some Genetic and Environmental Factors on Fertility in Beef Cattle," unpublished master's thesis, University of Missouri, 1958.
5. Burgess, J. B., N. L. Landblom, and H. H. Stonaker. "Weaning Weights of Hereford Calves as Affected by Inbreeding, Sex and Age," JAS, 13: 843, 1954.
6. Chambers, D., J. A. Whatley, Jr., and D. F. Stephens. "Growth and Reproductive Performance of Large and Small Type Hereford Cattle," Okla AESMP No. MP-34, 1954.

7. Chambers, D., G. O. Conley, and J. W. Whatley, Jr. "Reproductive Efficiency of Range Beef Cows," Okla AESMP MP-48, 22, 1957.

8. Crenshaw, D. B. "Measures of Reproductive Performance in Beef Cattle," unpublished master's thesis, University of Missouri, 1969.

9. Cundiff, L. F. and K. E. Gregory. "Improvement of Beef Cattle Through Breeding Methods," North Central Region Publication 120 (Revised), 1968.

10. Cundiff, L. F. "Crossbreeding Cattle for Beef Production," JAS, 30: 694, 1970.

11. Damon, R. A., *et al.* "Performance of Crossbred Beef Cattle in the Gulf Coast Region," JAS, 18: 437, 1959.

12. Dinkel, C. H., and J. H. Minyard. "Indexing Beef Cattle," SDAESC 144, 1958.

13. Dinkel, C. H., D. A. Busch, J. A. Minyard and W. R. Trevellyan. "Effect of Inbreeding on Growth and Conformation of Beef Cattle," JAS, 27: 313, 1968.

14. Ensminger, M. E. "Breeding Practices of the American Cattleman," NLP, 34: 8, 21, 1956.

15. Fuller, J. G. "Crossbreeding Types for Baby Beef Production," PASAP, p. 53, 1927.

16. Gerlaugh, P., L. E. Kunkle, and D. C. Rife. "Crossbreeding Beef Cattle; A Comparison of the Hereford and Aberdeen-Angus Breeds and Their Reciprocal Crosses," OhioAESRB 703, 1951.

17. Gregory, K. E. "Beef Cattle Breeding," USDA Agricultural Information Bulletin No. 286, 1964.

18. Harris, L. A., L. C. Faulkner, and H. H. Stonaker. "Effect of Inbreeding on the Estimated Breeding Soundness of Yearling Hereford Bulls," JAS, 19:665, 1960.

19. Hazel, L. N., M. L. Baker, and C. F. Reinmiller. "Genetic and Environmental Correlations between the Growth Rates of Pigs at Different Ages," JAS, 2: 118, 1943.

20. Ittner, N. R., H. R. Guilbert, and F. D. Carroll. "Rate and Efficiency of Gain, Carcass Value and Climatic Adaptability of Hereford, Hereford × Brahman and Hereford × Shorthorn Crossbred Cattle," PASAP, WS, 3: 1, 1952.

21. Knapp, B., Jr., and A. W. Nordskog. "Heritability of Growth and Efficiency in Beef Cattle," JAS, 5: 62, 1946.

22. Knapp, B., Jr., and R. T. Clark. "Genetic and Environmental Correlations between Growth Rates of Beef Cattle at Different Ages," JAS, 6: 174, 1947.

23. Knapp, B., Jr., R. C. Church, and A. E. Flower. "Genetic History of the Line 1 Hereford Cattle," MontAESB No. 479, 1951.

24. Knapp, B., Jr., A. L. Baker, and R. T. Clark. "Crossbred Beef Cattle for the Northern Great Plains," USDAC 810, 1949.

25. Knox, J. H. "Research for Cattlemen in the Southwest," AC, 10:18, 1955.

26. Koger, M., and J. H. Knox. "The Correlation between Gains Made at Different Periods by Cattle," JAS, 10: 760, 1951.

27. Koch, R. M. "Size of Calves at Weaning as a Permanent Characteristic of Range Hereford Cows," JAS, 10: 768, 1951.

28. Koch, R. M., and R. T. Clark. "Genetic and Environmental Relationships among Economic Characters in Beef Cattle, I, Correlation among Paternal and Maternal Half-sibs," JAS, 14: 775, 1955.

29. Kincaid, C. M. "Breed Crosses with Beef Cattle in the South," Southern Cooperative Series Bulletin 81, 1962.

30. Krehbiel, E. V., R. C. Carter, K. P. Bovard, J. A. Gaines, and B. M. Priode. "Effects of Inbreeding and Environment on Fertility of Beef Cattle Matings," JAS, 29: 528, 1969.

31. Lasley, J. F., and W. Pugh. "Some Genetic Aspects of Intrauterine and Post-uterine Growth in Beef Cattle," JAS, 15: 1218, 1956.

32. Lindholm, H. B., and H. H. Stonaker. "Economic Importance of Traits and Selection Indexes for Beef Cattle," JAS, 16: 998, 1957.

33. Lush, J. L., J. M. Jones, W. H. Dameron, and O. L. Carpenter. "Normal Growth of Range Cattle," TexAESR 409, 1930.

34. Mead, S. W., P. W. Gregory, and W. M. Regan. "Deleterious Recessive Genes in Dairy Bulls Selected at Random," G, 31: 574, 1946.

35. McCleery, N. B., and R. L. Blackwell. "A Study of the Effect of Mild Inbreeding on Weaning Weight and Grade of Range Calves," PASAP, WS, 5: 223, 1954.

36. Nelms, G., C. M. Williams, and R. Bogart. "A Completely Pelleted Ration for Performance Testing Beef Cattle," PASAP, WS, 4:XIV, 1–2, 1953.

37. Peacock, F. M., W. G. Kirk, and M. Koger. "Effect of Breeding of Dam on Weaning Weight of Range Calves," JAS, 12: 896, 1953.

38. Peters, H. F. and S. B. Slen. "Range Calf Production of Cattle × Bison, Cattalo, and Hereford Cows," C. J. A. S., 46: 157, 1966.

39. Quisenberry, J. R. "Livestock Breeding Research at the U.S. Livestock Experiment Station," USDAAIB 18, 1950.

40. Rhoad, A. O., and W. H. Black. "Hybrid Beef Cattle for Subtropical Climates," USDAC 673, 1943.

41. Rhoad, A. O. "The Santa Gertrudis Breed," JH, 40: 115, 1949.

42. Romo, A., and R. L. Blackwell. "Phenotypic and Genetic Correlations between Type and Weight of Range Cattle at Different Periods," PASAP, WS, 5: 205, 1954.

43. Roubicek, C. B., R. T. Clark, P. O. Stratton, and O. E. Pahnish. "Range Cattle Production, 3," "Birth to Weaning—A Literature Review," ArizAESR 136, 1956.

44. Roubicek, C. B., R. T. Clark, R. M. Richard, and O. F. Pahnish. "Range Cattle Production; 4; Post Weaning Performance—A Literature Review," ArizAESR 138, 1956.

45. Roubicek, C. B., R. T. Clark, and O. F. Panish. "Range Cattle Production; 5; Carcass and Meat Studies—A Literature Review," ArizAESR 143, 1956.

46. Roubicek, C. B., R. T. Clark, and O. F. Pahnish. "Range Cattle Production; 7; Genetics of Cattle—A Literature Review," ArizAESR 137, 1957.

47. Semple, A. T., and H. E. Dvorachek. "Beef Production from Purebred, Grade and Native Calves," USDATB 203, 1930.

48. Stewart, H. A. "Breeding Productive Beef Cattle," NCAESR 15, 1952.

49. Stonaker, H. H. "Colorado A. & M. Research on Improvement of Cattle through Breeding," ColoAESGSP No. 642, 1956.

50. Stonaker, H. H. "Hybrid Breeding Increases Gains of Beef Cattle," A. I. Digest, 8: 11, 1960.

51. Tallis, G. M., *et al.* "A Topcross Breeding Experiment With Outbred and Inbred Hereford Sires," JAS, 18: 475, 1959.

52. Turner, J. W., B. R. Farthing, and G. L. Robertson. "Heterosis in Reproductive Performance of Beef Cows," JAS, 27: 336, 1968.

53. Urick, J. J., *et al.* "Heterosis in Postweaning Traits Among Lines of Hereford Cattle," JAS, 27: 323, 1968.

54. Urick, J. J., A. E. Flower, F. S. Willson, and C. E. Shelby. "A Genetic Study in Steer Progeny Groups During Successive Growth Periods," JAS, 16: 217, 1957.

55. Warwick, E. J. "Performance of Brahmas in Crosses with British Breeds," FCLJ, April, 1953.

56. Warwick, E. J. "Crossbreeding and Linecrossing Beef Cattle Experimental Results," World Review of An. Prod. 4: 37, 1968.

57. Wiltbank, J. N., K. E. Gregory, J. A. Rothlisberger, J. E. Ingalls, and C. W. Kasson. "Fertility in Beef Cows Bred to Produce Straightbred and Crossbred Calves," JAS, 26: 1005, 1967.

58. Weber, A. D. "Medium Is the Size," AHJ, March 15, 1951.

QUESTIONS AND PROBLEMS

1. Why is information on beef cattle breeding less complete than with swine?

2. List the important economic traits in beef cattle and indicate how to measure them as accurately as possible.

3. Fertility in beef cattle is very lowly heritable. Does this mean that inheritance does not affect this trait? Explain.

4. Outline methods that could be used to improve fertility in beef cattle.

5. What do weaning weights indicate about the level of performance of beef cows and calves?

6. Why are adjustment factors necessary for adjusting weaning weights of calves to a standard age basis?

7. What adjustment factors should be used to adjust weaning weights of calves? Why should they be used?

8. When and why should weaning weights of calves be adjusted to the same sex? How could this be done?

9. What is meant by the MPPA of cows?

10. Assume that you own three cows in your herd with a different number of records. Cow A has weaned one calf with a 205-day weight of 500 lb. Cow B has weaned five calves with an average weaning weight of 440 lb. Cow C has weaned seven calves with an average 205-day weight of 435 lb. If the average weaning weight of all calves in this herd is 405 lb, which cow would you consider genetically superior based on her MPPA?

11. What procedure would you follow in improving weaning weights of calves in your beef herd?

12. How would you improve postweaning rate of gain in your beef herd?

13. Why aren't heifers performance-tested on a full feed as a general rule?

14. What practical problems are involved in measuring individual economy of gains in beef cattle? What is a good substitute for measuring individual economy of gain and why?

15. Assume that you own bull calves with the following records:

No. of bull	205-day weaning weight adjusted for age of dam	Actual age when weighed	Actual weight when weighed
065	400	380	1000
070	450	392	1050
090	390	340	890
025	410	360	1000

Calculate the adjusted 365-day weights of each of these calves.

16. Do type and size mean the same thing? Explain.

17. How important are type and size in beef cattle production?

18. Does an 1800-lb cow require as much feed for maintenance as two 900-lb cows? Explain.

19. What is the optimum size of beef cow in the United States?

20. Why is it important to develop methods of estimating carcass quality and quantity in the live beef animal? What methods have been investigated for this purpose?

21. Why is it especially important to use superior progeny-tested sires for artificial insemination purposes?

22. Why is longevity important in beef cattle?

23. What traits may be genetically correlated in beef cattle? What do genetic correlations mean from the practical standpoint?

24. Why is it especially important to know if genotype × environment interactions affect economic traits in beef cattle?

25. What effect does inbreeding have on important economic traits in beef cattle?

26. Define topcrossing and linecrossing. How and why could they be important in beef cattle production?

27. Why is crossbreeding important in commercial beef cattle production? What does crossbreeding do genetically?

28. What traits in beef cattle are most likely to show heterosis when crossbreeding is practiced?

29. Discuss the possible use of Brahmans and some of the dairy breeds in crosses with European breeds for commercial beef production.

30. Why were crosses of Herefords × American Bisons studied in a Canadian experiment? What results were observed?

31. What kinds of gene action affect each important economic trait in beef cattle?

32. If you were using a selection index for selection in beef cattle, what traits would you use in this index? Why?

33. Outline a breeding and selection program for the production of purebred beef cattle.

34. Outline a breeding and selection program for commercial beef production.

35. Why are so many different breeds being imported into the United States?

19

Systems of Breeding and Selection in Sheep

Sheep were among the first animals domesticated and have been raised by man for food and clothing for many centuries. Many references were made to the sheep industry in the Holy Bible many hundreds of years before the birth of Christ. The very nature and habits of sheep were so well-known in the time of Christ that he used them as examples in his parables and other teachings.

Sheep were brought to the American continent by the early settlers. In fact, it is said that Columbus brought sheep with him his on his second voyage to the new world in 1493. The number of sheep reached a peak of 56 million in 1942 in the United States. Since then, the number has declined to less than 20 million. The number is still large enough to make sheep one of the major sources of income from livestock in some areas.

19.1 TRAITS OF ECONOMIC IMPORTANCE

Sheep are produced for both mutton and wool. In general, mutton is not as popular as pork and beef, although it is the preferred meat in some countries.

Wool for use in the manufacture of clothing and other textiles is being replaced somewhat by various synthetic fibers, but there is still a good demand for it in many parts of the world.

Traits of economic importance in sheep are those related to the cost of production of the kinds of wool and mutton demanded by the consumer. These traits will be discussed separately as was done in the chapters dealing with other species of farm animals.

19.1.1. Fertility

Sheep are seasonal breeders, and in most areas of the United States the breeding season is in the fall. Some breeds, such as the Merino and Dorset Horn, may be bred under some conditions to produce two lamb crops per year. Most breeds produce just one crop per year, although rams produce sperm throughout the year. Some rams are susceptible to high temperatures, however, and may be infertile or of low fertility in late summer during the first part of the breeding season.

The number of lambs raised per ewe is one of the most important factors determining the efficiency of production. Lamb production varies a great deal under different conditions and with different breeds. Ewes under farm conditions are usually more prolific than those produced on the range. This is probably due to a higher level of nutrition generally found in farm flocks. Twinning in sheep often is desirable, because a ewe that weans twins produces from 30 to 40 pounds more lamb than the ewe that weans a single lamb.

Extensive studies of breeding and lambing records have been made at the U.S. Sheep Experiment Station at Dubois, Idaho [17]. Complete sterility of rams was rare, and low fertility was uncommon. The percentage of ewes bred that lambed varied between 85 and 88 percent for the different breeds. The percentage of lambs born per ewe lambing varied from 118 to 129. An average of 95 lambs was born per 100 ewes bred. Another study of records at the same station [22] covering a 14-year period showed that 90.43 percent of the ewes bred lambed, with 92.54 percent of the lambs born being alive at birth. The average percentage lamb crop weaned of the live lambs born was 82.52. In contrast to these figures, in a Missouri study, 1.61 lambs were born per ewe with 1.16 weaned [6].

Fertility in sheep is lowly heritable, with an average heritability and repeatability estimate of 7 to 13 percent. These estimates are in agreement with those for other classes of livestock. This indicates that fertility in sheep is not greatly affected by additive gene action and could be improved very little by selection. Most of the phenotypic variation, therefore, is due to environmental factors, and attention to these should improve the lamb crop.

Fig. 19.1 Ewes of excellent type with twin lambs.
A large lamb crop weaned should be the goal of all sheep
breeders. (Courtesy of the University of Missouri.)

19.1.2 Weaning Weight

The age at which lambs are weaned varies under different conditions, but the age of 120 days is often used for selection purposes.

Lambs can be weighed as they reach 120 days, or weaning weights can be corrected to this age. The correction is done by multiplying the average daily gain from birth to weaning age by 120 and adding this product to the birth weight. Weights of twins may be adjusted to a single-lamb basis by multiplying the adjusted 120-day weight by the factors 1.0529; for triplets the factor is 1.0923.

The age of the ewe may have considerable influence on the weaning weight of her lambs. Two-year-old ewes wean lambs that are from five to ten pounds lighter than those from mature ewes. Production of ewes usually increases to four or five years of age. Probably the most important adjustment is that for weaning weights of lambs from ewes that are two years of age. This adjustment can be made by comparing the production of the two-year-old ewes with that of mature ewes in the same herd and then adding the difference to the weaning weights of lambs from the younger ewes. Or, if this is not possible, an adjustment may be made by adding seven pounds to the weights of the lambs.

Weaning weight in sheep is about 43 percent repeatable (Table 19.1), which indicates that culling ewes from the herd on the basis of their first year's production is practical. Culling the poorly producing ewes will increase the average weaning weight of the entire herd thereafter if environmental conditions remain unchanged. Weaning weight in sheep is about 33 percent heritable. Thus, selection for this trait will result in some improvement, although it may be slower than for postweaning gains or some other traits.

Table 19.1 Estimates of the percentage of repeatability of economic traits in sheep

Trait	Number of reports	Average	Range
Fertility:			
Lambing percent	2	7	0 to 15
No. lambs born	6	13	5 to 24
Birth weight	2	32	27 to 36
Weaning weight	5	43	22 to 76
Yearling body weight	3	73	71 to 78
Body type	2	33	26 to 40
Condition score	2	29	20 to 37
Wool characteristics:			
Face covering	2	76	70 to 82
Body folds	1	69	
Neck folds	2	56	54 to 58
Clean-fleece weight	2	66	63 to 69
Grease-fleece weight	5	46	7 to 74
Staple length	3	71	68 to 75
Resistance to trichostrongyles	2	24	22 to 25

19.1.3 Postweaning Gains

Some work has been done in performance-testing of sheep after weaning as has been done with beef cattle. Rate of gain in the feedlot in sheep is highly heritable, as shown in Table 19.2, so mass selection to improve this trait should be effective. Yearling body weight is also highly heritable, averaging between 40 and 45 percent. It is also very highly repeatable. Heavier yearling ewes also wean heavier lambs and produce heavier fleeces, so selection for this trait would probably be effective and desirable.

19.1.4 Type and Conformation

Desirable type and conformation have also received attention in sheep as in other classes of farm animals. With this species, however, attention must also be paid to selection for wool production in addition to mutton quality and rate and efficiency of gains.

Animals possessing very obvious defects, such as overshot jaws, undershot jaws, black wool, wool-blindness, skin folds, shallow bodies, and poor

Table 19.2 Estimates of the percentage of heritability for economic traits in sheep

Traits	No. of reports	Average	Range
Number of lambs born	5	13	7 to 25
Birth weight	8	33	12 to 61
Weaning weight	14	33	7 to 77
Yearling body weight	7	43	29 to 59
Postweaning daily gain	2	71	58 to 84
Efficiency of gains	1	15	
Body type	5	12	6 to 20
Condition score	3	12	4 to 21
Carcass characteristics:			
Fat thickness over loin	8	20	0 to 44
Loin eye area	10	48	11 to 93
Marbling	3	30	12 to 39
Tenderness	3	33	23 to 37
Percent fat	4	39	0 to 42
Percent lean cuts	6	37	0 to 61
Wool characteristics:			
Face covering	6	43	32 to 56
Neck folds	4	25	8 to 39
Body folds	6	41	20 to 51
Clean-fleece weight	4	52	38 to 62
Grease-fleece weight	9	47	28 to 66
Staple length	8	45	22 to 73
Fiber diameter	1	57	
Crimps per inch	2	44	40 to 47

mutton qualities, should be culled from the flock. If animals with these defects are eliminated from the breeding flock and selections are made on the basis of body weight and quantity and quality of wool, especially in the selection of rams, perhaps this will be sufficient attention to type.

Neale [13] has successfully used a system of selection and breeding which he calls "corrective mating." Ewes which may be highly productive yet have some objectionable characteristic may be mated to a ram that is especially outstanding in the trait in which the ewes are inferior. Many times this corrects the fault in just one cross. This is a principle of mating that could be used to improve many traits and is effective, as shown by the history of the development of the many present-day breeds of farm animals.

19.1.5 Carcass

Carcass traits in sheep are medium to highly heritable, as in swine and beef cattle. This indicates that selection for these traits should result in their genetic improvement.

Carcass traits in sheep can be measured only after the animal is slaughtered, making it impossible to measure this trait in animals to be used for breeding purposes. The next best measurement is probably that made in the close relatives of an individual. Full sibs are more valuable than half-sibs for this measurement, for an individual has twice as many genes in common with full sibs as he has with half-sibs. Although testing carcass quality in full sibs is limited to those instances where at least twin births occur, this phenomenon is much more frequent than in beef cattle.

Carcass information may also be obtained on the progeny of a ram, but the same disadvantage of progeny tests applies in sheep as in other classes of farm animals. It takes so long to get a good progeny test that the ram may be dead before his worth is fully known. Once he is progeny-tested, however, and proved to be superior, whether he is dead or living, his offspring should be given preference when one is selecting breeding animals. If he is dead when the results of progeny tests are fully known, there is still the possibility of keeping the relationship of individuals in the flock closer to the outstanding ancestor through linebreeding.

19.1.6 Fleece Weight and Quality

The total yearly production of a flock is about 20 percent from the wool and about 80 percent from the lambs marketed. This will vary from year to year in different areas and in different flocks where lamb production per ewe is higher than the average. Nevertheless, this shows very clearly that wool production is very important from the economic standpoint.

Wool production is largely dependent upon fleece weight and staple length. In general, for each half-inch that staple length is increased, the weight of the grease wool is increased by three-quarters of a pound and clean wool by one-half pound [19]. Greater gains are obtained with increases in staple length in both the fine-wool and the coarse-wool breeds. With some breeds, selection may have to be practiced against coarseness of wool, but selection should be practiced for the grade of wool most desirable in a particular area.

Heritability estimates presented in Table 19.2 show that most wool characteristics are highly heritable. The average heritability of fleece weight in nine studies was 47 percent, and of staple length in eight studies it was 45 percent. These heritability estimates are high enough that the mating together of the best individuals for these two traits should result in genetic improvement, and because of their economic value, the traits should be given attention in a breeding program.

Face covering in ewes is an important economic trait [19]. Open-faced ewes produce more lambs and wean more pounds of lamb than those with wool-covered faces. In six studies, face covering was 43 percent heritable, so selection for open faces should be effective. Selection experiments also

Fig. 19.2 Wool blindness is a trait to be avoided. Open face ewes are more productive and selection for this trait is effective. (Courtesy of the University of Missouri.)

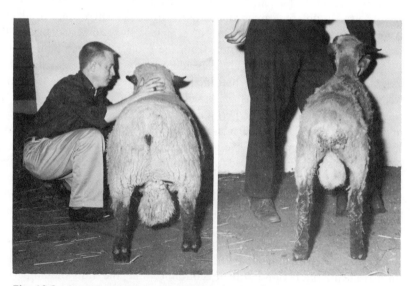

Fig. 19.3 A good and a poor type ram. A sire supplies half the inheritance for each of his offspring, which makes a good one of considerable value. (Courtesy of the University of Missouri.)

indicate that this trait can be eliminated in a flock in 10 to 15 years by selection within the group [19].

19.2 GENETIC CORRELATIONS AMONG TRAITS

Several studies have been made with sheep to determine the genetic correlations among the economic traits. In general, a large proportion of the same genes seem to be responsible for rapid gains in all phases of the growth period. Thus, there is no antagonism here, and the evidence indicates that selection for heavier body weight for any period in life should improve this trait for other periods. Since body weight at all ages seems to be medium-to-highly heritable, selection for this trait should be effective and very desirable.

Some of the same genes responsible for heavy body weights also seem to cause the production of heavier fleece weight. Although more information on these genetic correlations is needed, it seems that no important genetic antagonism exists between these two traits. Selection for body weight, for instance, should also cause some genetic improvement in fleece weight.

Many of the same genes responsible for greater staple length [3] are also responsible for heavier fleece weight. Or, at least, the studies do not suggest that there is an important antagonism between these two traits.

19.3 GENETIC ENVIRONMENTAL INTERACTIONS

Sheep are produced in many different environments throughout the world. Also, certain strains, or breeds, have been developed which appear adapted to a particular area. Because of these facts, it is desirable to know if important economic traits in sheep are affected by genetic × environmental interactions.

A review of the literature on this subject shows that, in general, genetic × environmental interactions have little influence on important economic traits in sheep (Pani, 1970) [14], unless the genotypes and environments in which they are compared vary widely.

19.4 INBREEDING IN SHEEP

Results from inbreeding in sheep are very similar to those with other farm animals. Inbreeding in some instances is followed by the appearance of defects that are due to the pairing of recessive genes. In one flock of sheep where inbreeding was practiced for ten years, fifteen blind lambs were born [16]. All traced back to the same sire, and the defect was found to be due to a single pair of autosomal recessive genes. In another study [17], mild inbreeding for four generations failed to uncover any detrimental recessive factors.

Inbreeding is usually accompanied by a decline in vigor [7, 10, 11]. Results summarized in Table 19.3 show that weaning weight and yearling body weight were decreased by three to four pounds for each ten percent increase in inbreeding. Body score and condition score were not affected adversely.

Table 19.3   Influence of inbreeding on various economic traits in sheep

Trait	No. of reports	Regression coefficient*	
		Average	*Range*
Weaning weight (lb)	2	−0.339	−0.302 to −0.375
Yearling body weight (lb)	5	−0.381	−0.055 to −0.585
Body type score	7	0.010	0.002 to 0.016
Condition score	7	0.007	0.002 to 0.013
Wool characteristics:			
Face-covering score	7	0.004	−0.004 to 0.016
Neck-folds score	7	−0.003	−0.009 to 0.001
Clean-fleece weight (lb)	3	−0.018	−0.008 to −0.025
Grease-fleece weight (lb)	5	−0.029	−0.013 to −0.057
Staple length (cm)	7	−0.007	0.000 to −0.015

*Unit change for one percent increase in inbreeding.

The weight of wool produced decreased with inbreeding, as did staple length. Possibly this was due to the decline in vigor usually associated with this system of breeding.

19.5   CROSSBREEDING IN SHEEP

Several hundred breeds of sheep are present in the world today. Ten or more breeds are represented by registry associations in the United States. However, the production of purebred sheep is in the hands of a relatively few breeders. Most commercial flocks are composed of grade or crossbred ewes, with the identity of the parent breeds not always clear and distinct. Systematic crossbreeding in sheep has the same great potential for the improvement of performance as it has with other species of farm animals.

Different breeds of sheep excel for certain economic traits. Some breeds are superior for the production of fine wool, others for the production of mutton. Some are highly fertile with multiple births the rule rather than an unusual occurrence. Some breeds are strictly seasonal breeders, whereas others tend to breed year round. Since existing breeds vary widely in their merit for different traits, it is possible that they can be combined in a systematic crossbreeding program for more efficient production of wool and mutton.

Results in which two-breed cross lambs are compared with purebred lambs are shown in Table 19.4. Most experiments in crossbreeding with sheep compare the crossbreds with only one parent breed. These are usually comparisons between the offspring of rams of different breeds mated to ewes of a single pure breed and the offspring of purebred rams mated to purebred ewes of the same breed. Very few experiments have been reported in which the crossbreds were maintained concurrently with the pure breeds used to produce the cross so that reciprocal crosses could be compared with the average of the two breeds.

Table 19.4 A comparison of the performance of crossbred and purebred lambs produced by purebred ewes*

Trait	As a percentage of purebreds
Fertility:	
Lambs born per ewe	103.2
Lambs weaned per ewe lambed	114.6
Lambs born per ewe bred	106.2
Percent twins	103.2
Weaning weight	106.6
Grease wool (lb)	114.6
Staple length	130.1
Mature ewe weight	112.6

*References 4, 5, 9, 12, and 20.

Intrauterine survival seems to be improved slightly in crossbred lambs. This is illustrated in the lambs born per ewe, which shows about 3.2 percent improvement. The greatest improvement from crossbreeding appears to come from increased survival rate from birth to weaning, as shown by 14.6 percent more crossbred than purebred lambs weaned per ewe that lambed. In these cases, all lambs were produced by purebred dams.

Studies of crossbreeding involving Border Leicester and Romney sheep have been made in New Zealand [2]. Data obtained to date show very clearly that the crossbreds have an advantage of 20 to 25 percent over the purebred Romney ewes in the percentage of lambs alive at 28 days per 100 ewes mated. The backcross of the crossbred ewes to Romney rams seems to give inferior results in this respect, as compared to the first-cross ewes. The results of this study, as well as from others, indicate that some breeds produce a considerably larger number of lambs than do others. The over-all performance of the crossbred ewes and lambs will be determined to a considerable extent by the prolificacy and mothering ability of ewes from the breeds involved. Thus, crossing two breeds noted for a high percentage lamb crop will give more lambs at weaning than will crossing two breeds that are known for producing a low percentage lamb crop. Even though the degree of heterosis is the same in both instances, the level of fertility may be considerably different because of the average production of the pure breeds used in the crosses. This is probably also true for other traits.

Fig. 19.4 Lambs from a two-breed cross. Crossbreeding improves the health and vigor of the lambs, resulting in a larger percent of lambs born weaned, and favors heavier weights. (Courtesy of the University of Missouri.)

Crossbreeding increases the weaning weights of lambs by about six to seven percent and the mature weights of ewes by 10 to 15 percent over the average of the pure breeds used in the cross. On the same basis, it also increases the weight of wool produced, which could be closely related to increased growth rate and viability of the individuals involved. Staple length was also increased by crossbreeding, but this 30 percent improvement could be out of proportion to the true heterosis, because some of the experiments summarized in Table 19.4 were those in which rams of long-wool breeds were mated to native ewes or ewes of the short-wool breeds.

Data summarized in Table 19.5 are from an experiment conducted at the Animal Research Service Center, Beltsville, Maryland. Hampshires, Shropshires, Southdowns, and Merinos were the four breeds used in this experiment. These data show about 13 percent heterosis in the two-breed

Table 19.5 The amount of hybrid vigor expressed in different systems of crossbreeding involving the Hampshire, Shropshire, Southdown, and Merino breeds*

	Crossbreds as a percentage of purebreds		
	2-breed cross	3-breed cross	4-breed cross
Ewes lambing of ewes bred	101	102	105
Lambs born of ewes lambing	93	108	109
Lambs born alive of total born	102	102	104
Lambs weaned of live lambs born	105	105	111
Lambs weaned of ewes bred	102	117	130
Weaning weight of lambs	110	119	120
Pounds of lamb per ewe bred	113	138	156

*Calculated from the report of Sidwell, G. Animal Research Center, Beltsville, Maryland. "Effective use of Breeds and Crossbreeding," *Proceedings of the Symposium for the Genetic Improvement of Wool and Lamb Production,* Texas A & M University, McGregor, Texas, April 18–19, 1968.

cross as compared to the purebreds in the pounds of lamb weaned per ewe bred. The three-breed cross showed a 38 percent improvement over the average of the pure breeds used in the cross with an advantage over the two-breed cross in pounds of lamb weaned per ewe bred of 25 percent. This 25 percent advantage of the three-breed over the two-breed cross is an estimate of heterosis in the crossbred ewe for maternal traits. It is interesting to note that the four-breed cross had an advantage over the three-breed cross of about 18 percent in this experiment. Certainly this experiment indicates a great advantage of the three- and four-breed crosses over purebreds in commercial lamb production.

A practical crossbreeding system in sheep should include the use of crossbred ewes for lamb production. These should be bred to a ram of a different breed. Crossbred ewes could be produced by combining breeds superior in reproductive efficiency, milk production, mothering ability, and wool quantity and quality. The rams could be purebreds that possess superior growthiness, carcass quality, sexual aggressiveness, and a high level of fertility (Whiteman, 1966) [21]. Hampshires and Suffolks are good examples of such ram breeds.

19.6 SELECTION STUDIES WITH SHEEP

A selection study with sheep in New Zealand has been in progress for several years [20]. In this study, selection has been practiced for high fertility by selecting dams with a high twinning record. In a second line, selection has been practiced for low fertility by selecting progeny from ewes that seldom, if ever, produced twins. A control flock was used in which no attempt was made to select for fertility in the ewes, with culling being done on the basis of phenotype and not fertility. The three flocks were handled as one except during mating. At the end of the tenth year, the high line produced an average of 148 lambs per 100 ewes lambing, the low line 117, and the control line 120. The results indicate that progress has been made in selection for this trait, even though its heritability is relatively low.

A study was made of the selection practiced and the progress made by selection in a flock of Rambouillet sheep [18]. Most of the selection differentials for the various traits were positive. Actual genetic progress was positive for staple length, type-score, and neck folds. Also some improvements in face covering and condition were observed. However, body weight and fleece weight declined slightly, in spite of positive selection differentials. Improvement in over-all merit of weanling lambs increased, as shown by a significant increase in the weanling index. However, the rate of improvement was slightly less than that expected from the amount of selection practiced.

Another trait that was used in a selection study was the multinipple trait in sheep. Usually, ewes have just two nipples, but some have more. Alexander Graham Bell selected for the improvement of this trait in sheep, beginning about 1890 [1, 15]. His objective was to develop a type of sheep that would produce two or more lambs and have the milking ability to raise them. He selected for extra nipples with the idea that selection for a large number of nipples would increase the milk supply. It is of interest, however, that in these experiments little or no association was found between increased nipple number and fertility or milk production.

Selection for a larger nipple number was highly effective at first and soon increased the average number in the flock to four. Nipple number seemed to become more or less stabilized at this level, however, and further progress in selection was slow or ineffective, although parents with six nipples were used for breeding purposes. The heritability estimate for nipple number was 14.4 percent when calculated from the intrasire regression of the offspring on their dams. This was determined in the population where stabilization of nipple number had already occurred, so perhaps the additive genetic variation for this trait had been exhausted or greatly reduced. Perhaps more than four nipples may have been the result of a combination of genes with non-additive effects, and this would make mass selection for more than four nipples ineffective.

19.7 SELECTION INDEXES

Several indexes may be used for selection purposes, depending upon the goals of the sheep breeder. An index based on the weaning weight of the lamb minus twice its face-covering score might be practical on the farm for some breeds [8]. For example, if a lamb weighed 55 pounds at weaning and the face score were 3, the index for that lamb would be 55 minus 6, or 49.

An index used for measuring ewe productivity [18] is the weight of the lamb at weaning plus 2.5 times the weight of wool the ewe produces each year, divided by the body weight of the ewe. Where the ewe weighed 100 pounds, weaned twin lambs weighing 90 pounds, and produced 8 pounds of wool per year, the index would be $[90 + (2.5 \times 8)]/100 = 1.10$.

Still another index, which is more complex and gives considerable weight to wool characteristics, is [19]

$$\text{Index} = 100 + (1.54 \times \text{clean-fleece wt.}) - (1.25 \times \text{staple length in centimeters}) - (0.01 \times \text{body weight in pounds}) - (0.75 \times \text{skin-fold score}) + (0.13 \times \text{crimps per inch of wool}).$$

19.8 BREEDING PLAN FOR SHEEP IMPROVEMENT

The following points are suggested in planning a sheep-improvement program:

1. Identify all individuals in the flock by means of a tattoo, ear-tag, or some other means.
2. Establish some system of record-keeping for all individuals in the flock. (See Fig. 19.5 for sample record forms.) Record the performance records on each lamb, ewe, and ram.
3. Adjust weaning weights of each lamb for age, sex, for twins, and age of dam where necessary and record these.

FLOCK RECORD EXAMPLE

No. of Ewe_____ Breed_____ Date Born_____

Birth Weight_____Weaning Weight_____ Yearling Weight_____

Face Covering Score_____

Lambing Record of Ewe

Date Lambed	Sex of Lamb	Birth Weight	Date Weaned	Adjusted Weaning Weight	Weaning Score	Remarks

Wool Record of Ewe

Date Sheared	Weight of Wool	Weight Scoured Wool	Staple length	Wool Grade	Index of Ewe	Remarks

Figure 19.5

4. Cull ewes that have a poor record of fertility, have obvious defects, or wean light lambs, and produce a light fleece at shearing. If desired, a productivity index may be calculated for each ewe in the flock for weaning weight of her lambs and her wool production, and then they may be culled on this basis. Both traits are highly repeatable. Replace ewes with ewe lambs that rank highest on the basis of one of the indexes mentioned earlier, or select those that are superior in body weight, mutton qualities, and wool.

5. Where more than one ram is used in the flock, compare the records of their offspring and cull those rams whose offspring are below average for the

desired traits. Replace them with ram lambs whose sires and dams have high production records for lambs weaned, mutton type, and heavy wool production. Ram lambs should also be selected on the basis of their own weaning weight, their rate and efficiency of gains after weaning, quality and quantity of wool, and mutton type. Or they could be selected on the basis of an index, as was suggested for ewe lambs.

6. In commercial lamb production, the use of crossbreeding may be profitable. It would be best to cross those breeds that are both known to be superior in fertility, performance, mutton type, and high quantity and quality of wool.

REFERENCES

1. Bell, A. G. "Saving the Six-nippled Breed," JH, 14: 99, 1923.

2. Clark, E. A. "Improving Fertility by Crossbreeding," ARRARS, 1957-58.

3. Doney, J. M. "Problems of Hill Sheep Improvement," PBSAP 3, 1956.

4. Foster, J. E., and E. H. Hostetler. "Changes in Meat and Wool Characteristics Resulting from the Use of Purebred Mutton Rams on Native Ewes," N.C. AESTB 60, 1939.

5. Gorman, J. A., F. S. Hultz, R. L. Hiner, O. G. Hankins, and D. A. Spencer. "Crossbreeding for Lamb and Wool Production," WyoAESB 254, 1942.

6. Guyer, P. Q., and A. J. Dyer. "Study of Factors Affecting Sheep Production," MoAESRB 558, 1954.

7. Hazel, L. N., and C. E. Terrill. "Effects of Some Environmental Factors on Weaning Traits of Range Rambouillet Lambs," JAS, 4: 331, 1946.

8. Karam, H. A., A. B. Chapman, and A. L. Pope. "Selecting Lambs under Farm Conditions," JAS, 12: 148, 1953.

9. Miller, K. P., and D. L. Dailey. "A Study of Crossbreeding Sheep," JAS, 10: 462, 1951.

10. Morley, F. W. H. "Selection for Economic Characters in Merino Sheep," ISCJS, 25: 304, 1951.

11. Morley, F. H. W. "Selection for Economic Characters in Australian Merino Sheep, III," "The Effect of Inbreeding," JAR, 5: 305, 1954.

12. Neale, P. E. "Production of Wool and Lamb from Different Types of Ewes and Breeds of Rams," N. M. AESB 305, 1943.

13. Neale, P. E. "Selection as a Method of Improving Sheep," NWG, 47: 26, 1957.

14. Pani, S. N. "Genetic × Environmental Interactions in Beef Cattle," Unpublished doctoral thesis, University of Missouri, 1970.

15. Phillips, R. W., R. S. Schott, and D. A. Spencer. "The Multinipple Trait in Sheep," JH, 37: 19, 1946.

16. Ragob, M. T., and A. A. Asker. "Effects of Inbreeding on a Flock of Ossimi Sheep," J, 45: 89, 1954.

17. Terrill, C. E., and J. A. Stoehr. "Reproduction in Range Sheep," PASAP 369, 1939.

18. Terrill, C. E. "Selection for Economically Important Traits of Sheep," JAS, 10: 17, 1951.

19. Terrill, C. E. "Sheep Improvement through Selection," NWG, 47: 14, 1957.

20. Wallace, L. R. "Breeding for Fecundity in Romney Ewes," ARRARS, 1957–58.

21. Whiteman, J. V. "Effective Use of Breeds and Crossbreeding of Sheep," *Proceedings of Symposium for the Genetic Improvement of Wool and Lamb Production,* Texas A & M University, McGregor, Texas, April 18–19, 1968.

22. Wiggins, E. L., C. E. Terrill, and L. O. Emik. "Relationship between Libido, Semen Characteristics and Fertility in Range Rams," JAS, 12: 684, 1953.

QUESTIONS AND PROBLEMS

1. When and by whom were sheep first introduced into America?

2. Are sheep numbers in the United States increasing or decreasing? Why?

3. Is twinning in sheep desirable?

4. What is the degree of heritability of fertility in sheep? How may it be improved in a flock?

5. What factors should be used to adjust weaning weights of lambs for selection purposes?

6. What is the degree of heritability of weaning weights in lambs? What does the degree of heritability of this trait indicate?

7. Would selection for rapid postweaning gains in sheep be effective? Explain.

8. What is meant by corrective mating?

9. What practical problems are encountered in selection for carcass traits?

10. Could carcass quality and quantity be improved by selection? Why?

11. Outline methods you would recommend in selection for carcass quality and quantity.

12. What percentage of the yearly production of the flock normally comes from the marketing of wool?

13. Are open-faced ewes more productive than ewes with wool-covered faces? How could the covered-face trait be eliminated from a herd?

14. What traits in sheep appear to be genetically correlated? What does this mean from the practical standpoint?

15. Are genetic × environmental interactions in sheep important to the producer?

16. What are the results of inbreeding in sheep?

17. Is crossbreeding in sheep recommended as practical? Explain.

18. Outline in detail a crossbreeding program you would recommend for commercial lamb production, naming the breeds to use, the crossbreeding system to follow, and how breeding stock would be selected.

19. Would selection for a high twinning rate in sheep be effective and practical? Explain.

20. Why was selection for a larger nipple number in sheep effective at first but ineffective in later generations?

20

Systems of Breeding and Selection in Dairy Cattle

More attention has been given to the improvement of performance of dairy cattle through breeding than to any other class of farm animal except poultry. One reason for this is that performance can be measured relatively easily and accurately in this species by weighing the milk produced and by testing for butterfat content at certain intervals during lactation. Another reason is that consumer demands for quality in dairy products are well known, and there has been no need to revise selection objectives over the years as there has been in swine, where the demand has shifted from fat hogs to lean hogs in recent years. In addition, there is a fairly constant demand for dairy products throughout the year, since they are so essential for health and they are used daily in most homes throughout the United States.

The same outline will be used in discussing dairy cattle breeding and selection as was used with swine and beef cattle.

20.1 TRAITS OF ECONOMIC IMPORTANCE

The traits of greatest economic importance in dairy-cattle breeding are fertility, milk production, butterfat production, type, and productive life span.

20.1.1 Fertility

Normal and regular reproduction in dairy cattle is of great importance, because the lactation period begins when a calf is born. The heritability and repeatability estimates for fertility are very low in dairy cattle, as in beef cattle. This is illustrated by the data summarized in Table 20.1. These low estimates indicate that most of the variations observed in fertility are due to environment and that selection to improve this trait would not be effective. The greatest improvement within a herd would come from proper attention to environmental factors such as nutrition, management, and disease control.

Table 20.1 Heritability and repeatability of fertility in dairy cattle*

Fertility Trait	No. of reports	Average	Range
Repeatability:			
Nonreturn to first service	4	10	3 to 27
Services per conception	2	7	6 to 8
Calving interval	2	7	0 to 13
Time to postpartum estrus	2	9	2 to 15
Heritability:			
Nonreturn to first service	3	7	0 to 11
Services per conception	3	-3	-15 to 3
Breeding efficiency†	1	32	
Time to postpartum estrus	2	7	6 to 8

*References 7, 13, 14, and 34.
†Calving interval as a percentage of 365 days.

The fact that heritability and repeatability estimates for fertility are low does not mean that genes do not affect this trait. It merely means that the amount of additive genetic variance affecting fertility is small, and it does not discount the possibility that single pairs, or at least a small number of pairs, of genes with nonadditive effects have an important influence on this trait. For instance, gonadal hypoplasia has been reported to lower fertility in cattle and is thought to be conditioned by a recessive gene [29]. Inherited sterility in bulls has been reported in which the sperm is abnormal, thus preventing normal fertilization [21]. In addition, inbreeding often causes a decline in fertility that is due to the uncovering of recessive genes [19] or the disruption of a nicking effect due to overdominance and/or epistasis.

20.1.2   Milk and Butterfat Production

Improvement of production of milk and butterfat has received the most attention by breeders through the years. Breeds have been developed which differ significantly in the amount of milk and butterfat they produce. Some breeds produce large amounts of milk with a tendency toward a lower percentage of butterfat, whereas the reverse is true of other breeds. These breed differences strongly suggest a genetic control of both milk and butterfat production.

Heritability and repeatability estimates for milk and butterfat production as well as butterfat percentage are summarized in Table 20.2. These estimates show that milk and butterfat production are from medium to high in heritability, so that selection for these traits should show improvement. Selection for butterfat percentage should be especially effective, since this trait is between 60 and 65 percent heritable.

Table 20.2   Heritability and repeatability estimates for milk and butterfat production in dairy cattle*

Repeatability	No. of reports	Average	Range
Milk yield	3	53	41 to 64
Butterfat yield	3	42	41 to 43
Butterfat percent	3	68	59 to 80
Nonfat solids	1	76	
Heritability Based on Parent-Offspring and Sib Resemblance			
Milk yield	30	36	5 to 71
Butterfat yield	17	40	20 to 84
Butterfat percent	13	62	33 to 83
Persistency of lactation	2	31	27 to 35
Peak milk yield	3	35	14 to 74
Total solids	2	36	34 to 37
Total nonfat solids	2	35	34 to 35
Longevity	1	37	
Type	4	25	14 to 31
Heritability Based on Resemblance Between Identical Twins			
Milk yield	4	89	80 to 90
Butterfat yield	1	86	
Butterfat percent	2	86	86 to 87
Sugar content	1	60	
Protein content	2	83	78 to 88
Persistency of lactation	1	84	

*Reports in the literature are so numerous that a complete list of references will not be given.

Heritability estimates derived from indentical twin data average considerably higher than estimates calculated from parent-offspring or sib resemblances that measure mostly the variation due to additive gene action but include some that is nonadditive. Heritability estimates based on identical-twin data have been obtained from split-twin and combined identical and fraternal twin records [6]. Several factors may be responsible for heritability estimates from identical-twin data being higher than those from nontwin data. One explanation is that twins may be more alike than nontwin relatives because their maternal and contemporary environment may be more alike. This would increase the size of the heritability estimates, because members of a pair of twins would resemble each other more closely. In addition, heritability estimates based on identical-twin data may include much of the nonadditive genetic variance, such as dominance and epistasis, in addition to that due to additive gene action, which is not measured to any great extent in nontwin data.

Heritability estimates based on nontwin data would seem to correspond more closely with the realized (or actual) heritability one might obtain in selection under practical conditions. These estimates are still high enough to indicate that mating the best to the best would be the system of choice for improved milk and butterfat production.

Some attention has been given in research to whether heritability estimates are higher in a low-producing or in a high-producing herd. In a study of records of 13,000 cows in Denmark produced by artificial insemination [31], the heritability estimates of milk fat yields were only slightly greater in cows of a high level of production than in those of a medium level, However, heritability estimates were considerably lower for both traits in cows of a low level of production. No evidence of a herd-sire interaction was found for either trait, with the true ranking of bulls tested on cows of a low, medium, or high level of production being the same in each case. It was concluded that selecting bulls on the basis of their daughters' records in a higher-yielding herd would be preferred in testing bulls for use in artificial insemination. A similar study in Sweden [24] showed that heritability and repeatability of milk yield and butterfat percentage were only slightly higher in high- than in low-producing herds. These two studies suggest that selection or progeny-testing in herds of medium-to-high production would be preferred, although the advantage is small.

20.1.3 Measurement of Milk and Butterfat Production

Several nongenetic factors are known to cause variations in the production records of dairy cattle. Adjusting records for factors known to cause variations would make selection more effective, because the superior animals would then be more likely to be superior because of inheritance. Some of these factors may be corrected for by recording production for a standard length of time or by using adjustment factors derived from a large body of data from many animals.

The Dairy-Herd-Improvement-Association [27] recommends that records of production be adjusted for length of lactation period, for number of milkings per day, and for age of the cows when they produce the records. Adjustments of records for these variables are necessary for a more valid comparison of sires as well as dams for selection and herd improvement programs. Several of the breed associations have correction factors of their own, but those presented here are recommended for standardizing Dairy-Herd-Improvement-Association records in proving sires.

Dairy Herd Improvement Associations (DHIA) have been formed in many states of the United States [26]. They are self-supporting, nonprofit, cooperative associations, organized and operated by dairymen for the purpose of obtaining and using information on breeding and production and management for the improvement of the efficiency of milk production. These associations are operated under the supervision of state extension specialists and county agricultural extension agents, who cooperate with the Federal Extension and the Dairy Cattle Research Branch of the USDA. Accurate feed and income records are an important part of the program, and they help dairymen cull the least profitable cows and feed excellent producers for the best possible production. Owners also keep accurate records on the milk production of each of their cows, and a supervisor tests samples of milk for butterfat content for herds enrolled in his area. Some associations operate a central testing laboratory where milk samples are tested.

Several types of organizations are found in Dairy Herd Improvement Associations. All of these have definite rules and plans for the most efficient operation. The Standard DHIA program operates under the procedures outlined in the *National Cooperative Dairy Herd Improvement Program Handbook*. Records from herds on this plan are used in the USDA sire evaluation and research program. Colleges and universities also use the records for research and educational purposes.

The effectiveness of DHIA in the United States is illustrated by the fact that cows enrolled in the Standard DHIA program have produced 60 to 80 percent more milk and butterfat than all other cows in the United States over a period of several years. Furthermore, there has been a steady increase in milk and butterfat production in cows enrolled in DHIA over the past 30 years.

DHIA records make it possible to make genetic improvement in milk and butterfat production because genetically superior sires and dams are recognized. However, the opportunity of selection among progeny of cows is limited because of their generally short productive life due to losses from injuries, diseases such as mastitis, and reproductive failure. As a result, they often leave few progeny in a herd. The greatest improvement in selection comes from the recognition and use of genetically superior sires.

In preceding chapters it was pointed out that both environment and heredity cause phenotypic variations in quantitative traits in a group of individuals. It was also shown that the genotype of the individual is fixed at conception and remains constant throughout the individual's life. Reducing variations caused by environmental factors results in a larger portion of the remaining variations being due to heredity. Thus, the productive level of the individual is more indicative of its genotype.

Records are usually adjusted to a 305-day (305D), mature equivalent (M.E.), two-times (2×)-a-day milking basis. Correction factors recommended in a particular area may be obtained from the local or state dairy extension specialist. Such correction factors will apply more appropriately in a specific area than those recommended on a national basis. Some adjustment factors are given here for purposes of illustration.

Lactation records are usually reported on a 305-day basis, because this reduces the variation in production records caused by varying lengths of lactation and because pregnancy has little or no influence on production during a lactation period of this length. A 305-day lactation period is also more desirable, because cows should calve each year and should have a dry period between two successive lactations.

It is not possible, nor practical, to weigh and determine the pounds of milk and butterfat produced each day of the lactation period for each cow in the herd. Therefore, methods have been developed whereby milk and butterfat production records are obtained one day each month and 305-day production records are calculated from these records. Monthly determinations are necessary because most cows decline in production as lactation progresses, and the calculation of a cow's total milk production for 305 days from consecutive monthly determinations would be more accurate than when calculated from only partial records. Sometimes it is necessary to calculate 305-day records from partial records, and projection factors for this purpose are presented in Table 20.3. To illustrate how these projection factors may be used, let us assume that a Holstein cow 32 months of age has a 270-day production record of 8000 pounds of milk and 320 pounds of butterfat. Appropriate projection factors given in Table 20.3 for cows under 36 months of age which have a 270-day lactation period are 1.08 for milk and 1.09 for butterfat. Thus, this cow's projected 305-day production record would be 8640 pounds of milk (8000 × 1.08) and 349 pounds of butterfat (320 × 1.09). When milk and butterfat production are measured each month for 305 days or more, these figures may be used to calculate each cow's 305-day

Table 20.3 DHIA 305-day projection factors

	Guernsey*								
Days in milk	<36 mo. of age		≥36 mo. of age		Days in milk	<36 mo. of age		≥36 mo. of age	
	Milk	Fat	Milk	Fat		Milk	Fat	Milk	Fat
15	15.73	16.98	14.07	14.39	165	1.62	1.69	1.50	1.54
30	7.89	8.51	7.06	7.25	180	1.50	1.56	1.40	1.44
45	5.29	5.71	4.75	4.90	195	1.40	1.45	1.32	1.35
60	4.00	4.31	3.61	3.74	210	1.32	1.36	1.25	1.27
75	3.24	3.48	2.93	3.05	225	1.25	1.28	1.19	1.21
90	2.74	2.93	2.49	2.58	240	1.19	1.21	1.14	1.16
105	2.39	2.54	2.17	2.26	255	1.13	1.15	1.10	1.11
120	2.12	2.25	1.94	2.01	270	1.09	1.10	1.07	1.07
135	1.91	2.02	1.76	1.82	285	1.05	1.05	1.04	1.04
150	1.75	1.83	1.61	1.66	300	1.01	1.01	1.01	1.01

*Factors for less than 36 months were based on 3667 records, and factors for cows greater than or equal to 36 months were based on 10,422 records.

	Holstein†								
15	16.67	15.88	14.83	13.64	165	1.62	1.64	1.51	1.52
30	8.32	7.99	7.42	6.89	180	1.51	1.52	1.41	1.42
45	5.54	5.39	4.96	4.69	195	1.41	1.43	1.33	1.34
60	4.16	4.10	3.74	3.60	210	1.32	1.34	1.26	1.27
75	3.35	3.33	3.02	2.95	225	1.25	1.27	1.20	1.21
90	2.82	2.82	2.56	2.52	240	1.19	1.20	1.14	1.15
105	2.44	2.46	2.22	2.21	255	1.13	1.14	1.10	1.11
120	2.16	2.18	1.98	1.97	270	1.08	1.09	1.06	1.07
135	1.94	1.96	1.79	1.79	285	1.05	1.05	1.03	1.04
150	1.77	1.79	1.64	1.64	300	1.01	1.01	1.01	1.01

†Factors for less than 36 months were based on 39,310 records, and factors for cows greater than or equal to 36 months were based on 92,763 records.

	Jersey‡								
15	15.24	16.39	14.22	14.46	165	1.60	1.65	1.51	1.55
30	7.65	8.22	7.14	7.27	180	1.48	1.53	1.41	1.44
45	5.14	5.51	4.79	4.90	195	1.39	1.43	1.33	1.36
60	3.89	4.17	3.63	3.73	210	1.31	1.34	1.26	1.28
75	3.17	3.38	2.95	3.04	225	1.25	1.27	1.20	1.21
90	2.68	2.85	2.50	2.58	240	1.19	1.20	1.15	1.16
105	2.34	2.47	2.19	2.26	255	1.13	1.14	1.11	1.11
120	2.09	2.19	1.96	2.01	270	1.08	1.09	1.07	1.07
135	1.89	1.97	1.78	1.82	285	1.05	1.05	1.04	1.04
150	1.73	1.79	1.63	1.67	300	1.01	1.01	1.01	1.01

‡Factors for less than 36 months were based on 2627 records, and factors for cows greater than or equal to 36 months were based on 6317 records. (Courtesy of USDA.)

record. The monthly production record of a cow, for example, would be determined by multiplying the test-day production by the number of days the cow was in milk during the test period. Totals for each month would then be added to give the 305-day production record. The 305-day production record calculated on this basis should be more nearly correct than the projection of partial records.

The age of the cow may have an important influence on the amount of milk she produces during a lactation period. Very young and very old cows do not give as much milk, as a general rule, as cows six to eight years of age, when they should be at their peak in production. Therefore, adjustment factors for age of the cow have been developed for both milk and butterfat production. These should be used when cows of different ages are compared. Research has shown that milk and butterfat production may vary with the season of the year, the region of the country where the records are made, and the breed to which the cow belongs. Therefore, age adjustment factors should take these factors into consideration. Persons interested in obtaining age adjustment factors for their area should contact their local extension director or state dairy extension specialist.

Age adjustment factors for Guernseys, Holsteins, and Jerseys, respectively, are given in Tables 20.4, 20.5, and 20.6. To illustrate how these factors may be used, let us assume that the 305-day lactation record of a Guernsey cow three years and five months of age, which lactates in the periods from November to June, is 5000 pounds of milk and 225 pounds of butterfat. Appropriate adjustment factors selected from Table 20.4 are 1.11 for milk and 1.07 for butterfat. Thus, the production of this cow adjusted for age (M.E. or mature equivalent basis) would be 5550 pounds of milk (5000 × 1.11) and 241 pounds of butterfat (225 × 1.07).

The number of times cows are milked per day is also an important source of variation in milk production, although most are milked only two times per day (2×). More frequent milkings result in the production of more milk, and therefore comparisons between cows, some of which are milked twice daily and some three or four times, are not valid. Since most cows are milked only two times per day, records are usually adjusted to this basis by conversion factors presented in Table 20.7. This is usually done after adjustments are made to a 305-day, mature equivalent basis.

Since breeds of dairy cattle differ in the amount of milk they give and in butterfat percentage, yields are sometimes reported on a four percent fat-corrected-milk basis. The formula for this conversion is

Fat-corrected milk (4 percent milk) = $(0.4 \times \text{milk}) + (15 \times \text{fat})$

For example, if a cow produced 12,000 pounds of 3.5 percent milk containing 420 pounds of fat, her four percent-equivalent record would be

Table 20.4 Guernsey age adjustment factors for milk and fat production: for cows that calve in Indiana, Illinois, Ohio, Michigan, Wisconsin, Minnesota, Iowa, Missouri, North Dakota, South Dakota, Nebraska, and Kansas*

Age in years and months	Adjustment factors by season of calving				Age in years and months	Adjustment factors by season of calving			
	November to June		July to October			November to June		July to October	
	Milk	Fat	Milk	Fat		Milk	Fat	Milk	Fat
1–9	1.36	1.32	1.26	1.23	5–3	1.03	1.01	1.02	1.00
1–10	1.30	1.26	1.22	1.19	5–4	1.03	1.01	1.02	1.00
1–11	1.26	1.23	1.20	1.17	5–5	1.02	1.01	1.02	1.00
2–0	1.24	1.21	1.19	1.16	5–6	1.02	1.01	1.02	1.00
2–1	1.22	1.19	1.18	1.16	5–7	1.01	1.00	1.01	1.00
2–2	1.21	1.18	1.18	1.15	5–8	1.01	1.00	1.01	1.00
2–3	1.19	1.17	1.17	1.14	5–9	1.01	1.00	1.01	1.00
2–4	1.19	1.16	1.16	1.14	5–10	1.01	1.00	1.01	1.00
2–5	1.18	1.15	1.16	1.13	5–11	1.01	1.00	1.01	1.00
2–6	1.18	1.14	1.15	1.12	6–0 to 6–2	1.01	1.00	1.01	1.00
2–7	1.17	1.14	1.14	1.11	6–3 to 6–5	1.01	1.00	1.01	1.00
2–8	1.17	1.14	1.14	1.10	6–6 to 6–8	1.01	1.00	1.01	1.00
2–9	1.17	1.14	1.13	1.09	6–9 to 6–11	1.00	1.00	1.01	1.00
2–10	1.17	1.14	1.12	1.08	7–0 to 7–2	1.00	1.00	1.00	1.00
2–11	1.16	1.13	1.11	1.07	7–3 –to 7–5	1.00	1.00	1.00	1.00
3–0	1.15	1.12	1.11	1.07	7–6 to 7–8	1.00	1.00	1.00	1.00
3–1	1.14	1.11	1.11	1.07	7–9 to 7–11	1.00	1.00	1.00	1.00
3–2	1.13	1.10	1.11	1.07	8–0 to 8–2	1.00	1.00	1.00	1.00
3–3	1.12	1.09	1.11	1.07	8–3 to 8–5	1.00	1.01	1.00	1.00
3–4	1.11	1.08	1.11	1.07	8–6 to 8–8	1.00	1.01	1.00	1.00
3–5	1.11	1.07	1.11	1.07	8–9 to 8–11	1.00	1.01	1.00	1.00
3–6	1.10	1.06	1.10	1.06	9–0 to 9–2	1.01	1.02	1.00	1.01
3–7	1.10	1.06	1.09	1.05	9–3 to 9–5	1.01	1.02	1.01	1.02
3–8	1.10	1.06	1.08	1.04	9–6 to 9–8	1.01	1.02	1.02	1.03
3–9	1.10	1.06	1.08	1.04	9–9 to 9–11	1.01	1.03	1.03	1.04
3–10	1.10	1.06	1.07	1.03	10–0 to 10–2	1.01	1.03	1.03	1.04
3–11	1.09	1.06	1.06	1.03	10–3 to 10–5	1.01	1.03	1.03	1.05
4–0	1.08	1.05	1.06	1.02	10–6 to 10–8	1.01	1.04	1.03	1.05
4–1	1.07	1.04	1.06	1.02	10–9 to 10–11	1.01	1.04	1.03	1.05
4–2	1.07	1.04	1.05	1.02	11–0 to 11–2	1.02	1.04	1.03	1.05
4–3	1.06	1.03	1.05	1.02	11–3 to 11–5	1.02	1.05	1.03	1.05
4–4	1.05	1.03	1.05	1.01	11–6 to 11–8	1.02	1.05	1.03	1.05
4–5	1.05	1.02	1.04	1.01	11–9 to 11–11	1.02	1.06	1.04	1.06
4–6	1.04	1.02	1.04	1.01	12–0 to 12–2	1.03	1.06	1.04	1.06
4–7	1.04	1.02	1.03	1.00	12–3 to 12–5	1.03	1.07	1.04	1.07
4–8	1.04	1.02	1.03	1.00	12–6 to 12–8	1.03	1.07	1.04	1.07
4–9	1.04	1.02	1.03	1.00	12–9 to 12–11	1.04	1.07	1.05	1.08
4–10	1.04	1.02	1.03	1.00	13–0 to 13–2	1.04	1.07	1.06	1.09
4–11	1.04	1.02	1.03	1.00	13–3 to 13–5	1.04	1.07	1.07	1.10
5–0	1.04	1.02	1.03	1.00	13–6 to 13–8	1.04	1.08	1.08	1.11
5–1	1.04	1.01	1.02	1.00	13–9 to 13–11	1.05	1.08	1.09	1.13
5–2	1.03	1.01	1.02	1.00	14–0 and up	1.05	1.08	1.11	1.15

*These factors were based on 57,520 records of cows that calved from November through June and on 32,450 records of those that calved from July through October. (Courtesy of the USDA.)

Table 20.5. Holstein age adjustment factors for milk and fat production: for cows that calve in Indiana, Illinois, Ohio, Michigan, Wisconsin, Minnesota, Iowa, Missouri, North Dakota, South Dakota, Nebraska, Kansas, Oklahoma, and Texas*

| Age in years and months | Adjustment factors by season of calving | | | | Age in years and months | Adjustment factors by season of calving | | | |
| | November to June | | July to October | | | November to June | | July to October | |
	Milk	Fat	Milk	Fat		Milk	Fat	Milk	Fat
1–9	1.40	1.38	1.32	1.30	5–3	1.02	1.01	1.02	1.01
1–10	1.35	1.33	1.28	1.26	5–4	1.02	1.01	1.02	1.01
1–11	1.32	1.30	1.26	1.24	5–5	1.02	1.01	1.02	1.01
2–0	1.30	1.28	1.25	1.23	5–6	1.02	1.01	1.02	1.01
2–1	1.29	1.27	1.24	1.22	5–7	1.02	1.01	1.02	1.01
2–2	1.28	1.25	1.23	1.21	5–8	1.02	1.01	1.01	1.00
2–3	1.26	1.24	1.23	1.21	5–9	1.02	1.01	1.01	1.00
2–4	1.25	1.23	1.22	1.20	5–10	1.01	1.01	1.00	1.00
2–5	1.25	1.22	1.22	1.20	5–11	1.01	1.00	1.00	1.00
2–6	1.24	1.22	1.21	1.19	6–0 to 6–2	1.01	1.00	1.00	1.00
2–7	1.24	1.22	1.21	1.18	6–3 to 6–5	1.00	1.00	1.00	1.00
2–8	1.24	1.22	1.20	1.17	6–6 to 6–8	1.00	1.00	1.00	1.00
2–9	1.24	1.21	1.19	1.16	6–9 to 6–11	1.00	1.00	1.00	1.00
2–10	1.23	1.21	1.18	1.15	7–0 to 7–2	1.00	1.00	1.00	1.00
2–11	1.21	1.19	1.17	1.14	7–3 to 7–5	1.00	1.00	1.00	1.00
3–0	1.19	1.17	1.16	1.13	7–6 to 7–8	1.00	1.00	1.00	1.00
3–1	1.18	1.16	1.15	1.13	7–9 to 7–11	1.00	1.00	1.00	1.00
3–2	1.16	1.15	1.14	1.12	8–0 to 8–2	1.00	1.00	1.00	1.01
3–3	1.15	1.14	1.14	1.12	8–3 to 8–5	1.00	1.01	1.00	1.01
3–4	1.14	1.13	1.14	1.12	8–6 to 8–8	1.01	1.01	1.01	1.01
3–5	1.13	1.12	1.14	1.12	8–9 to 8–11	1.01	1.02	1.01	1.02
3–6	1.13	1.11	1.13	1.11	9–0 to 9–2	1.01	1.02	1.01	1.02
3–7	1.12	1.11	1.12	1.10	9–3 to 9–5	1.01	1.02	1.01	1.03
3–8	1.12	1.11	1.11	1.09	9–6 to 9–8	1.02	1.03	1.02	1.03
3–9	1.12	1.10	1.10	1.08	9–9 to 9–11	1.02	1.04	1.02	1.04
3–10	1.11	1.10	1.09	1.07	10–0 to 10–2	1.03	1.04	1.02	1.04
3–11	1.10	1.09	1.08	1.06	10–3 to 10–5	1.03	1.04	1.03	1.05
4–0	1.09	1.08	1.07	1.05	10–6 to 10–8	1.03	1.05	1.03	1.05
4–1	1.08	1.07	1.06	1.04	10–9 to 10–11	1.03	1.05	1.03	1.05
4–2	1.07	1.06	1.06	1.04	11–0 to 11–2	1.04	1.06	1.03	1.05
4–3	1.06	1.05	1.06	1.04	11–3 to 11–5	1.04	1.06	1.04	1.06
4–4	1.05	1.04	1.06	1.04	11–6 to 11–8	1.04	1.06	1.04	1.06
4–5	1.05	1.04	1.06	1.04	11–9 to 11–11	1.05	1.07	1.04	1.07
4–6	1.05	1.03	1.05	1.03	12–0 to 12–2	1.05	1.07	1.05	1.07
4–7	1.05	1.03	1.04	1.03	12–3 to 12–5	1.06	1.08	1.05	1.08
4–8	1.04	1.03	1.03	1.02	12–6 to 12–8	1.06	1.08	1.06	1.08
4–9	1.04	1.03	1.03	1.02	12–9 to 12–11	1.07	1.09	1.06	1.09
4–10	1.04	1.03	1.02	1.01	13–0 to 13–2	1.08	1.10	1.07	1.09
4–11	1.04	1.03	1.02	1.01	13–3 to 13–5	1.08	1.11	1.08	1.10
5–0	1.04	1.03	1.02	1.01	13–6 to 13–8	1.09	1.12	1.09	1.11
5–1	1.03	1.02	1.02	1.01	13–9 to 13–11	1.10	1.13	1.10	1.13
5–2	1.03	1.02	1.02	1.01	14–0 and up	1.12	1.15	1.12	1.16

*These factors were based on 222, 032 records of cows that calved from November through June and on 153, 263 records of those that calved from July through October. (Courtesy of the USDA.)

Table 20.6 Jersey age adjustment factors for milk and fat production: for cows that calve in Indiana, Illinois, Ohio, Michigan, Wisconsin, Minnesota, North Dakota, South Dakota, Nebraska, Kansas, Iowa, and Missouri*

Age in years and months	November to June (Milk)	November to June (Fat)	July to October (Milk)	July to October (Fat)	Age in years and months	November to June (Milk)	November to June (Fat)	July to October (Milk)	July to October (Fat)
1–9	1.34	1.31	1.26	1.24	5–3	1.02	1.00	1.02	1.01
1–10	1.32	1.29	1.24	1.22	5–4	1.02	1.00	1.02	1.01
1–11	1.29	1.27	1.23	1.21	5–5	1.02	1.00	1.02	1.01
2–0	1.28	1.25	1.22	1.20	5–6	1.02	1.00	1.02	1.01
2–1	1.27	1.24	1.21	1.19	5–7	1.02	1.00	1.02	1.01
2–2	1.26	1.23	1.21	1.18	5–8	1.02	1.00	1.02	1.01
2–3	1.26	1.23	1.21	1.18	5–9	1.02	1.00	1.02	1.01
2–4	1.26	1.23	1.21	1.18	5–10	1.01	1.00	1.02	1.01
2–5	1.26	1.23	1.20	1.17	5–11	1.01	1.00	1.01	1.00
2–6	1.25	1.23	1.19	1.16	6–0 to 6–2	1.00	1.00	1.01	1.00
2–7	1.25	1.22	1.18	1.15	6–3 to 6–5	1.00	1.00	1.01	1.00
2–8	1.23	1.20	1.16	1.14	6–6 to 6–8	1.00	1.00	1.01	1.00
2–9	1.21	1.17	1.14	1.12	6–9 to 6–11	1.00	1.00	1.00	1.00
2–10	1.18	1.15	1.13	1.10	7–0 to 7–2	1.00	1.00	1.00	1.00
2–11	1.16	1.13	1.13	1.09	7–3 to 7–5	1.00	1.01	1.00	1.00
3–0	1.15	1.12	1.13	1.09	7–6 to 7–8	1.00	1.01	1.00	1.00
3–1	1.14	1.11	1.12	1.09	7–9 to 7–11	1.00	1.01	1.00	1.00
3–2	1.14	1.10	1.11	1.08	8–0 to 8–2	1.00	1.01	1.00	1.01
3–3	1.14	1.10	1.11	1.08	8–3 to 8–5	1.00	1.01	1.01	1.02
3–4	1.13	1.10	1.11	1.08	8–6 to 8–8	1.00	1.01	1.01	1.02
3–5	1.13	1.10	1.11	1.08	8–9 to 8–11	1.01	1.02	1.01	1.02
3–6	1.13	1.10	1.11	1.08	9–0 to 9–2	1.01	1.02	1.01	1.02
3–7	1.13	1.10	1.11	1.08	9–3 to 9–5	1.01	1.03	1.01	1.02
3–8	1.13	1.10	1.11	1.08	9–6 to 9–8	1.01	1.03	1.01	1.03
3–9	1.12	1.09	1.10	1.07	9–9 to 9–11	1.01	1.03	1.01	1.03
3–10	1.11	1.08	1.08	1.05	10–0 to 10–2	1.01	1.03	1.01	1.03
3–11	1.09	1.06	1.06	1.03	10–3 to 10–5	1.02	1.03	1.02	1.03
4–0	1.07	1.04	1.05	1.02	10–6 to 10–8	1.02	1.03	1.02	1.03
4–1	1.06	1.03	1.05	1.02	10–9 to 10–11	1.02	1.03	1.02	1.03
4–2	1.05	1.03	1.05	1.02	11–0 to 11–2	1.03	1.04	1.02	1.03
4–3	1.04	1.03	1.05	1.02	11–3 to 11–5	1.03	1.04	1.02	1.04
4–4	1.04	1.03	1.05	1.02	11–6 to 11–8	1.03	1.04	1.02	1.04
4–5	1.04	1.03	1.05	1.01	11–9 to 11–11	1.04	1.05	1.03	1.04
4–6	1.04	1.03	1.04	1.01	12–0 to 12–2	1.04	1.05	1.03	1.05
4–7	1.04	1.03	1.04	1.01	12–3 to 12–5	1.05	1.05	1.04	1.05
4–8	1.04	1.03	1.03	1.01	12–6 to 12–8	1.05	1.06	1.04	1.06
4–9	1.04	1.03	1.03	1.01	12–9 to 12–11	1.06	1.06	1.05	1.06
4–10	1.04	1.03	1.02	1.01	13–0 to 13–2	1.06	1.07	1.05	1.07
4–11	1.04	1.02	1.02	1.01	13–3 to 13–5	1.06	1.07	1.05	1.07
5–0	1.03	1.02	1.02	1.01	13–6 to 13–8	1.07	1.08	1.06	1.08
5–1	1.02	1.01	1.02	1.01	13–9 to 13–11	1.07	1.08	1.06	1.08
5–2	1.02	1.01	1.02	1.01	14–0 and up	1.08	1.09	1.07	1.09

*These factors were based on 22,601 records of cows that calved from November through June and on 13,438 records of those that calved from July through October. (Courtesy of the USDA.)

Table 20.7 Factors for reducing 305-day, age-corrected records to a twice-a-day milking basis

Number of days milked	Factor for 3-times-a-day milking			Factor for 4-times-a-day milking		
	2 to 3 years of age	3 to 4 years of age	4 years of age and over	2 to 3 years of age	3 to 4 years of age	4 years of age and over
5 to 15	0.99	0.99	0.99	0.98	0.99	0.99
16 to 25	0.98	0.99	0.99	0.97	0.98	0.98
26 to 35	0.98	0.98	0.98	0.96	0.97	0.97
36 to 45	0.97	0.98	0.98	0.95	0.96	0.96
46 to 55	0.97	0.97	0.97	0.94	0.95	0.96
56 to 65	0.96	0.97	0.97	0.93	0.94	0.95
66 to 75	0.95	0.96	0.96	0.92	0.93	0.94
76 to 85	0.95	0.95	0.96	0.91	0.92	0.93
86 to 95	0.94	0.95	0.96	0.90	0.91	0.93
96 to 105	0.94	0.94	0.95	0.89	0.91	0.92
106 to 115	0.93	0.94	0.95	0.88	0.90	0.91
116 to 125	0.92	0.93	0.94	0.87	0.89	0.90
126 to 135	0.92	0.93	0.94	0.87	0.88	0.90
136 to 145	0.91	0.93	0.93	0.86	0.88	0.89
146 to 155	0.91	0.92	0.93	0.85	0.87	0.88
156 to 165	0.90	0.92	0.93	0.84	0.86	0.88
166 to 175	0.90	0.91	0.92	0.83	0.85	0.87
176 to 185	0.89	0.91	0.92	0.82	0.85	0.86
186 to 195	0.89	0.90	0.91	0.82	0.84	0.86
196 to 205	0.88	0.90	0.91	0.81	0.83	0.85
206 to 215	0.88	0.89	0.90	0.80	0.83	0.85
216 to 225	0.87	0.89	0.90	0.79	0.82	0.84
226 to 235	0.87	0.88	0.90	0.79	0.81	0.83
236 to 245	0.86	0.88	0.89	0.78	0.81	0.83
246 to 255	0.86	0.88	0.89	0.77	0.80	0.82
256 to 265	0.85	0.87	0.88	0.77	0.79	0.82
266 to 275	0.85	0.87	0.88	0.76	0.79	0.81
276 to 285	0.84	0.86	0.88	0.75	0.78	0.80
286 to 295	0.84	0.86	0.87	0.75	0.78	0.80
296 to 305	0.83	0.85	0.87	0.74	0.77	0.79

From USDAP ARS-52-1, January 1955.

$$\text{Fat-corrected milk} = (12{,}000 \times 0.4) + (420 \times 15)$$
$$= 4800 \qquad + 6300$$
$$= 11{,}100 \text{ lb}$$

20.1.4 Productive Life Span in Dairy Cattle

This is another trait of economic importance. A study of 101 commercial dairy herds in Florida [4] showed that the average productive life of dairy cows in a herd maintained mostly by purchased replacements was 3.9 years after entering the herd at two years of age. In 14 herds where replacements were home-raised, this figure increased to 4.7 years. The reasons for the disposal of 58 percent of 2182 cows in these herds were udder trouble, low production, and reproductive disorders.

In a New Jersey [55] study, longevity in a Holstein-Friesian herd was found to be about 37 percent heritable, as calculated from the intrasire regression of daughters on dams. Breeding efficiency, expressed as a percentage derived from the actual calving interval in days to the ideal of 365 days, was 32 percent heritable. The association between productive life span and breeding efficiency was low and insignificant.

20.1.5 Type and Conformation

Type in dairy cattle has received much attention both in the show-ring and from breeders in selecting replacement animals. As shown in Table 20.2, type in dairy cattle is about 25 percent heritable, which indicates that

Fig. 20.1 Princess Breezewood R A Patsy 3816059 (VG) is one of those individuals that have good type as well as performance. Her record was 36,821 pounds of milk and 1866 pounds of butter-fat at five years of age. (Courtesy of the Holstein-Friesian Association of America.)

only moderate progress could be made in selection for the improvement of this trait.

Type and conformation are valuable because superiority in these traits should help the animal to maintain a long and highly productive life. The desirable items are size and development of the mammary gland, proper placement of the teats, soundness of feet and legs, and body capacity, which should give some indication of the animal's ability to consume large amounts of grains and roughages.

Differences between dairy- and beef-cattle type are very obvious. In a general way, there is a relationship between body form and milk production in dairy cattle, but within a herd or breed the relationship seems to be relatively small. Most studies between dairy type and production show a positive but low phenotypic correlation between these traits. The genetic correlations are also very low, which indicates that selection for type alone would result in little improvement in production. The two traits seem to be inherited independently, and to improve both, selection for both must be practiced.

20.2 GENETIC CORRELATIONS BETWEEN PRODUCTION TRAITS

Genetic correlations between the various traits in dairy cattle are summarized in Table 20.8. The data indicate very clearly that there is a strong genetic correlation between milk yield and butterfat yield, with an average value of 0.81. Thus, many of the same genes affect both traits, and intensive selection for one should bring about improvement in the other. This is a desirable correlation, since it is positive. The genetic correlation between milk yield and butterfat percentage was negative in all studies and averaged—0.41. This suggests that many of the genes responsible for high milk yields cause the production of a lower percentage of fat in the milk. These figures seem to be borne out by the association of the two traits seen in the different breeds of dairy cattle. Holstein-Friesians, for instance, give large amounts of milk but a lower butterfat percentage. Jerseys give a smaller amount of milk, and the percentage of butterfat is much higher. This negative genetic association is not of great economic importance, because most dairymen are probably more interested in the total fat and milk yield than they are in the percentage of fat in the milk.

The data presented in Table 20.8 show that there is a very low genetic correlation between fat yield and fat percent in the milk. This means that very few of the same genes affect these two traits, and selection for one should not cause a genetic change in the other.

Table 20.8 Genetic correlations between traits in dairy cattle

Traits correlated	No. of studies	Average	Range
Type and butterfat yield	3	−0.15	−0.52 to 0.08
Type and milk yield	3	0.05	0.00 to 0.08
Milk yield and butterfat percent	7	−0.43	−0.58 to −0.20
Milk yield and butterfat yield	6	0.81	0.62 to 0.92
Fat yield and fat percent	3	0.14	−0.03 to 0.26

References 17, 18, 22, 25, 42, 45, and 49.

20.3 SELECTION OF SUPERIOR DAIRY COWS

Dairy cows produce a limited number of offspring in their lifetime because their productive life span is short and their reproductive rate is slow, usually being limited to one calf per year. For this reason, it is not possible to make much improvement through selection over a period of years by placing emphasis on selection of cows for higher production. Nevertheless, it is important to determine the productivity of each cow in the herd for milk and butterfat yield by determining her record on a 305-day, M.E., 2× per day milking basis. Cows with poor records may then be culled, which tends to raise the average production record of the entire herd. Replacement stock may then be selected from the most highly productive cows that are genetically superior. Prospective herd sires should also be selected from superior dams.

20.4 SELECTION OF SUPERIOR DAIRY SIRES

Most of the improvement realized through selection for increased productivity in dairy cattle results from the identification and use of genetically superior dairy sires. This is true because a sire may produce hundreds, or even thousands, of progeny in his lifetime, especially when he is used for artificial insemination. A genetically superior dairy cow, however, usually produces no more than eight to ten progeny during her lifetime and for this reason as an individual she has a smaller influence on the amount of genetic progress made through selection than does a genetically superior sire.

The selection of superior dairy sires requires special methods, since the bull produces no milk. He does transmit genes for milk production to

his offspring, however. The most accurate method for selecting superior dairy sires is through the use of the *progeny test* and *herdmate comparisons* [26]. The progeny test, of course, refers to the estimation of the sire's transmitting ability for milk and butterfat production through the records of his daughters. Herdmate comparisons refer to the comparison of the records of the daughters of a sire with those of other sires which lactate in the same herd during the same season of the year. It is necessary that the records of all cows being compared within the herd be converted to a 305-day, $2 \times$, M.E. basis before comparisons are made.

In herdmate (or contemporary) comparisons, variations due to herd, year, and season are reduced, since cow records are compared within the same herd, during the same year and the same season. The average records of the daughters of a sire are compared with those of the daughters of all other sires within the same herd that calved in the same months as the daughters or in the previous two months or in the succeeding two months. Thus, the averages compared represent a five-month average.

20.4.1 Adjusted Herdmate Average (AHMA) [26]

The number of herdmates available for a comparison with the average of a particular sire's daughters often varies from herd to herd. Therefore, the herdmate average should be adjusted for the number of herdmates available, in order to place the herdmates' and the daughters' records on a more comparable basis, by using the following formula:

$$\text{AHMA} = \text{RBYS avg.} + \frac{\text{NHM}}{\text{NHM} + 1} (\text{HM avg.} - \text{RBYS avg.})$$

where

AHMA is the adjusted herdmate average.
RBYS avg. is the regional-breed-year-season average.
NHM is the number of herdmates.
HM is the herdmate average.

The RBYS averages are calculated from the $2 \times$, 305-day, M.E., DHIA records of cows calving from January 1, 1950, to May 1, 1967. These may be found in the DHIA Sire Summary Lists for the various breeds. These averages are adjusted from time to time as more records become available.

The RBYS helps reduce environmental differences that result from a comparison of records of cows located in different regions of the United States. The portion of the formula in which the number of herdmates is divided by the number of herdmates plus one tends to adjust for a different number of records, because the adjusted herdmate average increases toward the herdmate average as the number of herdmate records increases. This indicates the probability of a greater accuracy with a larger number of records used to calculate herdmate averages. The use of artificial insemination in dairy-cattle breeding makes it possible to compare records of a sire's daughters with those of their herdmates in many herds. These figures from many herds may then be averaged before comparisons are made.

When artificial insemination is used, it is possible to compare the progeny of several sires in more than one herd. However, it is possible that some sires will be used more often and produce more offspring than others. The average of a sire's daughters may be adjusted for a difference in number as follows:

$$\frac{\text{Adjusted daughter}}{\text{average}} = \frac{\text{daughter}}{\text{average}} - 0.9(\text{AHMA} - \text{RBYS average})$$

This adjustment is necessary because it has been shown that the records of the daughter average of a sire reflects about 0.90 of the difference between the herdmate average and the breed average. This formula adjusts for differences in herdmate levels in different herds, and helps correct for the problem that might arise when a sire has daughters produced by artificial insemination in a low-producing as compared to a high-producing herd.

To illustrate the calculation of the AHMA, let us assume that the RYBS average for cows of the Holstein breed is 14,000 pounds of milk, and the herdmate average is 15,000 for 20 herdmates. The calculations would be as follows:

$$\text{AHMA} = 14,000 + \frac{20}{20 + 1}(15,000 - 14,000)$$
$$= 14,000 + 0.95(1000)$$
$$= 14,950$$

20.4.2 Predicted Breeding Value or Predicted Difference (PD) [26]

The estimate of a sire's probable breeding value is referred to as the Predicted Difference (PD). This is the amount the bull's progeny would be expected to vary from their herdmates in breed average herds. The PD for a sire may be calculated from the following formula:

$$\text{PD} = \frac{\text{no. of daughters}}{\text{no. of dau.} + 20}\left(\frac{\text{adj. daughter}}{\text{average}} - \frac{\text{breed}}{\text{average}}\right)$$

As an example, let us assume that the adjusted daughter average for 50 daughters of a sire is 11,000 pounds of milk and the breed average is 9000 pounds. The PD of the sire would be

$$PD = \frac{50}{50 + 20}(11{,}000 - 9000)$$
$$= \tfrac{50}{70}(2000)$$
$$= 0.7143 \times 2000$$
$$= 1429$$

In other words, the daughters of this sire should have a 305-day milk production record 1429 pounds above their herdmates that are average for the breed.

The accuracy of the PD in estimating the true breeding value of a particular sire as compared to other sires in the same breed may be affected by many factors. These include (1) the number of daughters used in the comparison, (2) the number of herds in which his daughters are located, (3) how well these daughters of a sire are distributed through these different herds, and (4) the number of lactations per daughter. The USDA-DHIA use more sophisticated calculations and use a computer for calculating the PD for sires, but the formula given above gives a general idea of how it is calculated. Sire summary lists, including the PD of many sires in DHIA, are published periodically by the USDA-DHIA and may be purchased from the United States Department of Agriculture.

20.5 RESULTS OF SELECTION IN DAIRY CATTLE

Definite breed differences in milk and butterfat production leave little doubt that selection for these traits has been effective. Nevertheless, very few well-controlled selection experiments with dairy cattle have been reported.

In a USDA study [10], a Holstein herd was established in 1918 and a selection study was initiated. No females were culled until after they completed at least one lactation record. Environmental conditions were kept as constant as possible throughout the experiment. Ten proven sires were used during a period of 28 years. Sires were used whose unselected daughters were uniformly high producers and produced better than their dams.

The average yearly production of 16 foundation cows was 17,524 pounds of milk and 601 pounds of butterfat. This was a high level of production. The average production of 183 unselected daughters that were descendants of the ten sires was 17,491 pounds of milk and 629 pounds of butterfat. Some of the proven sires improved the average of their daughters' records over their dams, whereas others lowered them. The last three progeny-tested sires used produced unselected daughters whose milk production records ranged between 18,680 and 19,850 pounds of milk and between 683 and 711 pounds of butterfat. This study showed conclusively that superior, progeny-tested bulls were hard to find.

Several reports in the literature have attempted to assess the progress made by using artificial insemination in dairy herds [20, 39, 53]. Usually records of performance of cows produced by this method are compared with those of cows produced by natural matings in the same herd at the same time. Cows from artificial insemination have proved to be significantly superior in production in some instances.

A study of Holsteins produced in New York State [40] showed that in the period from 1951 to 1959 the gain from artificial insemination was 58 percent faster for butterfat yield and 28 percent faster for milk yield than gains realized from natural service.

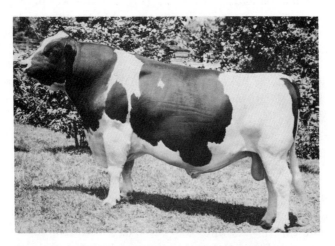

Fig. 20.2 WIS Captain (1738), a *Gold Medal Proved Sire.* Thirteen pairs of daughter-dam comparisons showed a difference of +1350 pounds of milk and 56 pounds of butterfat in favor of the daughters. In four years he performed 219,043 services by artificial insemination with approximately 145,000 offspring. At that time he was still going strong as a sire. (Courtesy of the American Breeders Service, Chicago, Illinois.)

The magnitudes of intraherd environmental and genetic trends in milk production of Holsteins in New York State [23] included records from December, 1956 to November, 1962. These data were from cows artificially sired and from the DHIA population. The improvement per year due to environment was estimated to be 282 pounds of milk and 11 pounds of butterfat. Yearly genetic improvement over the same period was estimated at 104 pounds of milk and three pounds of butterfat. Although most of the trend for improvement appeared to be due to environment, some improvement was genetic.

Research at the California Station [35] indicated that milk production in cows in DHIA herds in that state produced an average of 13,536 pounds of milk and 507 pounds of butterfat in 1968 as compared to an average production in 1930 of 8464 pounds of milk and 330 pounds of butterfat. No estimate was made of how much of this improvement was genetic or environmental. In 1968, the year of this study, the average milk production of cows on the DHIA test was 39 percent higher than for cows not on the DHIA program.

Results of these studies indicate that some genetic improvement can be made in milk and butterfat production through the application of proper mass selection programs.

20.6 INBREEDING OF DAIRY CATTLE

Many experiments have been conducted with inbreeding in dairy cattle. The main objective in most studies was to determine if pure lines could be formed in which individuals had a level of production as high as that of outbred animals. In addition, it was desired to learn what the effects of inbreeding might be in this class of farm animals.

20.6.1 Occurrence of Detrimental Genes

Several experiments have clearly shown that inbreeding in dairy cattle uncovers recessive genes if they are present in the foundation stock. A study of genetic and environmental factors in the development of the American Red Danish cattle [43] showed that 65 calves in 27 herds were born with paralyzed hind quarters. Forty-two calves in 11 herds were dead at birth and showed ankylosis and mummification. These defects are inherited, and they have been reported in Denmark. Both defects were traced to certain bulls of the breed. A gene frequency analysis showed that about 25 percent of American Red Danish cattle were heterozygous for the paralyzed condition and 11 percent were heterozygous for ankylosis.

In a study conducted in California [37], an increase in calf mortality that accompanied inbreeding was partly accounted for by two lethal genes, one of which controlled an anomaly of the liver and the other an anomaly of the heart. Neither of these defects could be determined by their external morphological appearance. In a USDA study of inbreeding [56] a few Guernsey calves were deformed at birth, and apparently a recessive gene was involved. In an inbreeding experiment at the New Jersey station [2], data on four foundation herd sires were studied. As inbreeding progressed, some of the descendants of one bull died at birth or shortly afterward of an inherited defect called "Bulldog." Various abnormalities of the reproductive tracts also occurred, and this sire family had to be abandoned. Another sire produced offspring that were undesirable and many of which also carried a factor for red spotting of the hair coat. Apparently, two of the four sires transmitted no apparent genetic defect, and one of these produced descendants of very satisfactory type and performance.

It is apparent from the literature that dairy cattle may carry several recessive genes that are uncovered by inbreeding. Most of these defects in the heterozygous state cannot be recognized by the morphological appearance of the individual and can be discovered only by an appropriate progeny test. Thus, any breeder who practices inbreeding risks an increase in the occurrence of genetic defects in his herd. The only sure way to determine if the breeding animals are carrying such genetic defects is to inbreed and progeny-test them. This takes time and is costly, but there are probably some sires of outstanding genetic merit that do not carry lethal or detrimental genes and that should be identified.

20.6.2 *Effect of Inbreeding on Growth*

Results of experiments do not all agree as to the effect of inbreeding on growth rate. In general, however, inbreeding seems to decrease birth weight and mature weight [1, 11, 33, 36, 41, 56]. In one study [33], it was found that inbreeding slowed the growth rate early in life but permitted the later rapid growth to continue longer so that mature size was not decreased but may have even been increased. At the New Jersey Station [30], Holstein-Friesians were inbred up to 20 percent without any decrease in weight at maturity as compared to outbred animals. When inbreeding was more than 20 percent, inbred females grew normally to approximately first calving age, then developed more slowly thereafter.

20.6.3 *Effect of Inbreeding on Fertility*

Inbreeding did not cause an increase in the number of services per conception in grade Holsteins [56] and seemed to cause little or no increase in abortions and stillbirths. However, in most experiments, an increase in inbreeding has resulted in an increase in calf mortality after birth. Part, but not all, of this increased mortality was due to lethal factors. Apparently, inbred calves were less able than outbreds to cope with environmental conditions during this stage of life.

20.6.4 Effects of Inbreeding on Production

Some of the early experiments on inbreeding in dairy cattle did not report the coefficient of regression for milk and butterfat production on inbreeding as calculated from Wright's coefficient of inbreeding. Results of later studies which reported such figures are summarized in Table 20.9.

Table 20.9 Effects of inbreeding on milk and butterfat production

Production trait	No. of reports	Regression coefficient*	
		Average	Range
Pounds of milk	4	−71.900	−209.800 to −0.074
Percentage of butterfat	2	0.006	0.003 to 0.008
Pounds of butterfat	4	−2.310	−4.880 to −0.300

*Decrease or increase for each 1 percent inbreeding.
References 28, 43, 50, and 52.

In some experiments, inbreeding seemed to result in an increase in production [2, 3, 56], but, in general, this was not true. At the New Jersey Station [3] experimental results indicated that inbreeding up to 20 percent, accompanied by rigid selection, could result in superior animals. It was concluded that the primary result of the inbreeding work at that station was the development of superior inbred sires with a marked prepotency for desirable growth, type, butterfat test, and production.

20.6.5 Conclusions on Inbreeding

Results of inbreeding work in dairy cattle show that inbreeding is often detrimental, as measured by the increased occurrence of recessive defects, greater calf mortality, and lowered production of milk and butterfat in inbred cows. Nevertheless, adverse effects have not always been observed. Certain inbred sires and dams seem to be very prepotent for high production. This suggests that a linebreeding program using outstanding animals within the breed could develop outstanding lines. The development of several such lines within a breed could produce some very superior animals if the best lines were crossed. The heritability of the traits suggests that additive genes have a moderate-to-great effect on production, whereas inbreeding and crossbreeding effects are only moderate. This would suggest that selection of high-

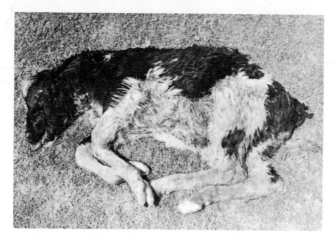

Fig. 20.3 A calf affected by a recessive lethal gene which causes pregnancy to be prolonged in some cases as long as 60 to 70 days with a birth weight of up to 150 to 160 pounds. The calves have to be delivered by Caesarian section in some cases. (Courtesy of Dr. L. W. Holm, School of Veterinary Medicine, University of California, Davis, California.)

producing stock and fixing the desired traits by moderate inbreeding would be of value in a dairy cattle breeding program.

20.7 CROSSBREEDING IN DAIRY CATTLE

Results of experiments in which the performance of the F_1 crossbreds are compared with that of the average of the two parent breeds are summarized in Table 20.10. In some of these experiments, numbers of animals involved are small, but they still give some indication of the effects of heterosis on milk and butterfat production.

Results indicate that there is some heterotic advantage in milk and butterfat yield, with little or no advantage for the percentage of butterfat in the milk. In three experiments involving crosses of the dairy breeds, from 3 to 12 percent heterosis was observed in milk yield and from 4 to 37 percent in fat yield. This means that there was a tendency for the F_1 average to be closer to the average of the more productive breed. It should be pointed out, however, that in none of the studies did the average of milk production by the F_1 crossbreds exceed that of the more productive parent breed.

In the crosses of the dairy breeds, the average butterfat yield of the F_1 crossbreds exceeded that of either parental breed in two out of three experi-

Table 20.10 Influence of crossbreeding on milk production in dairy cattle

| Breeds crossed | F_1 as percentage of P_1 average | | |
	lb milk	% fat	lb fat
Red Danish × Jersey	103	98	104
Friesian × Guernsey	107	97	109
Friesian × Jersey	112	115	137
Jersey × Angus	96	96	90
Friesian × Angus	98	97	98
Average	103	101	108

References 8, 9, 16, 38, 57, and 58.

ments. In two of the three reported experiments, the percentage of butterfat in the milk of the F_1 crossbreds was slightly lower than the average of the parental breeds. This trait shows that heterosis effects were slight, or even slightly negative. In none of the three experiments did the average percentage of butterfat in the milk of the crossbreds exceed that of the most productive parent.

Results of two experiments summarized in Table 20.10 involving the cross of a dairy breed with a beef breed (Angus) actually showed a slightly negative heterosis for all traits studied. In other words, there was a tendency for the average of the F_1 for milk yield, fat yield, and butterfat percentage to be less than the average of the two parent breeds. Numbers of animals involved in these experiments were small, however, and the results may not be conclusive.

A summary of the results of a crossbreeding experiment with dairy cattle by the USDA at Beltsville, Maryland, is given in Table 20.11. The experiment was initiated in 1939 with foundation cattle from the Holstein-Friesian, Guernsey, Jersey, and Red Danish breeds. These cows were mated in rotation to proven sires of the Holstein-Friesian, Jersey, and Red Danish breeds. All production records of crossbred females and their foundation dams, except the Red Danish dams, were made on the basis of three-times-per-day milking for a 365-day lactations period. In this experiment, there was a considerable advantage of the two-breed crosses over the foundation purebreds in milk and butterfat yield and, to a lesser extent, in the percentage of butterfat. The three-breed cross had only a slight advantage over the two-breed crosses. Crossbred cows had a higher degree of persistency of production than purebreds, which accounted for a part of their higher level of production.

In the Beltsville experiment, the average of the crossbreds exceeded or

Table 20.11 Results of crossbreeding experiments with dairy cattle at Beltsville, Maryland*

Groups tested	No. of cows	Mature-Equivalent Values		
		lb milk	% fat	lb fat
Foundation purebreds	55	13,799	4.30	594
		(100)	(100)	(100)
2-breed crosses	55	17,811	4.49	799
		(129)	(104)	(135)
3-breed crosses	58	18,240	4.39	801
		(132)	(102)	(135)

*Adapted from USDA TB 1074, 1954. Figures in parentheses are production figures as a percentage of the foundation purebreds, or an estimate of heterosis.

equalled the average of any of the parental foundation breeds on a mature-equivalent basis. This experiment shows a higher degree of heterosis for milk and fat yield than other experiments reported in Table 20.10.

An intensive crossbreeding study has been conducted by the Illinois Agricultural Experiment Station in cooperation with the USDA. It was initiated in 1949, using the Holstein and Guernsey breeds. One of the first reports from this study was concerned with the livability of purebred and crossbred cows [12]. Deaths due to hardware disease and the ingestion of poisons were excluded from the study. Over-all death losses in two generations of this study were 32.7 percent for purebreds and 13.4 percent for F_1 crossbreds and 3/4-breds. It was concluded that longevity in cows in this study was affected to a considerable extent by nonadditive gene action. A later report dealt with the effects of crossbreeding on services per conception, age at first calving, interval from first service to conception, gestation length, calving interval, calving date to first heat, and calving date to first service. Crossbreeding had no significant effect on any of these reproductive traits. Still later reports [45] from this study showed small increases in birth weight (3.4 lb) due to crossbreeding, but little or no effect on the length of gestation. Crossbreeding effects were significant early in life for size and body weight in females [46] but decreased linearly as age increased, and production traits were affected very little by heterosis in generations one, two and three. However, a later study [5], which included records on 519 first and 345 second lactations, showed the percentage of heterosis for yields of milk, butterfat, solids-not-fat, and protein of 7.3, 6.4, 4.7, and 7.4, respectively. Percentages of heterosis for weight at 6, 12, 18, and 24 months of age were 6.0, 6.5, 4.7, and 5.9, respectively. When all aspects of production were combined and considered [47], the crossbred cow had enough over-all advantage over the purebred cow to suggest that their use under practical conditions would be profitable.

A Florida Agricultural Experiment Station report presented information on the life span and livability of straightbred and crossbred dairy cattle [55]. About 85 percent were straightbred Jerseys or Holsteins, seven percent were Jersey $\times$ Holstein crosses, and eight percent were Jersey $\times$ Guernsey crosses. Although about 88 percent of the cows were culled for reasons suggesting a weakness in livability, no evidence of heterosis was detected for this trait.

Red Sindhis, Zebu cattle from India, have been crossed with Jerseys in an attempt to develop a type of dairy cattle adaptable to the southern United States [32]. Crossbred cows carrying varying proportions of Red Sindhi inheritance showed no advantage over straightbred Jerseys in dairy characteristics.

20.7.1 Conclusions on Crossbreeding in Dairy Cattle

From the standpoint of milk and fat yield, the results of crossbreeding indicate that selection for high performance within the existing breeds might be the breeding method of choice. Some established breeds are noted for high milk yield, whereas others are noted for the production of a high percentage of butterfat, which gives a creamy, yellow color to the milk. Keeping a herd containing cows of the two breeds and mixing the milk for sale purposes perhaps would be more advantageous to the dairyman than crossbreeding when only these traits are considered. On the other hand, when a dairyman has a herd whose production is average or below, crossbreeding might be of some advantage. In such a case, the transmitted high productivity of the sire used would be more important than the breed to which he belongs.

Crossbreeding seems to increase longevity and reproductive life span in dairy cows. If true, crossbreeding may have a greater advantage when all traits are considered than when milk and butterfat alone are considered. More experimental results are needed in which cows of two or more breeds are crossed that are similar, on the average, in milk and butterfat production. Crossbreeding may have more advantages for these traits in such experiments.

20.8 GENETIC-ENVIRONMENTAL INTERACTIONS

It is important to know if dairy characteristics are greatly affected by genetic $\times$ environmental interactions, because semen from the same dairy sire is often used in many different herds in different areas. A review of many studies indicates that, in general, genetic × environmental interactions are not important for milk production unless differences in environment are very large due to large differences in climate and other conditions. Studies show that there is a trend for butterfat percentage to be more highly heritable in herds where the production level is very high. This suggests that genetic × environmental interactions may be of importance when sires selected in one herd are used in others at different locations and under variable feed and management conditions. Variable results have been reported for the effect of genetic × environmental interactions on other economic traits (Pani, 1970).

20.9 ARTIFICIAL INSEMINATION IN DAIRY CATTLE BREEDING

Artificial insemination has found widespread use in dairy cattle breeding in the past few years. The first practical use of artificial insemination in cattle breeding was in Russia, where approximately 20,000 cows were artificially bred in 1931. In 1936, the first cooperative artificial breeding association was established in Denmark. The first cooperative of this kind was established in the United States in 1938, and almost 8,000,000 cows are now bred each year by this method. The average number of cows bred per sire in the United States by artificial insemination is about 3300 as compared to 30 to 50 if natural mating is practiced.

The development of methods of successfully freezing bull semen has resulted in its use for storing semen all over the world. Frozen semen may be stored for several years and possibly indefinitely. Frozen semen may now be shipped all over the world and its use has greatly reduced the wastage of semen from outstanding sires that occurred when the storage time was limited to only a few days.

Artificial insemination is of tremendous value in making the most possible use of dairy sires. The main problem is to identify outstanding sires on the basis of progeny tests. Progeny tests should be conducted to identify bulls with superior transmitting ability for milk and fat production. It is also necessary to progeny-test sires for any detrimental recessive genes they might possess in the heterozygous form. This can best be done by mating a sire to 35 or more of his own daughters, as described in Chapter 11. Any sire that is heterozygous for a detrimental recessive gene transmits this gene to approximately one-half of his offspring. Thus, if he sires 3000 calves in a year, 1500 of them would receive this gene from their sire. For this reason, it is desirable to use sires for artificial insemination that are free of detrimental recessive genes.

REFERENCES

1. Baker, G. A., S. W. Mead, and W. M. Regan. "Effect of Inbreeding on the Growth Curves of Height at Withers, Weight, and Heart Girth of Holstein Females," JDS, 28: 607, 1945.

2. Bartlett, J. W., R. P. Reece, and J. P. Mixner. "Inbreeding and Outbreeding Holstein-Friesian Cattle in an Attempt to Establish Factors for High Milk Production and High Fat Test," N.J.AESB 667, 1939.

3. Bartlett, J. W., and S. Margolin. "A Comparison of Inbreeding and Outbreeding in Holstein-Friesian Cattle," N.J.AESB 712, 1944.

4. Becker, R. B., P. T. D. Arnold, and A. H. Spurlock. "Productive Life-span of Dairy Cattle," FlaAESB Exp. Sta. Bull. 540, 1954.

5. Bereskin, B. and R. W. Touchberry. "Crossbreeding Dairy Cattle. III. First-lactation Production," JDS, 49: 659, 1966.

6. Brumby, P. J. "Monozygotic Twins and Dairy Cattle Improvement," ABA, 26: 1, 1958.

7. Carman, G. M. "Interrelations of Milk Production and Breeding Efficiency in Dairy Cows," JAS, 14: 753, 1955.

8. Castle, W.E. "Inheritance of Quantity and Quality of Milk Production in Dairy Cattle," PNAS, 5: 428, 1919.

9. Cole, L. J., and I. Johansson. "Inheritance in Crosses of Jersey and Holstein-Friesian with Aberdeen-Angus; 3; Growth and Body Type, Milk Yield and Butterfat Percentage," AN, 82: 265, 1948.

10. Dawson, J. R., and D. V. Kopland. "A Breeding Experiment with Dairy Cattle at the Huntley, Montana Field Station," USDATB 965, 1949.

11. Dickerson, G. E. "Effects of Inbreeding in Dairy Cattle (Progress Report)," JDS, 23: 546, 1940.

12. Dickinson, F. N., and R. W. Touchberry. "Livability of Purebred vs. Crossbred Cows," JDS, 44: 879, 1961.

13. Donald, H. P. and D. Anderson. "A Study of Variation in Twin Cattle, II, Fertility," JDR, 20: 361, 1954.

14. Dunbar, R. S., Jr., and C.. R. Henderson. "Heritability and Fertility in Dairy Cattle," JDS, 26: 1063, 1953.

15. Eldridge, F. E., and G. W. Salisbury. "The Relation of Pedigree Promise to Performance of Proved Holstein-Friesian Bulls," JDS, 32: 841, 1949.

16. Ellinger, T. "The Variation and Inheritance of Milk Characters," PNAS, 9: 111, 1923.

17. Farthing, B. R., and J. E. Legates. "Genetic Covariation between Milk Yield and Fat Percentage in Dairy Cattle," JDS, 40: 639, 1957.

18. Freeman, A. E., and R. S. Dunbar, Jr. "Genetic Analysis of the Components of Type, Conformation and Production in Ayreshire Cows," JDS, 38:428, 1955.

19. Gregory, P. W., W. N. Regan, and S. W. Mead. "Evidences of Genes for Female Sterility in Dairy Cows," G, 30: 506, 1946.

20. Hahn, E. W., J. L. Carmon, and W. J. Miller. "An Intra-herd Contemporary Comparison of the Production of Artificially and Naturally Sired Dairy Cows in Georgia," JDS, 41: 1061, 1958.

21. Hancock, J. L. "The Spermatozoa of Sterile Bulls," JEB, 30:50, 1953.

22. Hancock, J. L. "Studies in Monozygotic Cattle Twins. 7. The Relative Importance of Inheritance and Environment in the Production of Dairy Cattle," NZJSTA, 35: 67, 1953.

23. Harville, D. A., and C. R. Henderson. "Environmental and Genetic Trends in Production and Their Effects on Sire Evaluation," JDS, 50: 870, 1966.

24. Johansson, I. "The Manifestation and Heritability of Quantitative Characters in Dairy Cattle under Different Environmental Conditions," AGSM, 4: 221, 1953.

25. Johnson, K. R. "Heritability, Genetic and Phenotypic Correlations of Certain Constituents of Cows Milk," JDS, 40: 723, 1957.

26. King, G. J., and R. H. Miller. *The National Cooperative Dairy Herd Improvement Program Handbook.* Agriculture Handbook No. 278, U.S. Department of Agriculture.

27. Kendrick, J. F. "Standardizing Dairy-Herd-Improvement-Association Records in Proving Sires," USDAARAR, B.D.I.-Inf. 162, 1955.

28. Laban, R. C., P. T. Cupps, S. W. Mead, and W. M. Regan. "Some Effects of Inbreeding and Evidence of Heterosis through Outcrossing in a Holstein-Friesian Herd," JDS, 38:525, 1955.

29. Lagerlof, N. "Hereditary Forms of Sterility in Swedish Cattle Breeds," FS, 2: 230, 1951.

30. Margolin, S., and J. W. Bartlett. "The Influence of Inbreeding upon the Weight and Size of Dairy Cattle," JAS, 4:3, 1945.

31. Mason, J. L., and A. Robertson. "The Progeny Testing of Dairy Bulls at Different Levels of Production," JAgS, 47: 367, 1956.

32. McDowell, R. E., J. C. Johnson, J. L. Fletcher, and W. R. Harvey. "Production Characteristics of Jersey and Red Sindhi-Crossbred Females," JDS, 44: 125, 1961.

33. Nelson, R. H., and J. L. Lush. "The Effects of Mild Inbreeding on a Herd of Holstein-Friesian Cattle," JDS, 33:186, 1950.

34. Pau, J. W., C. R. Henderson, S. A. Asdell, J. F. Sykes, and R. C. Jones. "A Study of the Inheritance of Breeding Efficiency in the Beltsville Dairy Herd," JDS, 36: 909, 1953.

35. Pelissier, C. L., and F. D. Murrill. "Impact of Dairy Herd Improvement Association on Milk Production Efficiency," C. A., 24:4, 1970.

36. Ralston, N. P., S. W. Mead, and W. M. Regan. "Preliminary Results from the Crossing of Two Inbred Lines of Holsteins on Growth and Milk Production," JDS, 31:657, 1948.

37. Regan, W. N., S. W. Mead, and P. W. Gregory. "Calf Mortality in Relation to Inbreeding," JAS, 5:390, 1946.

38. Robertson, A. "Crossbreeding Experiments with Dairy Cattle," ABA, 17:201, 1949.

39. Robertson, A., and J. M. Rendel. "The Performance of Heifers Got by Artificial Insemination," JAgS, 44:184, 1954.

40. Spalding, R. W., C. R. Henderson, H. W. Carter, R. Albrechtsen, and A. M. Meek. "The Selection and Evaluation of Dairy Sires," Cornell E. B. 1118, 1963.

41. Swett, W. W., C. A. Matthews, and M. H. Fohrman. "Effect of Inbreeding on Body Size, Anatomy and Production of Grade Holstein Cows," USDATB 990, 1949.

42. Tabler, K. A., and R. W. Touchberry. "Selection Indices Based on Milk and Fat Yield, Fat Percent, and Type Classification," JDS, 38:1155, 1955.

43. Thompson, N. R., L. J. Cranek, Sr., and N. P. Ralston. "Genetic and Environmental Factors in the Development of the American Red Danish Cattle," JDS, 40:56, 1957.

44. Touchberry, R. W. "Genetic Correlations between Five Body Measurements, Weight, Type and Production in the Same Individuals among Holstein Cows," JDS, 34:242, 1951.

45. Touchberry, R. W., and B. Bereskin. "Crossbreeding in Dairy Cattle. 1. Some effects of Crossbreeding on the Birth Weight and Gestation Period of Dairy Cattle," JDS, 49:287, 1966.

46. Touchberry, R. W., and B. Bereskin. "Crossbreeding in Dairy Cattle. II. Weights and Body Measurements of Purebred Holstein and Guernsey Females and Their Reciprocal Crossbreds," JDS, 49:647, 1966.

47. Touchberry, R. W. "A Comparison of the General Performance of Crossbred and Purebred Dairy Cattle," JAS, 31:169, 1970.

48. Tyler, W. J., and A. B. Chapman. "A Simplified Method of Estimating 305-day Lactation Production," JDS, 27:463, 1944.

49. Tyler, W. J., and G. Hyatt. "The Heritability of Milk and Butterfat Production and Percentage of Butterfat in Ayrshire Cattle," JAS,, 6:479, 1947.

50. Tyler, W. J., A. B. Chapman, and G. E. Dickerson. "Growth and Production of Inbred and Outbred Holstein-Friesian Cattle," JDS, 32:247, 1949.

51. Verley, F. A., and R. W. Touchberry. "Effects of Crossbreeding on Reproductive Performance of Dairy Cattle," JDS, 44:2058, 1961.

52. Von Krosigk, C. M., and J. L. Lush. "Effect of Inbreeding on Production in Holsteins," JDS, 41:105, 1958.

53. Wadell, L. H., and L. D. McGillard. "Influence of Artificial Breeding on Production in Michigan Dairy Herds," JDS, 42: 1079, 1959.

54. Wilcox, C. J., J. A. Curl, J. Roman, A. H. Spurlock, and R. B. Becker. "Lifespan and Livability of Crossbred Dairy Cattle," JDS, 49: 991, 1966.

55. Wilson, C. J., and K. O. Pfau. "Longevity of Dairy Cows within a Holstein-Friesian Herd; An Estimate of Heritability and Its Relationship with Breeding Efficiency," JDS, 38: A9, 1954.

56, Woodward, T. E., and R. R. Graves. "Some Results of Inbreeding Grade Guernsey and Grade Holstein-Friesian Cattle," USDATB 339, 1933.

57. Wriedt, C. "The Inheritance of Butterfat Percentage in Crosses of Jerseys with Red Danes," JG, 22: 45, 1930.

58. Yapp, W. W. "The Inheritance of Percent Fat Content and Other Constituents of Milk in Dairy Cattle," PSCBC 328, 1923.

QUESTIONS AND PROBLEMS

1. Why has more attention been given to the improvement in performance in dairy cattle than in beef cattle?

2. Why is fertility so important in dairy cattle?

3. What is the heritability of fertility in dairy cattle? How may fertility be improved in dairy cattle?

4. Some breeds of dairy cattle are quite different in the amount of milk and butterfat they produce in a single lactation period. What do these breed differences indicate about milk and fat production?

5. What is the heritability of milk and butterfat yield in dairy cattle? What do these heritability estimates indicate about these traits?

6. Why are heritability estimates based on the resemblance among identical twins usually higher than when based on parent-offspring or sib resemblance? Which heritability estimate should more accurately predict the amount of progress made in selection?

7. How are DHIA associations formed and operated?

8. What are the main objectives of DHIA?

9. Has the DHIA program been effective in increasing the efficiency of milk and fat production over the years?

10. What is meant by $2\times$, M.E., and 305-day records?

11. Why should records be adjusted to a 305-day basis?

12. A Guernsey cow 30 months of age has a 210-day lactation record of 6000 pounds of milk and 250 pounds of butterfat. What would be her projected 305-day lactation record?

13. Why is a projected 305-day lactation record usually less accurate than one calculated from actual monthly records?

14. Assume that a Jersey cow three years and five months of age has a 305-day record of 5000 pounds of milk and 270 pounds of butterfat. What would her record be on a M.E. basis if she calved in December?

15. Does a cow give more milk when milked three times or two times daily? Why?

16. A Holstein cow 4.5 years of age has a 305-day lactation record of 13,000 pounds of milk and 520 pounds of butterfat when milked three times daily. What would her record be if converted to a twice-a-day milking basis?

17. What is the average productive life span of a dairy cow? Why are cows usually culled from a dairy herd?

18. Does dairy type have a close correlation with the level of milk production? Discuss.

19. What traits in dairy cattle appear to be genetically correlated?

20. Should the cow or the bull receive the most attention in selection for higher milk production? Why?

21. What are the recommended procedures for estimating a bull's transmitting ability for milk and fat production?

22. What is meant by the adjusted herdmate average? Why is it used?

23. Assume that the regional-breed-year-season average for Holsteins is 13,000 pounds of milk and the herdmate average for 30 herdmates is 14,000 pounds of milk. What is the adjusted herdmate average?

24. Assume that the daughters of a sire averaged 11,500 pounds of milk, whereas the adjusted herdmate average is 13,000 and the regional-breed-year-season average is 13,200 pounds. What is the adjusted daughter average?

25. Let us assume that the adjusted daughter average for 60 daughters of a sire is 12,000 pounds of milk, whereas the breed average is 9000 pounds of milk. What would be the predicted difference for this sire?

26. Explain what is meant by the PD of a sire.

27. Assume that in question 26, the sire had 100 tested daughters. What would be his PD with this many daughters? Of what importance is the number of daughters in determining a sire's PD?

28. What factors influence the accuracy of the PD for a sire?

29. Do experiments suggest that milk and fat production can be improved through selection? Discuss.

30. Outline in detail all procedures you would recommend for the improvement of milk and fat production in a dairy herd.

31. Why do cows enrolled in DHIA have a higher production record than cows not enrolled in DHIA?

32. Outline in detail what one might expect to happen in the most important economic traits if inbreeding is practiced in dairy cattle.

33. Outline in detail what one might expect to happen in the most important economic traits if crossbreeding is practiced in dairy cattle.

34. Does it pay to use crossbreeding in dairy cattle? Explain.

35. How important are genetic $\times$ environmental interactions in dairy cattle?

36. What problems may be encountered if a sire produces thousands of offspring each year by artificial insemination? How could some of these problems be avoided?

21

Systems of Breeding and Selection in Horses

The horse was probably first domesticated in the Old World (Asia or Africa) between 4000 and 5000 years ago. Men and horses have enjoyed a close relationship since that time, and the development of agriculture in some areas was due as much to the presence of the horse as it was to the initiative and ingenuity of man.

Horses were first used by man in the conduct of his wars, and the nation that developed superior horses was the one that ruled supreme for many years. The horse was the chief weapon used by the Norman army in its successful invasion of England. Later, the horse played a very important part in the development of that country. The heavy combined weight of the soldier, his weapons, and his armor required large, strong, and heavy horses. With the introduction of guns and gun powder, much of the heavy armor was discarded, and lighter, more nimble horses suitable for riding, hunting, and racing were developed. The size, appearance, and performance of the horse was changed when needed because these traits must have been highly heritable, since they responded readily to selection pressure applied by man.

The life of the American Indian was greatly changed by the introduction of the horse into the Americas by the Spaniards. Before the introduction

413

of the horse, the American Indian depended upon his legs as a mode of travel. His mode of transportation depended upon what he and his wife could carry on their backs or what could be carried on the backs of his dogs which were often used as pack animals. With the horse at his disposal, the Indian was able to travel much farther and faster in a day than ever before, and he could transport more material more quickly, using the horse either as a pack animal or to pull a makeshift sled.

The great cattle industry of the western United States could not have been developed without the western cow pony. Even today the horse is absolutely necessary to work cattle on much of the western range, even though airplanes, helicopters, and jeeps are available. The trust and mutual dependence between a cowhand and his favorite cow pony still exists today and is something to behold. The author can speak from personal experience in this respect, since he spent several years in the range country and many pleasant hours, and even days, on horseback working cattle in Arizona.

The development of agriculture in the early years of the United States could not have taken place without the horse as a draft animal. With the development of railroads, automobiles, and trucks, as well as the tractor, the need for the horse became less and less, and it looked as if the horse in the United States was headed for extinction. In recent years, however, the horse has made a startling comeback. Although he is no longer needed as a draft animal, the horse is becoming more and more popular for pleasure and recreational purposes. Young boys and girls all over the land care for and ride horses as one of their personal projects. Furthermore, horse science courses are again being taught in many colleges of agriculture, in spite of the fact that they were dropped from the curricula of these same colleges only a few years ago. This revived interest in horses makes it more important than ever that we include a chapter in this book on the breeding and selection of this very popular species of farm animal.

21.1   TRAITS OF ECONOMIC IMPORTANCE

21.1.1   Fertility

Lowered fertility of the brood mare is as serious a problem as it is in any class of farm animals. A study of 45 draft and 35 light mares during two breeding seasons showed that only 69 percent conceived and produced foals [1]. This low percentage colt crop seems to be similar to that obtained all over the United States. Perhaps one reason for the decreased breeding efficiency in mares is the extreme length of the estrus period, which averages between five and six days. Ovulation, as a general rule, occurs one to two days before the end of estrus. The time of mating in relation to the time of ovulation in the mare is of great importance because of the limited life of both spermatozoa and ovum in the female reproductive tract. If the spermatozoa are introduced into the female reproductive tract too long before or too long after the release of the ovum from the ovary, they may die and fail to fertilize the egg.

The heritability of fertility in horses is low, and one report in the literature gives a heritability estimate of five percent [5]. This low estimate seems reasonable, since fertility is also lowly heritable in other species of farm animals. Selection for improved fertility in horses would probably be ineffective. Attention to the improvement of fertility through such environmental factors as nutrition, management, and disease control is indicated. In management practices, treatment for certain pathological and/or functional disorders responsible for lowered fertility or sterility could be of value, providing such abnormalities are not inherited. Much work has been done in this area in the past several years.

21.1.2 Performance

Horses have been bred for many different purposes, and their ability to do the task for which they were bred has been an important factor in their popularity. The several different aspects of performance will be discussed separately.

Racing ability. This trait depends probably upon certain physiological and nervous qualities of the individual, as well as upon its anatomical structure. Since so many different factors are involved, it seems very probable that the trait is influenced by the action and interaction of many genes. Breeders believe that racing ability is highly heritable, and they are willing to pay high prices for an outstanding stallion or even thousands of dollars for his services.

Several schemes have been used to measure the racing ability of Thoroughbreds. The average earning index has been used by several investigators [3]. This index is computed by first calculating the average amount of money earned by the breed each year; this is determined by dividing the total amount of earnings for all animals by the total number of starters. The earnings of a horse are then compared with the average for the breed for that year. A horse that wins exactly as much money as the average is given an adjustment factor of 1.0. A horse that wins four times the average of the breed has an adjustment factor of 4.0. This adjustment factor is figured for each year that the horse is raced; the factors are summed and then divided by the number of years raced. The resulting figure is the average earnings index.

Studies of the relationship of stake winners to the earnings index of their dam [4] indicate that this trait is highly heritable. Thus, mares with the higher earnings index would have the higher racing ability. The earnings of the offspring of mares with higher indexes were also higher than those of the offspring of mares with lower indexes.

A heritability estimate of 60 percent for racing ability was found in a study at the Kentucky Agricultural Experiment Station [8]. This study, together with other evidence, indicates that racing ability is highly heritable and is influenced by many genes with additive effects. The heritability, however, may not be as high as indicated here.

Racing ability in the Standardbreds and Quarter horses has not been studied in detail from the genetic standpoint, but the beliefs of breeders that this trait "runs" in certain families and that certain sires produce better performing offspring than others indicate that this trait is highly heritable. Speed on the track varies in different individuals. Some horses can run very fast for a short distance, after which they falter. Others can maintain a rapid pace for much longer distances. This "staying" ability must also be heritable, since the breeds or types differ in this respect, the Thoroughbred being noted for racing ability at the longer distances and the Quarter horse at the short distance of a quarter-mile.

Trotters and pacers. The ability to trot or pace definitely depends upon the genotype of the individual. The trotting gait has been reported to be due to a dominant gene, whereas the pacing gait is due to the recessive gene. Natural pacers under this mode of inheritance would breed true and would be homozygous for the recessive gene. It is possible to teach a trotter to pace, but the natural inclination to trot or pace seems to be inherited in a simple Mendelian manner.

Speed in trotting and pacing could be influenced by several pairs of genes, many of which act in an additive manner. As far as is known, no actual heritability estimates for speed in trotting or pacing have been published.

Cow sense. The term "cow sense" as used here refers to the ability or aptitude of the horse to understand and work cattle. In working cattle, a cow pony has several different jobs to do; often, different horses are used for different jobs. One is to carry its rider over the range to round up cattle in an area of several sections. The range covered is often rocky and brushy, and the pony must have stamina and surefootedness. Another job, mostly limited to specially rough areas where cattle are sometimes wild, is to carry the rider close enough to a running steer so that it can be roped. In the rodeo, cow ponies must be fast enough to catch a speeding calf and then must keep the rope taut when the calf is caught so that the cowboy can make a quick tie. Still another job is reserved for the "cutting" horse, who must have the agility and the cow sense to cut a single animal from a herd with very little guidance by the rider. Although considerable training is necessary to make a good roping or cutting horse, training alone is not sufficient.

Little or no research has been done to determine if these traits are inherited. Some horses seem to have "cow sense"—others do not. Cow men have long been of the opinion that cow sense is inherited. In many cases they will travel long distances and pay high prices to get a colt from a particular line of breeding or out of a particular stallion.

Type and conformation. Within certain broad limits, proper type and conformation are necessary for the individual to excel in performance. A Thoroughbred built like a draft horse would not be expected to be a stakes winner. Similarly, draft type is avoided in selecting and breeding cow ponies or Quarter horses. Where speed and agility are the prime requirements, attention must be paid to the degree of muscling and size of bone, as well as to the soundness of feet and legs. Type differences within the breeds are probably of less importance than between breeds. The statement that horses can have good type and conformation without good performance, but cannot have good performance without good conformation seems to be true [9].

The various breed-registry associations have definite descriptions as to the ideal type of individual within their breed. Since many breeds of horses are known, a description of each will not be given here. If both type and performance are important, then selection for both should be practiced.

21.2 CORRELATIONS BETWEEN TRAITS

Genetic correlations between various traits have not been studied in horses as they have in other classes of farm animals. Some phenotypic correlations have been studied, however, especially between a trait that can be easily measured and the ability to perform.

It has been reported that experienced breeders and handlers of race horses believe that nervous mares, as a general rule, are more likely to produce speedier offspring than are mares that are less nervous or docile. At least, it is thought that the highly spirited horses are usually speedy racers because they have a tremendous will to win in races in competition with others. This has been studied, using 50 high-spirited mares that had 272 offspring and 50 phlegmatic mares that had 248 offspring [3]. The study showed no significant difference between the average earning indexes of the offspring of the two groups of mares. It was concluded that the temperament of the dam was not correlated with the racing ability of the offspring. It was noted, however, that there was a significant positive correlation between the average earning index of the dam and that of her offspring. This, again, demonstrates that racing ability is a heritable trait.

Some breeders of Thoroughbreds believe that coat color is related to racing ability. Leicester, in 1959, made a comprehensive study of such a possible correlation [6]. The results, based on several hundred horses, led to the conclusion that the number of winners of a certain coat color was proportional to the opportunity of horses of that color to win. For example, horses of the chestnut coat color won more races than those that were grey or roan, but this was because more chestnut horses were raced. Another investigator, however, did not agree with these findings and felt that there was a definite correlation between coat color and racing performance [7].

21.3 SELECTION IN HORSES

The existence of several different types and breeds of horses in this country indicates that selection in this species has been effective. One needs only to observe the extreme difference in size between Shetland ponies and draft horses to realize that this is true. Further evidence is found in the various types such as draft horses, pacers, trotters, Quarter horses and Thoroughbreds that have been successfully developed with a definite purpose in mind.

As far as Thoroughbreds are concerned, selection practiced has been based mostly on individuality and/or performance of the parents. Most of the attention has been given to the sire's performance, and an attempt has been made to use sires that have been outstanding winners if possible. Better results are obtained when both parents have excellent racing records, but even in such matings many of the offspring are disappointing. In one study [4], it was shown that less than two percent of the mares studied had an earnings index greater than four. It has also been estimated that 0.05 percent of the colts registered as a thoroughbred will be good enough to improve the breed.

Selection practiced in Thoroughbreds has been to identify the best and then mate the best to the best. This is still the recommended procedure for best results, as is indicated by the apparently high heritability of racing ability.

In general, mares are not performance-tested as often as are stallions. They also cannot be as superior on the average as stallions because more of them must be kept for breeding, which makes the selection differential smaller. At the King Ranch in Texas, however, attention is given to the selection of mares for "cow sense" and other traits [2]. The mares are broken and ridden at three, four, and five years of age and are given a chance to show their ability for working cattle. Those that show a good potential and have speed and endurance are checked for soundness, quality, and blood lines. Those that meet the rigid requirements are added to the mare breeding herd. Mares in the breeding herd that fail to produce superior offspring after two or three foals are culled. Such a system of testing and culling should result in progress for the traits of importance, especially in large herds, where rigid selection can be practiced.

Fig. 21.1 Algo, an excellent type Quarter horse. He was shown for about two and a half years and won 52 first prizes, 10 grand championships, and 19 reserve grand championships. (Courtesy of the King Ranch, Kingsville, Texas.)

21.4 INBREEDING IN HORSES

In many species of farm animals, inbreeding combined with selection has been used for improving and forming the different breeds and types. This has been especially true in purebred beef cattle. Inbreeding, however, has not played such an important part in the development of the different breeds and types of horses.

A pedigree analysis of the Thoroughbred breed was made to determine the degree to which inbreeding has played a part in the improvement and development of this breed [10]. The analysis was made of the pedigrees of stakes winners, losers, and "millionaire sires," or those whose offspring have won a million dollars or more in racing. Pedigrees were traced to the years of 1748 as a base for stakes winners and losers in the years 1935, 1940, and 1941. The average inbreeding for 556 stakes winners was 8.23 percent and that for the losers was 8.00 percent. Thus, there was little or no difference in the amount of inbreeding in the two groups. It was concluded that the genetic composition of the breed has been influenced very little by inbreeding and linebreeding.

A study of recent pedigrees indicates that, in the Thoroughbred, inbreeding is carefully and purposely avoided. Possibly inbreeding causes a decline in racing ability because of an associated decrease in the stamina and vigor of inbred individuals. Stamina and vigor as well as speed are necessary if horses are to be winners on the track.

An investigation of present-day Thoroughbred pedigrees will show that outstanding stallions are often imported from foreign countries, and the inheritance of these individuals is mingled with that of superior racing individuals in this country. This is a form of outbreeding, and should be of value, especially if the imported horses possess some plus genes for racing ability not present in our own racing stock. They should have these genes, for imported horses probably have descended from ancestors different from those of our racing stock, at least in the last several years.

Pedigrees of present-day Quarter horses will show that some linebreeding and inbreeding is being practiced. Such a pedigree is given in Fig. 21.2. On

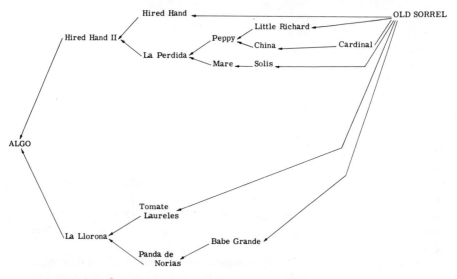

Fig. 21.2 Arrow diagram of the pedigree of Algo (P-63,952) illustrating how he is linebred to Old Sorrel.

the other hand, there are other pedigrees that show no inbreeding or linebreeding, at least in the three or four generations usually shown. Possibly in those instances in which other traits, such as "cow sense" and beauty of conformation are bred for, inbreeding and linebreeding may be of more value.

Although inbreeding and linebreeding have been used very little in the development and improvement of some of the present-day breeds, theoretically they could be used to advantage. For instance, linebred families could be formed in which the relationship was kept high to a particular stallion in the breed, with a minimum of inbreeding being involved. If several such distinct families were formed, crossing them might produce a larger proportion of desirable animals in the linecross individuals than is produced by present methods of breeding and selection. Even though individuals in the linebred families might not be outstanding winners themselves, they should breed truer than noninbred parents, and their linecross offspring should show increased heterosis for stamina and vigor. The continued use of linebred families for breeding purposes in a manner comparable to the use of inbred lines of corn could continue to produce a larger percentage of winners in future years. The same should also be true for gaited animals and Quarter horses. The disadvantage of such a system of mating, however, would be the time and systematic effort required to put it into practice. Whether or not this system of breeding would be profitable would have to be proved under actual practice.

21.5 CROSSBREEDING IN HORSES

Crossbreeding has not been studied in detail in horses as it has in some other species of farm animals. Most breeds of horses have been developed for a specific purpose which requires a particular form and function. Crossing two or more breeds that differ considerably in form and function would probably produce a vigorous crossbred progeny, but it would probably rank midway between the parent breeds in performance, and would not perform as satisfactorily for a particular purpose as one of the parent breeds. For example, crossing draft horses with Thoroughbreds would probably produce an offspring that would not perform as well as the draft parent for draft purposes or would not race as well as the Thoroughbred parent.

Crossbreeding of Thoroughbred and Quarter horses has been practiced in the United States in recent years. These are listed as two separate breeds in textbooks that discuss breeds of horses. The sale catalog of Quarter horses at the National Quarter Horse Congress in the fall of 1969 had 81 four-generation pedigrees listed. Of this number, 48 pedigrees showed some Thoroughbred breeding, usually in the top side of the pedigree. The Thoroughbred stallion Three Bars appeared at least 35 times in these 81 pedigrees. Very little inbreeding or linebreeding was detected in these pedigrees, and when it was found the inbreeding coefficient did not exceed 0.125 (12.5 percent inbreeding). These records indicate that some Thoroughbred blood in Quarter horses produced faster-running progeny and possibly improved type and conformation.

REFERENCES

1. Andrews, F. N., and F. F. McKenzie. "Estrus, Ovulation, and Related Phenomena in the Mare," MoAESRB 329, 1941.

2. Dinsmore, W. "Selecting Brood Mares," QHJ, 11(3): 215, 1958.

3. Estes, B. W. "A Study of the Relationship between Temperament of Thoroughbred Mares and Performance of Off-Spring," JGP, 81: 273, 1952.

4. Estes, J. A. "Dams of Stakes Winners," BH, 80: 99, 1960.

5. Hartwig, W., and U. Riechardt. "The Heritability of Fertility in Horse Breeding" (t.t.), Z, 30: 205. ABA, 27: 626, 1958.

6. Leicester, C. "Coat Color and Performance," TR, 172(2): 28, 1960.

7. Myers, R. "A Discussion of Hybrid Vigor," TR, 172(3): 28, 1960.

8. Pirri, J., Jr., and D. G. Steele. "The Heritability of Racing Capacity," BH, 63: 976, 1952.

9. Reynolds, F. "Type vs. Performance," WLJ, 38: 224, 1960.

10. Steele, D. G. "Are Pedigrees Important?" BH, 38: 574, 1942.

QUESTIONS AND PROBLEMS

1. When were horses first domesticated?

2. How did mankind first use horses?

3. Are horses natives of America?

4. How and why did the horse change the way of living of the American Indian?

5. What part did the horse play in the development of the cattle industry and agriculture in the United States?

6. Why did horses decline in numbers a few years ago? Why are horses so popular today?

7. What are the traits of economic importance in horses? How many of these traits appear to be heritable?

8. Are type and performance related in horses? Explain.

9. Are there any important genetic correlations among traits in horses?

10. Has selection in horses been effective? Why?

11. Why do horse breeders usually avoid inbreeding?

12. Is crossbreeding ever practiced in horses? Explain.

13. Could hybrid vigor be utilized in horse breeding? Explain.

14. What systems of breeding are used in the production of Thoroughbred horses today?

Index